SCIENTIFIC ADVICE TO THE
NINETEENTH-CENTURY BRITISH STATE

SCIENCE AND CULTURE IN THE NINETEENTH CENTURY

BERNARD LIGHTMAN, EDITOR

Scientific Advice to the Nineteenth-Century British State

Roland Jackson

UNIVERSITY OF PITTSBURGH PRESS

Published by the University of Pittsburgh Press, Pittsburgh, Pa., 15260
Copyright © 2023, University of Pittsburgh Press
Manufactured in the United States of America
Printed on acid-free paper
10 9 8 7 6 5 4 3 2 1

Cataloging-in-Publication data is available from the Library of Congress

ISBN 13: 978-0-8229-4790-5
ISBN 10: 0-8229-4790-0

COVER ART: Preliminary studies for the Big Ben clock tower, Houses of Parliament, Palace of Westminster, London. Courtesy RIBA Collections.
COVER DESIGN: Joel W. Coggins

Contents

Acknowledgments

MUCH OF THIS BOOK WAS WRITTEN DURING THE FIRST PHASES OF THE COVID-19 pandemic, when scientific and public health advice took center stage throughout the world. That advice was proffered and contested in the context of global politics, national systems of politics and government, religious beliefs, expert disagreement, and the hopes, fears, values, and beliefs of individual citizens everywhere.

I have been fortunate enough to experience aspects of policymaking involving science and technology quite closely in the United Kingdom, particularly through my involvement over nearly twenty years with the UK government's Sciencewise program. That experience came together with my interest in the nineteenth-century history of science in the gestation of this book. It was apparent to me when writing *The Ascent of John Tyndall*, a biography of one of the leading scientific figures of the time, that the scientific, engineering, and medical experts of the period had a far deeper involvement with Parliament, government, and administration than I had realized. I could find no book that offered any systematic overview, and this is my attempt to provide one. Those working at the interface of science and policy today will recognise the sorts of tensions and challenges experienced by their forebears, allowing for the distance of a couple of centuries.

The original manuscript of this book was too long for publishers. I am publishing much of the excised material as a set of complementary case studies, *Case Studies in Scientific Advice to the Nineteenth-Century British State*. These vary greatly in length, including a substantial section on vaccination, which sadly could not find a place in the main study. Taken together, I believe the examples in both books provide a comprehensive overview of the provision of scientific advice to Parliament and government in Britain at the national level. The *Case Studies* also include two appendices on my identification of 149 fellows of the Royal Society who were MPs (member of Parliament) during the nineteenth century, and of the relatively few MPs who had significant scientific, engineering, or medical expertise. While doubtless incomplete, I believe they are the most comprehensive such analyses to date.

Thanks go to many people and institutions who have provided advice and access to resources. I am particularly grateful to Bill Brock, who read the entire manuscript, even more grateful to Nicola Jackson, who read it twice, and also thank Lawrence Goldman and Frank James, who read substantial parts. I much appreciate the pertinent commentary and advice from three mystery readers solicited by the University of Pittsburgh Press, who substantially helped shape the book and its final arguments. Many others offered useful information and advice, including Michael Bailey, Daniel Belteki, Ian Blatchford, Gabriel Finkelstein, Laura Franchetti, Graeme Gooday, Bryan Gray, Rebekah Higgitt, Roger Knight, Roy McLeod, Elizabeth Neswald, Paul Ranford, David Sutherland, and Sophie Waring. Special thanks to Claire Batley, Katherine Emery, and Mari Takayanagi at the Parliamentary Archives, who went beyond the call of duty during lockdown, to Virginia Mills and Keith Moore at the Royal Society of London, to Kay Walters and Laura Doran at the Athenaeum, and to staff of the British Library, the London Library, The National Archives, the University College London and Institute of Education libraries, the Wellcome Library, and the Royal Institution of Great Britain. I thank the University College London Department of Science and Technology Studies for an honorary research fellowship, and the Royal Institution for a visiting fellowship.

It is a privilege to appear in the Science and Culture in the Nineteenth Century series, and I am grateful to Bernie Lightman and all at the University of Pittsburgh Press for the opportunity. Particular thanks to my editor, Abby Collier, to Amy Sherman, Jessica LeTourneur Bax, and to all others at the press, of whom I may be unaware, who have been involved in the production of this book.

List of Abbreviations

BAAS   British Association for the Advancement of Science

MP   Member of Parliament

PP   Parliamentary Papers. These are abbreviated in references by the number of the paper and the year of publication—for example, PP 290 (1814)—with the page reference if appropriate. The full or initial part of the title is given in the bibliography.

UCL   University College London

SCIENTIFIC ADVICE TO THE
NINETEENTH-CENTURY BRITISH STATE

Introduction

NINETEENTH-CENTURY BRITAIN WITNESSED AN ENORMOUS GROWTH IN the scope and size of central government. The same period saw major scientific advances, huge industrial development based on coal and steam power, and new medical understanding and procedures. To what extent are these great changes connected? This book brings together perspectives from the history of science and the history of the state to explore this question, and to focus on the influence of scientific, engineering, and medical expertise on national policy and administration.

While both politics and expert advice may represent the "art of the possible,"[1] their objectives, values, and practices are often in tension. As the state responded to the military, social, economic, and environmental challenges of the nineteenth century, the practitioners of science, engineering, and medicine, almost entirely men, were drawn into close involvement and contestation with the politicians.[2]

Several sets of questions arise: What were the social and political imperatives, and the scientific, engineering, and medical developments in this period that drove or offered opportunities for the government or Parliament to seek expert advice? How did the practitioners of science, engineering, and medicine achieve authority and influence with the British government and Parliament? What were the most significant political and cultural constraints within which they had to operate? To what extent was advice sought from individuals or from institutions? What were the roles in which individuals acted? How were they appointed, employed, and remunerated, or otherwise recognized? How did individual personal qualities, networks, and relationships influence their appointment, and the structures and modes of providing advice? Standing back, looking over the course of the nineteenth century, what factors shaped the system of advice that developed? How different was the system at the end of the nineteenth century, leading up to World War I, compared to its nature at the beginning?

The intent of this book is to take a broad approach, across both military and civil policy areas. That allows connections to be made and trends perceived across an entire century, with glances back to eighteenth-century roots and glimpses forward to the twentieth century. It means that nuances and detail will be lost in individual policy domains. I have left a trail of references to enable anyone to explore particular aspects in more depth, and can only regret any omissions of work that other scholars might have expected to see acknowledged.

SCIENTIFIC ADVICE AND THE STATE

In twenty-first-century Britain, scientific advice to government is highly organized, integrated across government departments, and led by a chief scientific adviser who reports to the prime minister. Each individual government department has a chief scientist, formally a civil servant, who acts as the broker of advice within that department. Some of these may be engineers or social scientists, and there is a chief medical officer at the Department of Health. The beginning of the nineteenth century reveals a very different picture. The word *scientist* was not coined until 1833,[3] and was barely used in Britain until the twentieth century. Likewise, the terms *expert* and *expertise* did not come into colloquial use until the 1860s.[4] The meanings of *science*, *scientific advice*, and *expertise* will be treated here in a fluid manner.[5] They will often encompass as a shorthand both engineering and medicine, although the specialties will be distinguished where appropriate. What counts as *scientific advice* or *technical expertise* to inform policy and its implementation is as much determined by who is prepared to listen and confer validity on it, as it is by who claims to offer it, and by the actual basis of such claims. From the point of view of practical politics, *expert advice* is a combination of what its protagonists claim it to be and what its recipients accept as valid.

The men of science, or savants, of the day were those who derived their knowledge of the natural world from experience, observation, and experiment, and include those who sought to understand human behavior and society in an analogous manner. They encompass mathematicians, natural philosophers, physicists, chemists, geologists, natural historians, biologists, and statisticians.[6] Engineering, growing out of practical and artisanal skills and knowledge, developed further as mathematics and the theoretical under-standing brought by physics and chemistry offered greater rigor and further opportunities for innovation. The medical profession, a career with a status rather below that of the law and the church at the turn of the nineteenth century, became increasingly scientific, suffering constant tensions between its different elements, and between medical practitioners and the advocates of new scientific research. The developing social sciences are not treated in detail here.[7] The "state" is envisaged as the whole system of institutions and processes that constitute the direction and management of Britain as an entity, national and local. Britain emerged from the eighteenth century as a "fiscal-military" state.[8] It had powers to raise money to finance wars that enabled it to increase its colonial possessions, and to control trade and commerce that allowed it to benefit from those conquests. In 1811, during the Napoleonic Wars, one half of all government expenditure, £43 million out of £85 million, was spent on the navy, army, and Board of Ordnance.[9] This fiscal-military state of the late eighteenth century, with its aristocratic ethos, evolved into the liberal but more centralized and bureaucratic state of the nineteenth century, now rooted in free trade.[10] The emphasis of this book is on advice to the executive and the legislature—in other words, to government and Parliament, alongside the civil service and the military, and in the context of other institutions such as private corporations, charities, and social and professional associations. This complexity makes the concept of a single identifiable "state" problematic. The responsibilities and powers of different public bodies were less coordinated at the beginning of the nineteenth century than they were at the end. Courts, towns, parishes, and other local bodies had substantial autonomy, and there were strong political pressures tending to favor local over central control. This tension between the local and central elements of the state runs throughout this book. There were also differences, many of which still exist, between England and Wales, Scotland, and Ireland. Ireland only became part of the United Kingdom in 1801, with representation in Parliament at Westminster.

DRAWING BOUNDARIES

Some boundaries must be drawn. Policies on primarily social issues such as education, crime, policing, prisons, and labor relations are not addressed here in any detail.[11]

This book does not tackle government support for the scientific research community nor government furtherance of scientific, technical, or medical education. Throughout the nineteenth century, many people within the scientific community lobbied for funding from the state for the systematic support of scientific research and of individual researchers, without great success.[12] British governments left fundamental scientific research to private initiative, but encouraged practically oriented science where it offered benefits to the state. The scientific community was split on this. Although many sought the endowment of research by government, others considered it best done by private means, as they would retain independence. Men of science also lobbied for parliamentary legislation and money to improve scientific and technological education, with somewhat more success. Both these dimensions have been extensively investigated, through numerous case studies and broader assessments.[13] By contrast, this is a book about science for policy, not policy for science, although the two are inevitably linked. It is the technical content and relevance of advice that is emphasized.

There are many individual studies of the nature and impact of scientific advice on particular policy areas. But although there has been recent research in specific domains, relating to both civil and military policies and practices, there is little synoptic overview.[14] Such a synthesis is attempted here. Nevertheless, the emphasis is on Britain's domestic policies rather than on its colonies and extending empire. They would deserve a volume to themselves. But it seems likely that the manifest usefulness of the scientific experts in colonial expansion reinforced their authority in domestic affairs too.[15]

I have taken a policy-led approach to writing this account. Two introductory chapters, presenting context on the state of science and scientific institutions in the early nineteenth century, are followed by twelve chapters on different policy areas. Within each chapter the treatment is for the most part chronological. The consequence of this approach is that common issues recur with different emphases in different domains. I start with two chapters on military policy. That is partly because the fiscal-military state of the late eighteenth century already had a strong demand for scientific and technical expertise but also because the factors affecting military policy prove to be distinct from those affecting civil policy. The ten following chapters address these civil policy areas, concentrating on social and industrial issues. While some analysis is offered in individual chapters, they are primarily descriptive of what occurred, highlighting the research evidence. The final chapter seeks to bring out the key themes arising from these narratives, and sets out my overall analysis and conclusions.

In considering the existing literature on these various policy areas, it becomes evident that historians have taken different approaches depending on the topic.[16] For example, with respect to the military and to transport systems, there is much focus on technological developments connected to the state. In public health, by contrast, theories of disease interact with disputes about appropriate local and central state relationships. In the area of utilities, such as gas provision, economic issues and the role of the private sector may be stressed over the provision of technical expertise. Studies have also been made from the point of view of the legal system—for example, in the case of river pollution seen from the perspective of the nuisance laws. I do not engage critically with these contrasting approaches, but my method of building this account upward from the Parliamentary Papers (described below) enables me to some extent to cut across these differences.

THE EVIDENCE BASE

The main thesis of this book is that the expanding administrative system of the state at the central level, across multiple policy areas, was piecemeal and pragmatic, in contested interaction with several major constraints. Throughout the nineteenth century, the driving force was the development of scientific, engineering, and medical knowledge, expertise and authority, in conjunction with the increasing range and impact of social and political pressures for action on issues to which technical advice could be relevant. In the civil policy areas, the constraints are rooted in the liberal culture that prevailed during the nineteenth century, almost regardless of party politics.[17] Four overarching principles and beliefs, widely held throughout nineteenth-century society, emerge from the evidence presented here to set the primary constraints within which politics and expertise were contested. They are the sanctity of private property, the laissez-faire approach to capitalist private industry,[18] the emphasis on individual freedom and responsibility, and the importance of local government. These political forces will be met repeatedly as the chapters unfold.

I take as my starting point the deliberations and actions of central government and Parliament, the executive and the legislature. In legislative terms, Parliament can pass public acts, which change the general law, as well as private acts, also known as local and personal acts, which affect the powers of individual groups such as local authorities or private companies. These might encompass services such as public health provision in specific towns, or the development of individual railways. Local authorities could also acquire powers under public acts. They could incorporate clauses from public clauses acts—for example, to provide gas services or to acquire land, and adopt permissive clauses or general powers, such as making bylaws, under specific public acts.

This book focuses on central government and administration, and hence on public acts affecting the nation as a whole, or at least England and Wales. The scrutiny of public bills by select committees of Parliament prior to legislation was thorough, generally involved a range of expert advice, and had parliamentary confidence. Private bills, by contrast, received no independent expert evidence on technical aspects.[19] Instead, they were contested by affected local interests, who might commission experts to bolster their opposing cases. Most reports on private bills were unpublished.[20] The nature and operation of private acts inevitably informed national discussion and legislation, especially in areas such as infrastructure, transport, and public health.[21] The endemic tension between the local and the national led to iterative processes throughout the century, in which both national and local systems of government and administration evolved together. While acknowledging that, I concentrate here on the national perspective.

The reports of select committees on public bills, of royal commissions, of boards and commissioners, and of committees of central government departments, form the primary evidence base for this book. These reports were generally published, and in the case of select committee and royal commission reports in particular, they contain effectively verbatim transcripts of all the oral evidence. Allowing for the fact that these are constrained by formal processes, they nevertheless enable the authentic voices of parliamentarians and experts to be heard. It is this national level that is the major locus of diverse and independent scientific advice even if, in some policy areas such as public health, local initiatives preceded and strongly influenced later central government action. A detailed analysis of expert advice to committees scrutinizing local bills, and to local authorities themselves, would make an interesting complement to this study.

I have consulted more than seven hundred Parliamentary Papers, and list those I have referenced in a separate bibliography at the end of the book. These include reports of committees and commissions, and the evidence presented to them, as well as administrative papers. In making such extensive use of government publications, published and often sold in the "blue books" that avalanched out of the government's printers, I am following the lead of so many members of the Victorian public.[22] These publications were widely read and quoted, in newspapers, periodicals, and by private individuals, as were parliamentary speeches, captured also in *Hansard*. They were a means of informing the public and stimulating debate, as well as of influencing legislation or of improving administration, and for parking politically difficult problems.

Select committees and royal commissions prove particularly important.[23] Select committees, of either the House of Commons or the House of Lords, are established by Parliament. Their membership consists solely of parliamentarians, and their role is to scrutinise bills, or other matters that Parliament considers important. Witnesses, including technical experts, can be invited to give evidence under questioning, and the committee presents this evidence and its report to Parliament. Royal commissions, by contrast, are tools of government designed to explore particular issues in detail, often over many months or even years. Their purposes may include making recommendations on legislative or administrative changes, or kicking difficult issues into the long grass. The memberships of committees and commissions are chosen under political pressures to advance particular agendas. Those invited to give evidence may be selected to put forward specific views, or to buttress arguments that members wish to advance. Equally, witnesses to committees and commissions would know that they could put their views on the record to quote them later with the authority of a parliamentary paper. Their views would need to be acceptable, or at least to represent substantial and unavoidable opinion, if they were to be heard at all. Radical views threatening the political order would find purchase difficult. It is unsurprising to find that so many influential advisers were moderate in their politics, broadly sharing the agenda of the ruling class. The radical socialists, Chartists, and trade unionists, with limited representation in Parliament, had to make their views known through other means such as pamphlets, protests, and strikes.

The treatment here is in the British context rather than by comparison with other countries, although reference is made at times to international issues and constraints. The position of the scientific community in Britain, in relation to government, was different to that in places such as France, Germany, and the United States, as the political systems were so distinct. Compared to forms of the state in, for example, autocratic Prussia and revolutionary "statist" France, Britain was more politically liberal and less centralized.[24] The revolutionary upheavals on the continent and in the United States, the formation of countries such as Italy and Germany in the second half of the century, and the different national scientific cultures, give unique contexts to policymaking. Educational institutions were predominantly private in Britain, unlike the case in France and Germany. Many influential members of the British governing class shared a broad scientific interest and outlook, exemplified by their membership of the independent Royal Society of London for Improving Natural Knowledge (see chapter 2). There was therefore a receptive culture toward scientific expertise in general, which lasted throughout the nineteenth century under a stable political culture that avoided revolution. In France, the Académie des Sciences and the major educational institutions were part of the state.[25] Scientific figures were prominent advisers to King Louis XVI prior to the French Revolution. They were also members of the elite under Napoleon Bonaparte, either in government or as trusted advisers, but within a highly centralized system that persisted.[26] Later still, during the various French political upheavals, some became ministers. It has been argued that scientists in Germany, who tended to come from lower social backgrounds than in France or Britain,[27] had little influence on government before the end of the nineteenth century,[28] although advice was provided within individual cities and states. German universities, and later research institutes, had a research orientation tied to the emerging state.[29] Academic science, rather than applied research, was much later to take off in the United States compared to Europe.

The importance of individual states in the United States, with some similarities to the federal system in Germany, colors the context within which advice could be offered and influential. Comparative studies have been made in some policy areas, such as public health,[30] and in terms of approaches to higher education and scientific research.[31] A broader comparative analysis of how different countries accessed and used scientific advice for military and civil policy would be of great interest, but is not attempted here. Britain's unique features are examined in their own context.

GOVERNMENT GROWTH

The period chosen for this study is broadly from the end of the Napoleonic Wars (1803–1815) to the end of the century, with glances back to eighteenth-century roots and glimpses forward to the twentieth century. That choice follows from analysis that has identified three broad phases of government growth in the nineteenth century. These provide useful markers when reading the subsequent chapters, even if the growth looks in many respects more like a continuum.[32]

The first period runs from around 1815 to the Reform Act 1832. This overlaps with the presidency of Joseph Banks at the Royal Society until 1820, a time during which elite science was primarily the domain of gentlemen of independent means, like Banks.

The second period runs from the 1830s until around 1870. From the 1830s, reflecting a widespread move for reform, there was a dramatic increase in the number of parliamentary select committees, royal commissions, and consequential legislation across many areas of social policy. In addition, the line between political minister and permanent official became more distinct after 1830 than before.[33] This mid-century period, straddling the Crimean War (1853–1856), saw the founding of the British Association for the Advancement of Science,[34] a reformed Royal Society,[35] consolidation in the medical community, and the formation of the Social Science Association.[36] The men of science and other experts became increasingly involved with the reforming activities of government, with regulation, and with a civil service that was starting to become professionalized. The later mid-century years, after the initial burst of reform in the 1830s and 1840s, are sometimes known as the "age of equipoise." From 1852 and the coalition government of the fourth Earl of Aberdeen to the 1860s, there was a period of relative social calm and balance.[37]

The third period runs from the 1870s until the end of Queen Victoria's reign in 1901. The Reform Act 1867, brought in by the Conservative fourteenth Earl of Derby and Benjamin Disraeli, roughly doubled the electorate. It heralded William Gladstone's first Liberal government and a further burst of royal commissions and legislative activity in social policy. Legislation from around 1870, by both Liberal and Conservative governments, appears more consolidating than that of previous decades when so many major changes were introduced. The civil service became larger, more bureaucratic, and more capable. Gladstone's government of 1884 completed the electoral legislation of the nineteenth century by bringing in the Representation of the People Act 1884, which increased the franchise to the majority of men but still excluded women.

As the nature and organization of the state changed, so did the manner in which members of the scientific community projected and justified their activities in the public domain, especially to government.[38] Through much of the century, most scientific research was seen to be private and voluntary, independent of the state in the manner of private businesses. Exceptions were areas of direct relevance to state interests, such as navigation, surveying, and mapping, for domestic benefit and colonial expansion, and the armed forces. From the 1840s to the 1870s there was a move by a vocal minority of practitioners to create a professional scientific community and to establish cultural authority for a naturalistic approach to science separate from religious or other traditional authorities. Left to their own researches, the men of science would, they argued, make new discoveries that would lead to countless applications to benefit society. Science, which was a marginal activity in the early part of the nineteenth century, had become a part of general intellectual culture by mid-century. That created its tensions, in particular with respect to arguments about who could speak with scientific authority.[39]

From the 1870s, when the British government still neither funded fundamental scientific research to any significant extent, nor wholeheartedly supported science and technology education, the rhetoric of the scientific community changed. That coincided with a growing and separate scientific culture among its practitioners.[40] Now, with scientific knowledge increasingly specialized, the arguments of those promoting science shifted toward science in the service of the state, emphasising military security and economic growth.

The organization of political parties, of government, and of public institutions had developed further by this point. The political parties were establishing policymaking capabilities, changing the relationship of the government and Parliament to individuals or organizations that sought to influence policy. The civil service directly employed an increasing number of staff with technical knowledge, although political nomination was still the normal route into the civil service in the 1850s.[41] Change was accentuated by the Northcote-Trevelyan Report of 1854,[42] leading to examinations and open competition, but that took several decades to embed. The concept of the neutral civil servant slowly crystallized by the late century, consigning the operations of the likes of Edwin Chadwick and John Simon to the past.[43] In these last decades more formal structures, a larger government, and the growth of professional scientific, engineering, and medical disciplines and roles laid the foundation for the organizational and institutional changes in relation to science that took place in the early years of the twentieth century.

This book is a study of the interaction of scientific expertise with politics and administration. While technical advice might on the face of it seem paramount in policy areas such as weights and measures, vaccination, or the railways, in reality, questions including the extent of local determination or central control, individual rights, religious beliefs, and economic values set the social and political context within which technical aspects were debated and contested. I aim to examine the interconnections and mutual influences, to bring them to view, and to expose some of the underlying contingencies. Things happened as they did, but they might have been different. Many of these possible counterfactuals are concerned with the personal characteristics of individual advisers and the fortunes of politics, as ministers moved and governments changed or fell at unpredictable moments. Complex interactions of different scientific, commercial, religious, political, and personal interests and beliefs shaped the outcomes in each policy area. But exactly what happened when is contingent on many factors. I have attempted to bring these contingencies to light, and to place them in the context of broader trends across the nineteenth century.

PART I

THE RISE OF SCIENCE

1

FOUNDATIONS

ANY STARTING POINT FOR A HISTORY OF SCIENTIFIC ADVICE TO THE British state will be arbitrary. The founding in 1660 of the Royal Society of London for the Improvement of Natural Knowledge, generally known as the Royal Society, is as good a place as any to begin. That is because this organization, surviving today as Britain's elite scientific institution, had strong links to the state from the outset.[1] Both institutionally and through the advice of its individual fellows it played a central role throughout the nineteenth century, and it included engineers and medical men among its fellowship. Physicians and surgeons made up more than half the scientific fellowship in 1800, reducing to about a third by 1860. Engineers, surveyors, hydrographers, and instrument makers made up around 10 percent in this period.[2]

Most of the scientific advisers mentioned in this book were fellows of the Royal Society. Given its charter by King Charles II, the Royal Society's early membership included politicians, diplomats, and government officials, alongside the savants, the emerging cadre of natural philosophers such as Robert Boyle and Isaac Newton.[3] It was a body that brought aristocrats with an interest in knowledge about the natural world into contact with the savants who shared those interests. Even so, examples of substantial policy influence in the early days are limited. The system of administration in the king's name, a minimal state, lacked the demand or capacity for seeking or absorbing extensive advice.

The armed forces, and particularly the Admiralty, were an exception.

Samuel Pepys, the reforming secretary of the Admiralty, was elected a fellow in 1665 and was president from 1684 to 1686. He helped persuade Charles II to establish the Royal Mathematical School in 1673 and the Royal Observatory at Greenwich in 1675. Both were aimed at supporting the Royal Navy with astronomical charts for navigation and scientifically trained recruits, respectively. The Royal Observatory was the first scientific institution established by the British state. A generation later it was Newton, as president of the Royal Society in 1714, whose advice to Parliament was significant in establishing the Board of Longitude (see chapter 3).

DRIVERS OF CHANGE

The philosophy that drove the explorations of the savants of the eighteenth century was strongly influenced by Francis Bacon's ideas of the previous century.[4] The Baconian method, based on the premise that to command nature one must understand and obey her laws, was to accumulate and test knowledge, by experiment and the generation of general theories, from which material improvements for humanity were expected to follow. It proved difficult, however, to put this knowledge into practical use. Indeed many improvements in technologies required no new natural philosophy, although this did not stop people subscribing to Baconian rhetoric. Instead, the focus of the Royal Society moved toward natural philosophical investigations alone, and the discovery of universal laws. The applications of this knowledge were entrusted to others. That left a gap for a means of collecting and sharing knowledge about inventions, arts, and manufactures.

Following initiatives in Scotland and Ireland, an organization was established in England to encourage the development of improvements based on science. The impetus came from William Shipley, an artist and collector of natural objects and curiosities.[5] He was concerned that private interests were increasingly detrimental to the public good. The upper classes sought to defend their wealth and position, often wasting money on gambling and luxuries, while the lower classes were left in poverty and hunger. The rising manufacturing class guarded its inventions through commercial secrecy and patenting, driving up prices in a manner that, as Shipley saw it, effectively conspired against the public. His solution was the creation in 1754 of the Society for the Encouragement of Arts, Manufactures and Commerce, known as the Society of Arts.

The organization still exists, now known as the RSA. Shipley promoted the idea of medals and prizes as rewards to support improvements of social benefit. Prizes were given to women as well as men, and prize-giving ceremonies were attended by a mostly female audience.[6] Upper-class women also made up a substantial proportion of the audiences at the Royal Institution of Great Britain from its founding in 1799. They helped integrate science, and particularly chemistry, into fashionable culture and in the service of the country.[7]

Although the Society of Arts soon established specialist groups on topics such as agriculture, chemistry, mechanics, and manufactures, the emphasis was on the encouragement of practical improvements. It did occasionally seek legislation in Parliament. For example, members who were MPs introduced legislation that reduced the price of imported madder for the dyeing and calico printing industry, boosting production and benefiting the economy.[8]

Toward the end of the eighteenth century, having suffered defeat in the American Revolutionary War in 1783, and facing the consequences of the French Revolution (1789–1799), of growing industrialization and of more complex trade, public administration in Britain started to become more extensive and centralized. This continued throughout the Napoleonic Wars, leaving government by 1815 larger, more organized, and gradually losing its ethos of personal patronage and sinecure positions.[9] The nascent more professional civil service was emerging; the term *civil service* first appears in a Treasury letter of 1816.[10] Whereas senior administrative appointments had been political and based on patronage, they became increasingly depoliticized during the government reorganizations of the period in response to fighting the wars of the day.

JOSEPH BANKS

The colossal scientific figure in this period was Joseph Banks. As a natural historian, Banks took part in James Cook's voyage aboard the HMS *Endeavour* to Australia and New Zealand from 1768 to 1771. On his return, his relationship with King George III led to his appointment as effective director of the Royal Botanic Gardens at Kew in 1773. Through the position of Kew in a network of botanical gardens worldwide, Banks sought to develop botany and agriculture to the economic benefit of Britain.[11] Banks was soon an influential figure at the Royal Society, becoming president in 1778 until his death in 1820. He is a clear forerunner of all that came afterward, both in his primary interest of using science to promote Britain's interests abroad, and in his contributions to matters of internal policy.

Banks came from a landed background with strong mining interests, giving him status and conduits within the gentlemanly ruling elite. This differentiates him from most of the scientific advisers who will be encountered later, who have predominantly middle-class origins. His position reflects an eighteenth-century political culture that was oligarchical, based on voluntary public service, and in which success depended on patronage. A further difference from most later scientific advisers is that Banks's advice was wide-ranging, across different policy areas and government branches. In a sense he acted like a "chief scientific adviser" to government. Indeed Lord Auckland, perhaps regarding this state of affairs as somewhat continental, referred to Banks in 1791 as "His Majesty's Ministre des affaires philosophiques."[12] The formal post of chief scientific adviser to the government, reporting to the prime minister, would not be established until 1964.

Banks supervised the expenditure of the royal grant of £3,000 for the geodetic survey of England between 1783 and 1787, giving him contacts with the army and the master-general of the Ordnance. As president of the Royal Society, he held ex officio positions at the Board of Longitude and at the Royal Observatory.

Banks's connection with Lord Hawkesbury, later the first Earl of Liverpool, was significant for his rise, starting in 1787 when Hawkesbury was president of the Privy Council Committee for Trade and Plantations. Hawkesbury was appointed to the committee at the outset and to its presidency in 1786. He was advised by Banks on the plan to transfer breadfruit from Tahiti to the West Indies, which became the infamous *Bounty* voyage. Banks's status was recognized when he was appointed a privy councillor in 1797, joining the Committee for Trade and the Committee for Coinage.

Government departments sought his advice: from the Excise Office in 1791 on the proportion of duty to be paid on alcoholic drinks; from the Victualing Office of the Navy in 1809 on the effectiveness of iron water storage vessels at sea; and from the home secretary in 1814 on the safety of the Gas Light Company's reservoirs. The Committee for Coinage sought Banks's knowledge and connections with respect to problems of the wear of metals and the production of coins. For the latter, Banks drew in the engineer and industrialist Matthew Boulton to help. Banks also received many general requests for advice, including specifying species of cotton and tea that might grow in British lands; dealing with the threat of Hessian fly infection in wheat; means of producing saltpeter for the manufacture of gunpowder, alkalis, and extracting gold; and growing and supplying wheat, cinnamon, and hemp. Much of this was concerned with advancing military causes and with the colonial trade, using advice to help secure useful raw materials for home benefit. It meshed with Banks's desire as a British landowner to see, on the one hand, colonial territories used to supply Britain and, on the other, the promotion of self-sufficiency and the protection of the interests of British landowners from foreign imports.

Banks was no disinterested adviser, but an advocate for particular policies and for the use of science to advance them.

TOWARD THE NINETEENTH CENTURY

While he was president of the Royal Society, Banks was instrumental in founding the Royal Institution of Great Britain in 1799. Like the Royal Society, the Royal Institution has survived into the twenty-first century, and is still in its original home in London. In response to declining access to Continental markets during the recent wars, the original aim of the Royal Institution was to provide scientific and technical knowledge to a lay audience through lectures, to encourage innovation for agricultural and industrial self-sufficiency. Or, as the official version ran, "a Public Institution for diffusing the Knowledge and facilitating the general Introduction of useful Mechanical Inventions and Improvements, and for the teaching by courses of Philosophical Lectures and Experiments, the application of Science to the Common Purposes of Life."[13] The aim was to promote the practical value and the use of science. An explicit public policy remit was not part of the plan, even if Banks himself was active in political circles.

The Royal Institution's most significant early figure, Humphry Davy, was appointed professor of chemistry in 1802. He introduced scientific research into the organization, an activity that flourished there throughout the nineteenth century alongside the lectures. But he also ventured into the policy arena, particularly after becoming president of the Royal Society in 1820. Several of Davy's successors at the Royal Institution contributed scientific advice to government during the nineteenth century and feature later, among them Thomas Brande, Michael Faraday, John Tyndall, and Edward Frankland. Politicians would attend Royal Institution discourses on occasions, offering opportunities for private conversations and lobbying. However, Davy's advice to the navy of means for preventing the corrosion of hulls, based on his research on electrochemical action, was not an unqualified success (see chapter 3). Despite this sort of setback, the Baconian ideology retained an influence after the end of the Napoleonic Wars in 1815. Indeed, the early nineteenth century was a time of extensive discussion about the place and purpose of science and its practitioners in society. Davy, and many others, argued for the practical and theoretical importance of science, and thought that both aspects should be more widely understood and supported.[14]

Banks's death in 1820 can be taken to mark a transition between eighteenth-century modes of institutional governance and politics and the nineteenth-century approaches that form the subject of this book. In personal terms, influence and power was starting to pass to those with demonstrable knowledge rather than connection by birth. Reform was underway in both science and politics. Important changes to scientific institutions are often traced to the stimulus of Charles Babbage's polemic *Reflections on the Decline of Science and Some of Its Causes*, published in 1830. Babbage's attack was primarily a critique of the Royal Society, which he saw as failing to improve British science because its fellowship contained so many nonscientific figures. Although it had no immediate impact on the Royal Society, its arguments were instrumental in the founding of the British Association for the Advancement of Science in 1831. The British Association (described in the next chapter) soon entered the policy field alongside an eventually reformed Royal Society. Reform of the Royal Society had to wait until 1847, at which point fellowship was largely restricted to men of scientific achievement. Women were not elected to fellowship until 1945.

Several other scientific bodies were formed in the early nineteenth century, and these all helped buttress the credibility and authority of individual advisers. Preceding the British Association was the Geological Society in 1807, and the Astronomical Society of London in 1820, which became the Royal Astronomical Society in 1831. In medicine, the Royal College of Physicians had been founded in 1518, and the Royal College of Surgeons of London, formed in 1800, could trace its origins to forerunners in the fourteenth century. The engineers did not arrive substantively on the scene until the formation of the Institution of Civil Engineers in 1818, a reflection of the impact of industrialization and the building of canals, roads, and bridges. The purpose of the professional institutions was to mark out territory, to control the ability to practise through qualification and disciplinary processes, to protect their members' income and authority, and to share developing knowledge. They sought to influence the government for their own ends and to contribute their expertise to addressing the issues of the day.[15] In its turn, the government looked to them for advice, and from time to time would be stimulated to intervene to impose education, training, and qualification requirements, in a constant tension between professional autonomy and national regulation.

One further significant institution founded in this period was the Athenaeum, a private gentlemen's club formed in 1824 to attract the intellectual elite from the sciences, the arts, religion, law, and politics.[16] Women were not admitted as members until 2002. Banks's protégé, Davy, was the first chairman, and Faraday, who had been Davy's assistant, was the first secretary. Most of the scientific advisers featured in this book were members, and the club offered a place within which they could interact with politicians, civil servants, and other important social figures. Science was an enterprise of linked individuals, and those individuals were linked to others in their social circles, including in Parliament and government.[17]

This independent enterprise, ingrained in British culture, contrasts with the more centralized, state-supported activities and academies developing in Germany and France at the time. The British approach embraced voluntarism and self-help rather than state control. The role of the state was to remove barriers rather than to force particular action, although this would gradually change as dire social conditions became more manifest. Against this background there was no obvious need for state support for science, except where it was immediately useful for state purposes. Even in 1851 George Airy, as president of the British Association, and despite being an employee of the state, could hope that "in all cases the initiative of Science will be left to individuals or to independent associations." It was a prevalent view that the original researches of the practitioners of science should be unrestricted by government or institutional constraints, although Airy did propose that "when any branch of Science has been put in such a form that it admits of continuous improvement under a continued administrative routine, that administration should be undertaken by the Government."[18] At the Royal Observatory, he presided over a magnificent administrative machine funded by the Admiralty.

The next chapter outlines the development of significant scientific institutions. Institutional developments in engineering are introduced in chapter 7, and in medicine in chapter 12. Together, they set the backdrop for understanding the ways by which specialist advice would be shaped and communicated to influence policy.

2

THE ROYAL SOCIETY AND THE BRITISH ASSOCIATION

THROUGHOUT THE NINETEENTH CENTURY, THE ROYAL SOCIETY OF LONdon for Improving Natural Knowledge was a lynchpin of scientific advice to the state, as an institution and through the individual activities of its fellows. Its connections with the British government and Parliament were strong. At least 149 people who were MPs during the nineteenth century were fellows.[1]

As the century opened, Joseph Banks had been president since 1778 and would remain so until his death in 1820. Running through the early decades of the nineteenth century, and not resolved until 1847, was the tension between the fellows who were of a scientific mind, in that their interests and activities lay in the natural sciences, and those with broader interests in areas such as antiquities. The men with scientific expertise were a minority, making up less than a third of the fellowship in 1800 and a similar proportion of the council. Of the 149 scientific fellows in 1800, forty-four were mathematicians and physical scientists, eight were biological scientists, eighty-four were medical, and thirteen had engineering or technical expertise.[2] Not until about 1820 did the scientific fellows achieve a majority on the council, and not until 1860 in the fellowship as a whole. The majority in the earlier period tended to be richer and of higher social status. That set the image of the society, and was important financially. But it ran counter to the views of the reformers among the scientific fellows. They envisaged an organization of which membership was an assurance of serious scientific capability, and hence that the Royal Society could be seen unambiguously as the nation's premier scientific body, independent of government but the natural body to be consulted.[3] It would take several decades for the Royal Society to move to a position in which most fellows were elected on scientific merit alone.

In November 1820, after a brief interregnum by William Wollaston, and although opposed by figures such as John Herschel and Charles Babbage, Humphry Davy was elected president. Famous for his researches into electro-chemistry, and for his design of a miners' safety lamp, he had been knighted in 1812, the first man of science to receive the honor since Isaac Newton, and created a baronet in 1818, the first chemist to be so honored.[4]

Davy sought to continue where Banks had left off, seeking a closer relation-ship between the Royal Society and government. During his presidency, sev-eral Royal Society committees were established to advise government. Davy also played a central role in a committee established by the government in 1817 to advise on whether Dr. Sickler should be paid for his work in devel-oping a process for unrolling and deciphering the Herculaneum papyri, an initiative instigated by the prince regent. The committee found his method to be useless.[5]

Wollaston and Davy turned the Royal Society in the reformers' direc-tion, but Davy's character did not help. While his scientific credentials were respected, he was seen as arrogant and impatient, in many respects a baronet in the mold of Banks, although without the same tact. He was not able to prevail over the conservative elements among the fellows.[6] However, he did want to be seen to move the Royal Society away from criticism about patron-age, which would help him to strengthen its relationships with government. Change came gradually to the governance of the Royal Society. By 1824, both secretaries were men of science, and from 1827 one was expert in biological science and one in physical science, a situation that pertains to this day. In parallel, Davy and John Croker, secretary to the Admiralty, founded the Ath-enaeum Club. That offered a place in which the practitioners of science could continue to meet men of wealth and power, as they had in the Royal Society.[7]

The Royal Society and its fellows were repeatedly called upon to offer advice to different parts of government. Some were requests for which the Royal Society would establish a committee. Others were approaches to indi-vidual fellows, such as the Treasury request to James South in 1836 to report on the state of optical glassmaking.[8] The responsibilities undertaken by fellows included responding to direct commissions and requests, acting as commis-sioners for particular inquiries, and giving evidence to select committees and royal commissions. Although many other individuals proffered advice out-side the aegis of the Royal Society, the scale and scope of the Royal Society's involvement is plain to see. The emerging scientific community was strongly engaged with the state's institutions. The most active and influential members formed a small group, especially in the first half of the century. They were mostly fellows of the Royal Society, knew each other well, and their personal networks extended into the organs of state and the political class, many of whom were also fellows. The government departments with which they interacted included the Admiralty, the Board of Ordnance, the War Office, the Home Office, the Board of Trade, and the Treasury. This linkage into policy and practice is exemplified by the fact that most of the committees established by the Royal Society in this period, particularly in the 1820s during Davy's presidency, addressed newly emergent practical issues. Examples of those established include the Coal Tar on Ships Committee in 1822, the Meteorological Instruments Committee in 1822, the Copper Sheathing of Ships Committee in 1823, the Lightning Conductors on Ships Committee in 1823, the London Bridge Granite Committee in 1823, the Optical Glass Committee in 1824, the Prevention of Contagion Committee in 1824, the Gas Establishments Committee in 1825, the Lightning Conductors Committee in 1827, the Babbage Engine Committees of 1822 and 1829, and the Thames Levelling Commission in 1830. These intimate connections with governments of the day are reflected in public honors bestowed on them in recognition of services, many by Robert Peel. South received a knighthood in 1830, when he threatened to emigrate to France. David Brewster was knighted in 1831, the year in which he was involved in founding the British Association, as was Herschel.

ROBERT PEEL'S PATRONAGE OF SCIENCE

One personal connection of Davy was with Robert Peel. Peel was elected a fellow of the Royal Society in 1822, when he became home secretary. Despite his keen interest in scientific ideas, and his belief in the Baconian ideology of the potential of scientific knowledge to improve the human condition, he did not believe in the role of the state to promote scientific projects. These were primarily matters for private initiative and capital.[9] While that culture remained throughout the nineteenth century, there were significant exceptions. They include voyages of exploration, the funding of national institutions important to the state such as the Royal Observatory at Greenwich, and some commissioned research in areas like public and animal health as the century advanced. Initiatives like these supported practical state objectives of colonialization, trade, and the improvement of social conditions.

What Peel did do was to recognise individuals through the awarding of honors and pensions. His first significant act was to recommend to King George IV in 1825 that he create Royal Medals, to be awarded by the Royal Society, to "reward and encourage men distinguished by their literary talents and scientific attainments."[10] The medals are still awarded annually, but have acted primarily as a means of recognition rather than an incentive.[11]

Peel's short term as prime minister in 1834 and 1835 saw a baronetcy for John Barrow, the Admiralty figure who had promoted the extensive voyages of Arctic discovery. Peel also awarded civil list pensions to George Airy, just appointed Astronomer Royal in 1835; Mary Somerville, the mathematician and scientific author; and Michael Faraday. Pensions were mostly a reward for military, diplomatic, or political service, but Peel suggested that scientific and literary pursuits should be recognized too.[12]

Peel's second term as prime minister, 1841–1846, saw more pensions and honors bestowed. He awarded pensions to people, including James Forbes, Richard Owen, and William Rowan Hamilton.[13] Roderick Murchison received a knighthood in 1846.

Peel's principles, to reward merit and recognize scientific service, were contested politically by those who regarded the proper role of pensions as relieving financial distress, essentially applying a means test. That argument continued throughout the century. But by then Peel was dead, following a fall from his horse in 1850. He did not live to see the Great Exhibition of 1851, which his exertions with Prince Albert had done so much to bring about, and Britain lost one of its few leading politicians with a strong commitment to science.

DAVIES GILBERT, SCIENTIFIC PARLIAMENTARIAN

When Davy resigned as president due to ill health in 1827, he was succeeded by the treasurer Davies Gilbert. Over the course of the nineteenth century, only a handful of people who could be described as practitioners of science served as members of Parliament. Fewer still held ministerial responsibilities. In the first three decades the major figure is the Cornishman Gilbert, who sat as an MP between 1804 and 1832. He was a landed gentleman, with some underlying but suppressed reforming inclinations, traceable to his youth.

Gilbert was a strong supporter of scientific and engineering activity. His two most notable protégés were Davy and the engineer Richard Trevithick. Gilbert's interest and knowledge in mathematics and engineering, and his involvement in the Cornish mining industry, led to his connection with Trevithick. Over many years he sought to support Trevithick, his inventions, and his family.[14] As a landowner, Gilbert had strong agricultural interests. In 1808 he became a member of the Board of Agriculture (see chapter 5). He was one of just five members present when it was dissolved in 1821.[15]

As a dedicated parliamentarian, Gilbert served on committees, including finance and the Poor Laws, where his social beliefs and principles were in evidence. For example, he was a member of the select committee on the Poor Laws in 1817, with its widely held assumption that the poor rate tended to diminish the "natural impulse by which men are instigated to industry," effectively creating a class of undeserving poor dependent on the rate, and increasing the financial burden on the righteous. It was also said to increase the very evil of poverty the law was intended to counter,[16] as Robert Malthus had argued in 1798 in his pamphlet *Principles of Population*. Gilbert believed in a natural social order that it was the job of Parliament to reflect and help preserve. Nevertheless, he shared a common view that improvements to social conditions could be made through advances in science and engineering.

Gilbert's involvements with scientific and engineering applications started early in his parliamentary career. In 1805 he sat on the committee to consider the Thames Archway project to build a tunnel from Rotherhithe to Lime-house. The tunneling was beset by problems, and Gilbert was instrumental in recommending Trevithick to complete the project.[17] Trevithick failed just two hundred feet short, as water flooded the shaft. A tunnel under the Thames was not built until Isambard Kingdom Brunel completed one in 1843. Gilbert's engineering knowledge was also influential in the construction of the Menai Suspension Bridge. In 1824 he sat on the royal commission for the road from London to Holyhead,[18] which included examining plans for the bridges, and his most significant contribution was a result of his mathematical and engineering ability. Based on his understanding of the catenary curve, the shape a chain adopts when suspended between two points, he suggested to Thomas Telford, the engineer for the bridge, that he should raise the height of the towers and increase the length of the chains.

Gilbert sat on several committees of the Royal Society—for example, on the Coal Tar for Ships Committee at the request of the Admiralty in 1822, and on the Optical Glass Committee, which ran from 1824 to 1828 (see chapter 3), and he chaired the Admiralty-funded Thames Levelling Commission from 1829 to 1830.[19] He was instrumental in obtaining the initial government funding of £1,500 for Babbage's calculating machine, after approaching Peel. Peel referred it to Croker, and at the request of the Treasury a Royal Society committee was established which recommended funding.[20]

Two parliamentary enquiries demonstrate the close connection between Gilbert as a parliamentarian and as a fellow of the Royal Society. The first, the Select Committee on Weights and Measures in 1814, eventually culminated in the Weights and Measures Act 1824 (see chapter 14). The second was the Royal Commission for Inquiring into the Mode of Preventing the Forgery of Bank Notes.[21] The commission was chaired by Banks, and also included the technical and chemical knowledge of William Congreve, Charles Hatchett, and Wollaston among its members. The commissioners examined 108 schemes to counter forgery, and backed a method by Messrs Applegarth and Cowper, which the Bank of England had introduced just before the formation of the commission.[22] Gilbert was also a member of the Select Committee on Steam

Carriages in 1831 (see chapter 7). When he left the House of Commons in 1832 he was the last significant scientific expert to serve as an MP until Lyon Playfair in 1868.

TOWARD REFORM

As Gilbert's presidency drew to a close in 1830, a storm hit the Royal Society, and the forces of conservatism and reform were engaged in contest. Babbage's *Reflections on the Decline of Science and Some of Its Causes* was joined the same year by Augustus Bozzi Granville's *Science without a Head: Or the Royal Society, Dissected*, and South's *Charges against the President and Councils of the Royal Society.* Men like Herschel and Babbage considered that science was in decline because the Royal Society was neither exclusively scientific nor a government department.[23] They believed that the government was not supporting or rewarding science and its practitioners in the manner that they saw on the continent, although they themselves were independently wealthy and did not think that government should interfere in their researches. They wanted the removal of patronage for scientific positions. After a fraught election contested between Herschel and Prince Augustus Frederick, Duke of Sussex, sixth son of King George III, Sussex was elected president. His emollient style aided gradual reform. Peace was made with the astronomers, who had broken away in 1820 and criticized the Royal Society ever since. After the Astronomical Society gained its royal charter in 1831, it was recognized as a possible advisory body to government.[24] It took over supervision of the *Nautical Almanac* and joined the Royal Society in providing people to serve on the Board of Visitors to the Royal Observatory, its governing body. Specialist scientific societies, as they developed, would be important conduits of advice, although the Royal Society managed to retain pole position. Significant also in 1831 was the creation of the British Association for the Advancement of Science. Although founded primarily by men outside the circle of fellows of the Royal Society, it was not long before the fellows flexed their muscles and effectively took it over. The scientific community, in the shape of these two institutions in particular, bound together more strongly as it grew, and as distinct disciplines emerged. This stronger structure and organization was able to match and work with the growth of government as the century advanced. It underpinned the credibility and authority of individual advisers as they battled to influence the politicians.

When the Duke of Sussex resigned in 1838 he was succeeded by another gently reforming aristocrat, although one with greater scientific interests. This was the Marquess of Northampton, who held the presidency until 1848, through the last stages of reform. Northampton had a particular interest in geology, and was active in the Geological Society. He had been visible in his support of the British Association in the 1830s, which he would serve as president in 1848. Northampton was succeeded by the third Earl of Rosse. He was a notable astronomer, who discovered and studied spiral nebulae with his own giant telescope, and had been president of the British Association in Cork, Ireland, in 1843. The fact that he was also an aristocrat and member of the House of Lords neatly bridged the scientific, political, and social worlds. One more aristocrat would follow him, and after that the commoners. The savants were now gathering sufficient standing that the Royal Society no longer required an aristocrat at its head to signal its status.[25] At this point, after reforms in 1847, the Royal Society was firmly in the hands of the scientific experts, where it has stayed ever since. There developed nevertheless a tension between those who sought to paint the scientific enterprise as a disinterested seeking after truth and those who wished to make money from commercial activity and consultancy. Toward the end of the nineteenth century an increasing number of commercial men were elected as fellows, despite their lack of high attainment in science itself.[26]

PARLIAMENT AND THE ROYAL SOCIETY

This period of change in the Royal Society during the 1830s and 1840s is reflected in the number of parliamentarians who became fellows. Between 1800 and 1840, more than sixty members of the House of Commons were elected as fellows. Only about half a dozen had any significant scientific expertise and few were senior politicians. After 1840 the picture is different. Elections were less frequent, around forty in the remaining sixty years of the century, but the MPs elected as fellows were mostly senior politicians at the time of their election, including all the prime ministers of the period. They were elected because they shed luster on the Royal Society and created relationships with Parliament and government.

As with the engineers and medical fraternity, the influence of the scientific experts was mostly brought to bear from outside Parliament. Just a handful of significant scientific figures became MPs. In the first half of the century, only Gilbert and William Parsons stand out. Parsons served as an MP from 1821 to 1834, became the third Earl of Rosse in 1841, an Irish representative peer in 1845, and president of the Royal Society from 1848 to 1854. In the second half of the century the major scientific figures in Parliament were the Liberals Lyon Playfair, John Lubbock, and Henry Roscoe, and the Conservatives George Stokes and Nevil Story Maskelyne.[27]

BRITISH ASSOCIATION FOR THE ADVANCEMENT OF SCIENCE

The abolition of the Board of Longitude in 1828 (see chapter 3), and with it a ready channel into government support for exploratory voyages, was one challenge to the Royal Society. A second was the barrage from Babbage. Babbage took aim at many targets. He attacked the education system, founded on classics, claiming that "scientific knowledge scarcely exists among the higher classes of society."[28] He bemoaned the lack of a professional body of people devoted to science, unlike the clerics or lawyers, and unlike the situation he saw in Germany and France. He argued that scarcely anyone could pursue abstract science in Britain, which he thought essential for profitable exploitation but unprofitable in itself, without a private fortune. Babbage saw the Royal Society as failing to improve British science because its fellowship contained so many non-scientific figures. The system relied on personal contacts and patronage, a charge that Babbage laid at the door of the Royal Society in the form of its then-president, Gilbert. Babbage also railed against the opposition of Banks, as late president, to the formation of scientific societies with specific interests. He saw the formation of societies such as the Linnean, Geological, and Astronomical as essential for nurturing what became disciplinary and professional expertise.[29] In so many respects the Royal Society, he argued, was failing in its duty.

Reform of the Royal Society would not come until 1847. Its delay, and the Royal Society's lack of leadership of the scientific enterprise as a whole, was one cause of the formation of the British Association for the Advancement of Science in 1831, the year after Babbage's onslaught. Babbage had attended the meeting of the Gesellschaft Deutscher Naturforscher und Ärzte (Society of German Natural Scientists and Physicians) in Berlin in 1828. This gathering of people from across different areas of science suggested to Babbage the idea of a larger European gathering. That did not come to fruition, but it sowed the seed for a British counterpart. So too did the growth of local scientific and philosophical societies, in towns and cities across the nation, keen to share and discuss the latest findings and ideas. A small group of people took the initiative to establish the British Association for the Advancement of Science, to bring a wide range of people interested in science together.[30] Prime among them was the Reverend William Vernon Harcourt, who took up the idea of a scientific gathering advocated by the natural philosopher David Brewster, prompted by Babbage, and led the organization of the first meeting in York.

The British Association vowed to be a nonmetropolitan institution, moving from town to town each year. The new association settled on the following aim: "To give a stronger impulse and a more systematic direction to scientific inquiry; to promote the intercourse of those who cultivate Science in different parts of the British Empire with one another and with foreign philosophers; to obtain more general attention for the objects of Science and the removal of any disadvantages of a public kind which impede its progress."[31] In other words, it was concerned with promoting the activity of science itself, and it succeeded beyond all expectations over the next few years in putting science on the cultural map. People flocked to the annual meetings, with women attending in significant numbers and occasionally speaking.[32] Newspapers and journals reported the latest discoveries and ideas in detail. Although the initiators came mostly from outside the orbits of London and of the premier universities of Oxford and Cambridge, the eminent figures of the day soon saw the way the wind was blowing and joined in. In London, Roderick Murchison, president of the Geological Society, was an early activist. For Oxford and Cambridge, William Buckland and William Whewell, both friends of Peel, led the way.

The rise of the British Association brought to the fore the relationship between science and religion, with implications for the manner in which science informed policy and politics. The "gentlemen of science" who shaped the early British Association were preponderantly liberal Anglicans. For example, Harcourt, Buckland, and Whewell were ordained ministers. That led to difficulties with the literal interpreters of scripture on the one hand, with their belief contrary to geological evidence that the earth was created in six days, and with high churchmen such as John Henry Newman on the other. They perceived a danger in the ways by which the leaders of the British Association were able to decouple the spheres of authority between science and religion. The men of science, while placing their studies of the "Book of Nature" in the context of their religious beliefs, and seeing both in ultimate harmony, were becoming an authoritative body and a cultural force in their own right.

In the same way in which it managed to steer a course between religious extremes, the British Association pulled off that feat in politics. The underlying philosophy, expressed by Lord Francis Egerton in his presidential address at the 1842 meeting, was that it was "surely desirable that, under any form of government, the collective science of any country should be on the most amicable footing with the depositories of power; free indeed from undue control and interference . . . uncontaminated by the passions and influences with which statesmen have to deal, but enjoying its goodwill and favour, receiving and requesting with usury its assistance on fitting occasions, and organized in such a manner as to afford reference and advice on topics with respect to which they may be required."[33] Science should be separate from politics, even if that would inevitably prove difficult in practice. The association's founders believed in measured reform, rather than extreme democracy, which did not disturb treasured institutions.

Many early figures—such as Brewster, Airy, and presidents Lord Milton and the Marquess of Northampton—were Whigs. Others—such as Buckland, Whewell, and Murchison—were Peelite Conservatives, open to careful reform. Indeed, throughout the century the majority of the scientific experts were liberal in their politics. Perhaps this helped give scientific advisers a fair wind with the predominantly Liberal administrations of so much of the nineteenth century. So, despite the underlying liberal and reforming tendencies of many early advocates, the British Association nevertheless carved out a position in which science, as a search after knowledge, was separate from party politics. The leaders took care to balance the choice of presidents and vice presidents politically as far as possible. Their success is indicated by the position in 1844, when life members who were also members of Parliament were split equally between liberals and reformers on the one hand and conservatives on the other. In addition, the relationship between scientific advice and politics was rarely construed in religious terms. An exception, although it was not part of formal advice, is John Tyndall's public criticism of the government in 1865 for instituting a day of prayer to counter cholera, a challenge to the cultural leadership of the clergy.[34]

THE BRITISH ASSOCIATION AND PARLIAMENT

The British Association was not set up to engage the growing scientific community with issues of public policy, except where they concerned the support and funding of scientific activity and of science education. It was concerned with policy for science, not science to inform policy, and hence its direct activities for science, like those of the Royal Society, are peripheral to the focus of this book. In the first dozen or so years of its existence, up to 1844, the British Association lobbied government on thirty occasions for support.[35] The majority were successful, but they largely concerned money for scientific observations and publications. Most related to astronomy and the earth sciences, but their linkage to practical implementation for state purposes was limited, beyond the fact that many involved persuading government to spend additional money on extending work already administered by state-supported institutions, such as observatories, the Ordnance Survey, and the British Museum. One early success was the establishment of the Mining Record Office in 1838, under Henry De La Beche at the Museum of Economic Geology.[36] Equally telling is the range of suggested lobbies of government by individuals that the British Association did not take forward.[37] Many of these had a more practical policy focus: legislation to conserve salmon in 1832, following a parliamentary enquiry in 1824 and 1825;[38] large-scale experiments in naval architecture in 1832 and 1833;[39] the use of civil servants to collect agricultural data in 1839;[40] and an inquiry into the Mining Record Office in 1844.[41]

A further case in point relates to patent law reform. Requests to extend patents or to gain financial reward from the state required an approach to Parliament, which established a select committee to examine the issues in 1829. Gilbert and Marc Brunel were among those giving evidence, although

the committee failed to come to a conclusion and the inquiry was not continued.[42] Reform was advocated particularly by Babbage, who attempted to
have the question formally discussed at the 1832 meeting in Oxford, soon
after a select committee had advised against extending the patent for Thomas
Morton's slip for shipbuilding.[43] But Harcourt and others declined to take this
forward, so that the association sought no influence on legislation for patents
for two decades.[44]

After the Patent Law Amendment Act 1852, which established the Patent
Office and a simplified system for obtaining patents, a committee was eventually convened in 1854 under William Fairbairn and reported in 1858.[45] Fairbairn would serve a few years later on the royal commission on patent law.[46]
The decision not to pursue many of these possible initiatives exemplifies the
organization's desire not to seek to interfere in political matters other than
support for science itself. The association may have called itself "Her Majesty's
Parliament of Science,"[47] but it had no "loyal Opposition." The name betrays its
constitution as a deliberative body for advancing science by the practitioners
of science themselves. The politics was internal to the scientific community,
although the leaders took care to maintain relations with the scientifically
minded aristocrats who were also important in the Royal Society. It was not
until 1849 that the British Association established a parliamentary committee,
and then its primary role was as a channel to lobby for government support
for science and its practitioners.

This parliamentary committee was formed at the Birmingham meeting in
1849. Membership was open to all parliamentarians. The committee was to
be set up by Lords Northampton, Rosse, and Wrottesley, with the MPs Lord
Adare, Philip Egerton, and Charles Lemon, "to watch over the interests of Science, and to inspect the various measures from time to time introduced into
Parliament likely to affect such interests."[48] Such a large body proved unwieldy
and it was reconstituted in 1851, with only thirteen members, "to watch the
course of legislative measures that may affect the progress of science."[49]

It was Lord Wrottesley who took the leading role, as the committee finally
met in 1852. When he died in 1867, the committee died with him, implying
that it had depended on his initiative. Wrottesley is one of the many forgotten
people of scientific history.[50] An astronomer with his own private observatory
at Wrottesley Hall near Wolverhampton, he was a founding member of the
Astronomical Society and its president from 1841 to 1842. He was president
of the Royal Society from 1854 to 1858, and president of the British Association in 1860. Wrottesley's emphasis in public life, in his own words, was to
encourage the state "to regard with favour and respect, to cherish and foster, to appreciate and reward the labours of the cultivators of science."[51] The

Parliamentary Committee therefore had this focus, only occasionally straying into questions of the influence of scientific knowledge on public policy.

The committee's first report in 1852 advocated better pensions for men of science and, in support of the Royal Society, abolition of the postage tax on international scientific publications.[52] The members lobbied the Earl of Derby and the Earl of Malmesbury respectively, the relevant ministers, without immediate effect. Wrottesley was instrumental, with the support of the Parliamentary Committee, in persuading the government to make accommodation available in the new Burlington House in 1857 for the Royal Society and many other scientific organizations. But it was government support for science, international cooperation, and science education that most occupied the committee. Unlike some in the scientific community, such as Airy, Wrottesley wanted to see a more interventionist role for government.

The committee's report in 1855 proposed a Board of Science to advise government on scientific questions and to distribute government grants. Two members of the committee attempted to promote the idea in Parliament early in 1856, but their attempt was quashed by Prime Minister Viscount Palmerston. The following year, with the support of the Royal Society and other institutions, Wrottesley took a set of twelve resolutions based on the 1855 report directly to Palmerston. They covered science education, support for research, and the proposed Board of Science.[53] Again, the time was not ripe. Palmerston's government fell the next year and the surgeon Benjamin Brodie, with less interventionist views than Wrottesley, succeeded him as president of the Royal Society.

Further representations to government, and a major statement by Alexander Strange at the Norwich meeting in 1868, "On the Necessity for State Interventions to secure the Progress of Physical Science," helped stimulate the establishment in 1870 of the Royal Commission on Scientific Instruction and the Advancement of Science. Known as the Devonshire Commission, it was chaired by the seventh Duke of Devonshire, who had been a member of Wrottesley's parliamentary committee for ten years. This commission produced eight huge reports between 1870 and 1875. They were concerned primarily with science education and scientific institutions, but the final report concerned the advancement of science. It recommended the appointment of a minister of science or of a minister of science and education. That recommendation too fell on fallow ground and a minister of science, by that title at least, was not appointed until 1959.

The Parliamentary Committee did play a role, along with the Royal Society, in persuading the government into one particular commitment. The Admiralty had maintained the Hydrographic Office since 1795, which was responsible for coastal surveys, charts for navigation and, from 1832, for publishing tide tables.[54] In 1853 the committee had recommended a plan of the American lieutenant Matthew Maury for taking hydrographical and meteorological measurements at sea. Although James Graham, the first lord of the Admiralty, had given his support to navy ships taking part, he balked at recommending the creation of an office to coordinate the data.[55] But the following year Wrottesley was able to report that the government had now agreed to establish a new Meteorological Department in the Board of Trade, headed by Robert Fitzroy.[56] His position was unusual in that half his salary of £600 came from the Admiralty and half from the Board of Trade. Fitzroy produced his first report in 1857 for the department that eventually became the Meteorological Office, the United Kingdom's national weather service.[57] In 1861 Fitzroy's department produced the first weather forecasts, which were printed in newspapers and became highly contentious.[58]

Senior political figures engaged with the British Association. For example, Peel held a social gathering before the Birmingham meeting in 1839,[59] as his house, Drayton Manor, was nearby. Peel's name was discussed as a possible president or vice president on two occasions, though he declined to be a vice president in 1839.[60] Thomas Spring Rice was at the meeting too, and at the subsequent event in Glasgow in 1840 where, as Lord Monteagle, he gave a toast to "The Commercial and Manufacturing interests of the Country, which owe so much to Science for their advancement."[61] Spring Rice, educated at Trinity College Cambridge, was a friend of Whewell and George Peacock, two of the prime movers in the association, which was a useful connection when he was chancellor of the exchequer from 1835 to 1839 during Viscount Melbourne's Whig government.[62]

THE BRITISH ASSOCIATION'S SCIENTIFIC REPORTS

The British Association was a pressure group for science, and for the interests of its leaders. The desire to further scientific activity, and not to engage in party politics, is reflected in the disciplinary structure of "sections" that quickly developed. These formed a hierarchy by 1836, starting with the elite mathematical and physical sciences, then chemistry and mineralogy, geology and geography, zoology and botany, medical science, statistics, and finally mechanical science. It was the statistical section, introduced unofficially by Babbage in 1833, that proved controversial, given the intense political reform interest in alleviating the conditions of the poor. The large audiences attracted by the scientific programs of these sections soon generated a substantial financial surplus. Much of that was spent on research projects by favored people, including reports commissioned from individuals or committees. Until the advent in 1850 of the government grant fund of £1,000, made at the suggestion of Lord John Russell and administered by the Royal Society, the association was the major provider of funds for individual scientific research in Britain, apart from that directly supported by government departments such as the Admiralty.[63] William Wollaston had established a donation fund for the Royal Society in 1828 to be spent on promoting experimental researches, but it was not used extensively until the second half of the nineteenth century.[64]

Each year, the British Association published reports of work undertaken, and appointed people to carry out further activities. At the beginning these reports focused mostly on basic science, as those interested in individual emerging scientific disciplines sought to pull developing knowledge and understanding together, and give direction to further exploration. So at the second meeting at Oxford in 1832, reports were presented on topics such as astronomy, tides, meteorology, chemistry, optics, heat, thermoelectricity, mineralogy, and geology. At Cambridge in 1833, in addition to the reports on basic science, Peter Barlow produced a report on the strength of materials and George Rennie on hydraulics as a branch of engineering. In 1834, at Edinburgh, came a report on contagion from William Henry.

For much of the rest of the nineteenth century, reports on the applications of science were initiated and published alongside a much larger number on basic science. These offered a useful means of drawing public attention to questions of possible public policy, and a way of influencing discussion in scientific and political circles about significant issues. Sometimes, direct representations were made to Parliament or government. For example, in 1862 the association lobbied the Home Office on mortuary statistics, and the Board of Trade to encourage experiments on fog signals.[65] Association committees held a watching brief on patent law for much of the century.

The content and trajectory of reports on applied science from 1832 to the end of the century reveal a changing picture of the topics considered most salient, or at least most urgent in the minds of those who had the influence to put them on the association's agenda. A selection of topics gives a flavor: the properties of cast iron, the corrosion of iron rails, growing plants under glass, the vitality of seeds, manures for cultivated crops, the shape and construction of ships, steamships, steam boiler explosions, the prevention of smoke, chemical analysis of air and water, statistical inquiries and improving the census, guncotton, the cultivation of salmon and herring, weights and measures, underground water supplies, sewage treatment and utilization, fire damp in mines, and lighting using coal gas.

This shows the Victorians seeking to put science to use for the benefit of society and to ameliorate the conditions of life for the people of Britain. This voluntary activity came up time and again against governments generally disposed not to take centralizing action, unless there was irresistible pressure to do so.[66] The scientific community frequently railed against this lack of support, and the apparent disinterest shown by politicians in the potential of science. For example, in 1870, in the context of the demonstrable superiority of the Prussians over the French in the Franco-Prussian War (1870–1871), the recently formed magazine *Nature*, mouthpiece of advocates for government attention to science thundered: "Where is our science? At the Admiralty and the War Office, partisan placemen preside over technical administrations. Is that science? Under pressure of the newspapers or of private influence a ship of war is built by an amateur in spite of the demonstration of our professional adviser that she must be unsafe, and she goes accordingly to the bottom with 500 souls in the first gale of wind. Is that science?"[67] Yet slowly, throughout the century, the scientific contribution to standardization, inspection, and central administration grew, when public pressure allied to increased scientific knowledge and technological capability forced governments to act. The men of science, and a small number of women, in their published reports, their public statements, and behind the scenes, played an instrumental role.

STATISTICS

As the government, the medical community, and private organizations were confronted with growing social challenges, it became clear that effective responses needed to be based on the ability to measure and count more aspects of people's lives. That required statistics.

The history of statistics can be traced in Britain at least as far back as the seventeenth century, when William Petty used bills of mortality to examine the structure of the population.[68] This "political arithmetic" was later applied to developing life insurance, then to individual studies such as William Black's on smallpox in 1781. The origin of the word *statistics* derives from comparing the key features of states, such as their population, agriculture, and trade, which gradually became more quantitative. An example is John Sinclair's *Statistical Accounts of Scotland*, published from 1791, which first introduced the term *statistics* into the English language.

In the eighteenth century, governments had sought specific information—for example, on poor relief, emigration movements, and agricultural concerns.[69] The Home Office started publishing criminal statistics from 1810 in relation to concerns about the extent of capital punishment, and some educational surveys were also made around this time.[70] The growing number of these returns to Parliament, and the need for reliable information about provincial trade and manufactures, resulted in the creation of the Statistical Department of the Board of Trade in 1832.[71] Lord Auckland, president of the Board of

Trade, realized that savings could be made if his department published general annual returns of relevant statistics for many parts of government.

One political driver behind this, for the Whigs in government, was the view that having accurate information publicly available on social conditions might check misrepresentation and exaggeration that could lead to social and political unrest.[72] It is no coincidence that statistics became visible as an emergent discipline in the 1830s, just as society was grappling with the "Condition-of-England" question and political reform.[73] Was society progressing, or heading toward social strife and revolution? How would one know? The approach taken was to seek generalizations founded on empirical data, analogous to the approaches of the natural sciences.

It is arguable that statistics was formed in Britain as a collaborative scientific endeavor in 1833. At the British Association meeting in Cambridge that year, the Belgian statistician Adolphe Quetelet was a visitor. There was no suitable section of the association for the presentation of his work, so Babbage proposed creating a new section for statistics on the spot.[74] It was made clear that the new statistics section was to restrict itself "to facts relating to communities of men which are capable of being expressed by numbers, and which promise, when sufficiently multiplied, to indicate general laws."[75] The association tried hard to steer clear of politics, even if the establishment of "general laws" would seem inevitably to lead to political questions about their implications and operation in practice. The choice of which "facts" to examine would be made on particular grounds, and those facts would then be wielded in political argument.

The new statistics section recommended the creation of a permanent statistical organization. The result, after a committee meeting in Babbage's house early in 1834, was the formation of the Statistical Society of London. Its aims were stated as "procuring, arranging, and publishing Facts calculated to illustrate the Conditions and Prospects of Society."[76] In effect it would contribute to developing a natural science of society, in four categories of "economical," "political," "medical," and "moral and intellectual" statistics.[77] It was envisaged that these would generate predictive laws of social behavior.

Quetelet's views were particularly influential, including on significant figures such as William Farr and Florence Nightingale. He argued that statistical analysis could show the regularity of social phenomena, that these were equivalent to physical facts, and that they could be attributed to social conditions.[78] The improvement of human society, in its moral aspects such as crime and suicide, was therefore manageable.[79] This was a new social science beyond the narrow focus of political economy on the question of wealth. Indeed, throughout much of the nineteenth century, statistics was more a social than a mathematical science. But over time, the science of statistics evolved from a body of knowledge about important characteristics of the state to become a mathematical discipline by which such knowledge was reliably produced. It became a tool of social analysis rather than a science of social reform.[80]

The emerging disciplines that dealt with humans and societies were in an uneasy position alongside the natural sciences. On the one hand, such approaches sought to establish the existence of fixed laws of social life by rational and scientific approaches, analogous to the laws seen to govern matter. But on the other hand, studies in areas such as statistics, ethnology, or anthropology could lead too easily to heated political debate, which risked disturbing the calm, objective world of scientific discourse. The British Association sought to manage these tensions by trying to confine the scope of these sciences to "facts."

The difficulty of separating statistics from politics explains why the British Association was so wary of the statistical section. In principle statistics gave a scientific basis for understanding pressing social problems such as health, poverty, and education. Babbage saw practical benefits developing from such analysis, as did Herschel.[81] But Adam Sedgwick, in his presidential address at Cambridge in 1833, sought to separate statistics and science in general from politics. If not, in the "dreary wild of politics," the "demon of discord" would find its way into the "Eden of philosophy."[82] The gentlemen of science tried to deny that science is so often political. Proponents of agriculture came up against this challenge too, but were less successful. Attempts to introduce an agricultural section in the late 1830s failed because of concerns about the political controversy that might be engendered between agricultural and industrial interests, both important to the association, in the context of conflict over the Corn Laws. Aspects of agriculture were admitted in 1843, but firmly within Section B, chemistry and mineralogy, which was to include "their application to Agriculture and the Arts."[83]

CHEMISTRY, PHYSICS, AND BIOLOGY

The separate disciplinary and professional aspects of chemistry, physics, and biology emerged in parallel with the development of the British Association, although at different speeds.

The Chemical Society of London, a forerunner of the present Royal Society of Chemistry, was formed in 1841 with an academic and research emphasis, with Thomas Graham as its first president. It was followed by the Royal College of Chemistry in 1845, where Wilhelm Hofmann, a student of Justus Liebig, became the director. The college had an emphasis on teaching practical chemistry and on research. In its first few years it was the basis for the careers of several chemists who would become important advisers, including Frederick Abel and William Odling.

The Royal College in turn was absorbed into the School of Mines in 1853 under the government Department of Science and Art. The term *art*, or sometimes *mechanical arts*, meant "practical applications" rather than "the visual and graphic arts." Here, Edward Frankland succeeded Hofmann in 1865. Chemistry was seen as the most practically relevant science in the early nineteenth century, important in relation to agriculture; medicine; the analysis of pollution in water, air, and food; and to brewing and the chemical industries.[84] Indeed, the first significant institution devoted to the professionalization of science in the nineteenth century was the Institute of Chemistry, incorporated by the Board of Trade and founded in 1877 with Frankland as the first president.[85] It was in part a response to the formation of the Society of Public Analysts in 1874, as those who considered themselves serious chemists sought to claim their domain.[86]

In this professionalization, the natural scientists followed well behind the engineers and the medical men, both of whom had a longer tradition of industrial or private work and the need to carve out and protect their territories of employment. By 1877 there were already six major professional engineering institutions. The Institute of Physics was not formed until 1920, again incorporated by the Board of Trade, and the Institute of Biology not until 1950. The mind-set of so many of the leading savants, that science was a vocation rather than a career and required complete freedom for investigation by individual researchers, delayed their professionalization. Even so, it did not noticeably diminish their potential influence. Indeed, in some respects, their apparent disinterestedness may have aided their credibility. Concerns about the potential tainting of evidence by advisers who had financial interests and could be seen as biased became a significant issue from mid-century.

PART II

EMPIRE AND WAR

3

ADMIRALTY AND NAVY

THE DEVELOPMENT OF THE ROYAL NAVY, BRITAIN'S PREEMINENT WAR machine, was crucially dependent on new scientific and engineering knowledge that built on, and sometimes conflicted with, the craft expertise of centuries.

For the most part, the interactions of scientific experts and engineers with naval affairs took place directly with officials and ministers. That reflects the executive nature of many of the challenges and opportunities, which governments argued were not a matter for Parliament, even if new technologies would inevitably lead to operational, strategic, and political judgements and decisions. As a result, parliamentary inquiries into the navy throughout the nineteenth century were mostly concerned with questions of funding, manpower, management, and discipline, not with scientific and technical matters, although there was frequent criticism about these from individual MPs.[1] At the beginning of the nineteenth century the civil administration of the navy, for matters such as shipbuilding, dockyards, supplies, and pay, was the responsibility of the Navy Board. The Board of Admiralty was the ultimate executive body. The Lords Commissioners, who made up the board, were all active politicians with a preponderance of civilians alongside some serving officers.

In the early part of the century, the two significant figures who effectively acted as gatekeepers to scientific expertise were John Croker and John Barrow. Croker, an Irish Tory MP, became secretary to the Board of Admiralty in 1809 and held this political post until 1830. He was a powerful figure.

Although he described himself as a "servant of the Board," one member of the board observed that it was "precisely the other way round."[2] Barrow held the second secretaryship, which was non-parliamentary, for most of the period from 1804 until 1845. Barrow was elected a fellow of the Royal Society of London in 1805, at the instigation of Joseph Banks, as Banks sought to cement relationships with the Admiralty. Croker was elected a fellow in 1810, and it is no coincidence that both were elected soon after their appointments to Admiralty posts.[3] Barrow helped found the Royal Geographical Society in 1830. This interest exemplifies his commitment to furthering scientific exploration, for strategic imperial and economic ends. The large number of extensive and expensive voyages of exploration, which were important to the scientific community, made the relationships between them and the Admiralty particularly important, as did the proximity of the Royal Society in Somerset House to many naval offices. But the Admiralty, in Whitehall, had additional reasons to seek scientific advice.

THE BOARD OF LONGITUDE

By the time Croker and Barrow came to influence in the Admiralty, the Board of Longitude had been in existence for nearly a century.[4] Founded in 1714, the purpose of the Commissioners for the Discovery of the Longitude at Sea, as the members of the board were called, was to do just what their name said. It was vital for accurate navigation, and critical to a sea power such as Britain, to have a precise means of fixing longitude. The solution to the problem required an accurate timekeeper that would work at sea. Although it had effectively been solved by the 1770s when John Harrison had invented his chronometers, the method of lunar distances was mostly used until cheap, accurate chronometers became available in the mid-nineteenth century.

Navigational errors still resulted in huge numbers of casualties during the Napoleonic Wars, and the Board of Longitude remained sufficiently valuable to all parties to continue in existence long after its original purpose had been achieved in principle. It supported improvements to chronometers over its subsequent life. Its remit was broadened in the Longitude Act 1796 to include "making other useful discoveries and improvements in navigation, and for making experiments relating there to." In this sense, one might argue in modern terms that the Board of Longitude was established as a body of scientific experts to provide the government with scientific advice. But that would be a mischaracterisation, as the board contained powerful naval members, in addition to scientific figures such as the astronomer royal and the president of the Royal Society.[5]

Those in power in the state, and in the Admiralty in this case, identified individuals and groups as trusted advisers through their personal connections.

Those men of science who sought to have influence with state institutions used their networks in a complementary manner. The Royal Society straddled these camps, as it contained at this time not just the scientific experts but also members of the governing class, including Croker and Barrow.

Individual societies, such as the Astronomical Society and the Geographical Society, could do the same, leading to inevitable tensions with the Royal Society. Banks had sought to suppress these tensions by trying to prevent the creation of such specialist groups, but failed. Many more would emerge over the next few decades. Despite this, the Royal Society managed to preserve its position as the most important body to be consulted by government on scientific matters, a position it has retained to this day.

The cessation of the Napoleonic Wars in 1815 led to reviews of expenditure and policy across government, and the Board of Longitude was not immune. Further changes followed the Longitude Act 1818. The board was restructured, and here Banks was able to play a significant role to enhance his influence and that of the Royal Society, at least for a short period before his death.

Banks's connections with the Admiralty were deeply rooted, and he had been a member of the Board of Longitude since 1772.[6] As the nineteenth century opened, Banks developed a friendship with Barrow. The two extended support for voyages combining naval exploration and scientific discovery, a policy Barrow continued after Banks's death in 1820.[7] Banks had sought to increase the connections between the government and the Royal Society, not least to prise out more government funding for science. The Admiralty was a focus of interest, particularly for its support for voyages of exploration and their opportunities for scientific investigations in areas such as astronomy, geology, and natural history. The Royal Society already had a strong influence on the Royal Observatory at Greenwich, as it had acted since 1710 as the Board of Visitors to oversee the institution. That made the Royal Society the intermediary between the astronomer royal, the Board of Ordnance, and the Admiralty, a situation that lasted until the 1830s.[8]

The opportunity arose in 1818 for Banks and the Royal Society to make closer links when John Pond, the astronomer royal and head of the Royal Observatory from 1811 to 1835, failed to produce a suitable *Nautical Almanac*.[9] Its annual publication was critical for navigation. At this time, in a curious and unsatisfactory arrangement, the Royal Observatory was funded by the Board of Ordnance, although its head was appointed by warrant from the Board of Admiralty. At Croker's instigation, the Royal Observatory was transferred to the Admiralty, and the *Nautical Almanac* to the Board of Longitude, which was funded by the Navy Board under the control of the Admiralty. Davies Gilbert played a role here as chairman of the select committee on finance, which in its report on the Board of Ordnance recommended this transfer of responsibility.[10] The publication of the *Nautical Almanac* had the effect of giving a useful task to the Board of Longitude, along with the aim of discovering a northwest passage. That helped justify its existence and retained, advantageously to fellows of the Royal Society, one of the few government bodies that funded scientific activity.

Thomas Young, who had been foreign secretary of the Royal Society since 1804, became secretary of the Board of Longitude in 1818, holding both positions until his death. Originally medically trained, he is best known today as a natural philosopher for his foundational work on the wave theory of light. Young had worked for the Board of Admiralty since 1810, having been spotted by Barrow after publishing an account of ship carpentry and architecture. He had given advice on the ship-construction improvements of the leading shipwright Robert Seppings of Chatham Dockyard,[11] as well as serving on the Royal Society Pendulum Committee from 1816, and had consulted on the *Nautical Almanac*, for which he was now given responsibility. Young's critique of Seppings's proposals illustrates the difficulties of informing technical practice through scientific knowledge.[12]

In 1810 Seppings had developed a new way to strengthen ships, which was initially opposed by the Navy Board. Seppings went over their heads by writing to the Admiralty. Charles Yorke, first lord of the Admiralty, gave his support to trialling Seppings's method, encouraged by Barrow. After a successful trial the Navy Board was still resistant, so Barrow sought validation from his scientific advisers, including Banks and Young. Banks ensured that Seppings was elected a fellow, and that his key paper was read to the Royal Society in 1814 and published in the *Philosophical Transactions*. Unfortunately, Young's paper, read immediately afterward, did not offer unalloyed support. His report on the basis of abstract scientific principles did not unequivocally validate Seppings's practical engineering approach. It was a clash of practical craft experience with academic science, as if Young were saying that the technique seemed to work in practice but that theoretical understanding was lacking. Barrow, disturbed at the idea of what he saw as crucial changes being held up by half-hearted support, anonymously attacked Young in the press.[13] Banks, in support of the Admiralty, then nominated Barrow to the Royal Society's council, along with Croker, giving the Admiralty a stronger influence on Royal Society affairs.

Seppings's system was adopted as standard design practice by Admiralty Order. It saved money, enabled ships to increase in size by 50 percent, and to bear the weight and vibration of early steam engines. This particular episode illustrates the tensions between those with applied knowledge of immediate practical use to government, and the increasingly pure science focus of the Royal Society. Arguably it was the Royal Society's position and reputation that the Admiralty valued, rather than its specific scientific opinion, to validate what Yorke and others could see with their own eyes.[14] Humphry Davy's work would soon exemplify the same challenge of using abstract science and limited experimentation to inform a complex practical undertaking. When he and Banks had died, and the Navy Board was abolished in 1832, relationships between the Royal Society and the Admiralty would be different.

Under the Longitude Act 1818, three paid resident commissioners were appointed to the Board of Longitude. They were Young as secretary, William Wollaston, and Captain Henry Kater. Kater was a military surveyor with particular knowledge of pendulums, and an astronomer. According to Barrow, these were the first such salaried posts open to men of science, although those salaries were later withdrawn "in a subsequent fit of economy."[15] Despite the fact that Young, Wollaston, and Kater were all fellows of the Royal Society, three more commissioners formally represented the society, in addition to Banks as president. They were Charles Abbott, a Tory MP who had been the Speaker of the House of Commons until 1817; Gilbert; and Colonel William Mudge, an artillery officer and surveyor. All six were selected by Barrow and Banks.[16] Other Admiralty officials and men of science completed the board, though they rarely attended.

Without any formal system of employment, social networks and relationships shaped the manner in which scientific activities were organized. Official channels of responsibility were unclear, and personal relationships and patronage were crucial to making appointments and decisions on expenditure. In addition, given the individualistic nature of scientific activity, systematic coordination and standardisation were difficult. A case in point was the use of pendulums to measure the shape of the earth through gravitational variance between the poles and the equator. This question was connected to the use of a pendulum to establish the length of a yard, both being ultimately relevant to improving astronomy and hence navigation. Young was central to the selection of personnel and equipment for pendulum experiments supported by the Board of Longitude, but consistent agreed results could not be obtained in the early decades of the nineteenth century.

Over the years, the Board of Longitude established several subcommittees to examine matters requiring scientific knowledge. They included the Committee for Examining Instruments and Proposals, which might then require a report from an individual scientific expert, and a Tonnage Committee, to establish the best way of estimating a ship's weight (see chapter 8). Young played a central role for the final decade of his life, effectively acting as a scientific civil servant who brokered relationships between Croker, Barrow, and the Admiralty, and the individual scientific experts who served on the board or sought to influence it. With an annual budget of £1,000, the Board of

Longitude was the main source of public money for scientific activity in this period, apart from grants given to Babbage for his difference engine.[17]

In 1821 John Herschel was appointed by the Royal Society as a resident commissioner of the Board of Longitude, replacing Wollaston. An astronomer and natural philosopher, he was a man of independent means and a rising star. In the same year, aged just twenty-nine, he received the Royal Society's premier award, the Copley Medal, for his mathematical contributions. He had been one of the founders of the Royal Astronomical Society in 1820, and in 1824 he became secretary of the Royal Society. Although appointed by the Royal Society, his position was seen by the Royal Astronomical Society as strengthening their position.[18]

Correspondence between Young and Herschel exemplifies the contested boundaries between scientific advice and executive authority. Young was of the view that the Admiralty retained full control over the Board of Longitude. It was, in effect, a committee of the Admiralty. Herschel argued that the board, given the scientific experience it contained, should have some power over the operation of the Admiralty and the decisions of Barrow, particularly with respect to what he saw as scientific matters of judgement.[19] Yet the Board of Longitude, although it contained many scientific members, was essentially a political body. Its constitution included the first lord of the Admiralty, and Croker, the politician, generally presided. Decisions on expenditure were political, and could be scrutinized by Parliament. The navy and Admiralty, just like the army, War Office, and Board of Ordnance, had many men with technical expertise, so external advice was in any case often unnecessary.[20]

HUMPHRY DAVY'S "PROTECTORS"

When Davy became president of the Royal Society in 1820, it was the Admiralty that offered the most fruitful ground for influence. The navy's ships had increasingly been sheeted with copper in recent years. That was found to reduce the growth of weed, and made the vessel faster and more maneuverable. But it seemed to have the unfortunate side effect of corroding the ship's iron nails, although means were introduced to try to counter that.[21] Then the corrosion of the copper itself appeared to be a factor in the sinking of ships, although no trials were made for long enough or carefully enough to establish the facts. In 1822 Davy was asked by the Navy Board for advice.[22] Asked because he was the premier chemist of the day, with a reputation for successful practical application of science through his invention of a coal miners' safety lamp, Davy brought the request to the Royal Society. The council formed a committee to investigate, with Davy as chairman and with members including Herschel and the chemists Wollaston and Thomas Brande.[23] Brande, from a family of wealthy apothecaries, was another protégé of Davy. With the support of Charles Hatchett he had succeeded Davy as professor of chemistry at the Royal Institution in 1813, and became superintendent of the house, securing a post also as professor of chemistry to the Society of Apothecaries.

Davy's committee met only once, and Davy undertook the investigations himself in the laboratory at the Royal Institution, with Michael Faraday carrying out some analysis. He sent his report personally to Croker at the Admiralty, rather than to the Navy Board, perhaps seeking to ensure that the Navy Board would be instructed by the Admiralty to support practical tests. Davy was close to Croker at this time, while they were both involved in establishing the Athenaeum Club. Davy had found corrosion of the copper by seawater, and put it down to electrochemical action, a concept that he had proposed fifteen years earlier. He reasoned that attaching zinc to the copper would prevent its corrosion, a process now known as "sacrificial protection." After what looks in hindsight to be cursory tests in 1824, the Navy Board instructed all ships to be fitted with pieces of zinc, Davy's "protectors."

Within a few months it became clear that all was not well. A torrent of complaints was received from captains, who found that while the copper bottoms of their ships might be preserved, they were extensively fouled by barnacles and weeds. That was an unanticipated consequence of applying science to the problem, as the products of corrosion were toxic to barnacles and weeds. The Admiralty was eventually persuaded to allow the Navy Board to rescind the order to fit the "protectors," except in harbor where they appeared to be effective.

This was a stumbling attempt to apply scientific advice to naval practice, bringing scientific research and knowledge to bear directly on a technological problem. The simple Baconian rhetoric that scientific knowledge would be found unproblematically useful and applicable was found wanting. The time for extensive practical testing was neither sought by the Navy Board nor demanded by Davy. Understanding the relationship between laboratory science and its application to practical problems and policy would take time to develop, as other examples in this book attest.

Several smaller commissions came Davy's way at this time. In 1822 the Admiralty asked him to investigate the safety of using coal tar on ships. A committee including Wollaston, Young, Gilbert, Brande, and Hatchett found through experiments that coal tar would preserve timbers and would not emit any vapor flammable at temperatures below 200 degrees Fahrenheit, far higher than the highest temperature to which a ship would be exposed.[24] The following year, after a request from the Navy Office, a committee chaired by Davy, including Wollaston, Young, and Kater, approved the idea of copper lightning conductors designed by William Snow Harris, fixed to the mast and attached to a thick plate in the keel.[25] It would take twenty years for the

Admiralty to agree to implement the idea.[26] A few years later, in November 1826, the Board of Ordnance sought Davy's advice on lightning conductors for a building in the West Indies. He recommended several copper conductors given the amount of iron in the building. That request was followed up a year later by questions from the Board of Longitude to Gilbert, now the president, John Children, Edward Sabine, Kater, and Young.[27]

This was a period, in the 1820s, when the physical sciences had reached a point of usefulness to public policy, which required some state funding. The same was not true of the biological sciences—for example, with respect to agriculture—which in any case was not an activity for which the state yet took any responsibility, and corresponding requests for advice were not made.

HUMPHRY DAVY, MICHAEL FARADAY, AND OPTICAL GLASS

In April 1824, just a month before the Navy Board issued its order for all its ships to be fitted with Davy's "protectors," the Board of Longitude met to discuss optical glass.[28] Davy observed that optical glass, for vital instruments such as telescopes, was not of good enough quality. The Board of Longitude asked the Royal Society to convene a committee to agree with it a series of experiments to improve its quality, to be funded by the board. The Royal Society proposed a joint committee chaired by Davy and including Brande, Wollaston, Gilbert, and the optician George Dolland. Kater, Young, and Pond represented the board. They were also all fellows of the Royal Society.[29]

Faraday was asked to carry out the experimental work. A protégé of Davy, he had remained at the Royal Institution when Davy left in 1812, working mainly with Brande. Rapidly developing his own connections and reputation, he became superintendent of the house at the Royal Institution in 1821 and director of the laboratory in 1825. In those years, while carrying out his own research, he offered professional scientific advice on alloy steels and on gunpowders for the East India Company. He was already visible in the world of expert scientific advice, and would soon come to the notice of government bodies, aided by men such as Davy and Young.

Davy had prevailed on Faraday to act in the onerous role of unpaid secretary to the newly formed Athenaeum Club in the first three months of 1824. Faraday was able to escape when the post was made permanent, but soon found himself drawn into the glass work that Davy was leading. In 1825 he was appointed to the joint committee, and to a subcommittee consisting of himself, Dolland, and Herschel. His role was to supervise and make the glass with the firm Pellatt and Green, Dollond's to ground it, and Herschel's to test its optical properties.

Progress was difficult, and Davy dropped out as his health gave way. Seeking to make headway, the joint committee funded a furnace at the Royal

Institution in 1827, forcing Faraday to spend the next two years working with it. Davy's decline in health, perhaps as a consequence of these failures, led to his resignation from the Royal Society in 1827. The Board of Longitude, which he had so wanted to preserve under his strong influence, was abolished while he was away. He died in Geneva, Switzerland, less than a year later, in 1829. Faraday was thus able to escape also from the work on glass, which proceeded no further.

Released from the drudgery, Faraday discovered electromagnetic induction in 1831. It is difficult now to imagine life without it, although it took forty years before electric generators became commercially viable. But the work on glass was not wasted, even if Faraday thought so at the time. In 1845 he used the "heavy glass" he had created to make two significant discoveries. The first was the rotation of plane polarized light by a magnet, the "magneto-optical effect." The second was the weak repulsion of substances by magnets, or "diamagnetism." It was this latter discovery that inspired the initial researches of the young natural philosopher John Tyndall in late 1849, and brought him to the notice of Faraday, Sabine, and others.[30] It would lead to a life in scientific research and scientific advice. From such contingencies are careers made.

THE RESIDENT COMMITTEE OF SCIENTIFIC ADVICE

In July 1828, in the middle of the problems with optical glass and with Davy abroad having suffered a stroke, the Board of Longitude was abolished. In a sense it is surprising it had lasted so long, as its original purpose, to find an accurate means of measuring longitude at sea, had been largely accomplished by the 1770s. Those who ran it had found it sufficiently useful to their relationships and mutual ends to keep it alive. Financial constraints after the ending of the Napoleonic Wars in 1815 contributed to its demise.

The Admiralty appointed instead the Resident Committee of Scientific Advice, consisting of three members of the council of the Royal Society: Faraday, Young, and Sabine. The role of the committee was to advise "on all questions of discoveries, inventions, calculations and other scientific subjects."[31] They earned £100 a year each, although Young died not long after his appointment. He supervised the *Nautical Almanac* until his death, when it passed back to the Royal Observatory, before being handed to the Royal Astronomical Society in the early 1830s. Young had been unreasonably criticized throughout by astronomers for concentrating on its naval purposes and neglecting their needs. Sabine was appointed because of his surveying and pendulum knowledge.[32] He had done work for the Board of Longitude in both areas,[33] but does not appear to have done much for the committee,[34] not least because he was soon posted to Ireland. At that point he was one of the secretaries of the Royal Society and an engineer officer.

The creation of the committee was a move by Croker, who had attacked the Board of Longitude in Parliament. It is an example of a politician both seeking to save money, in the context of the financial pressures after the Napoleonic Wars, and to gain more control over independently minded experts, by making it an internal committee. Gilbert had assisted Croker in drafting the Longitude Acts Repeal Bill. Speaking before Croker in the debate before its third reading he tried to defend the board, and the value of what it had done, while acknowledging the argument for parsimony. But Croker argued trenchantly that dissolution of the Board of Longitude was not just about saving money. The board, he said, was a waste of time, as its meetings were "wholly occupied in reading the wild ravings of madmen, who fancied they had discovered perpetual motion and such like chimeras," and sought public money for their ridiculous notions.[35] Its abolition, and the creation of a more manageable internal committee, would take its function out of the public eye and potentially reduce the attention and criticism to which it was subjected. It would make his life easier. The Longitude Repeal Act 1828 became law. It soon became apparent that there were no savings to be had anyway.

Unlike the Board of Longitude, which was a channel of communication between the men of science, the Admiralty, and other interested parties, the members of the Resident Committee of Scientific Advice were given specific tasks by the Board of Admiralty, generally through Barrow. Herschel was so offended by this turn of events that he refused his final salary payment as a commissioner. As an independently wealthy gentleman he could afford to do so. He and Kater both declined to accept membership of the committee when the Board of Longitude was abolished, resulting in the invitations to Faraday and Sabine. In Herschel's view, the state needed to support scientific activity properly, and it was the scientific experts who had the knowledge as to how that money should be spent.

That tension, about the proper degree of funding and of control by politicians and the scientific community, would be fought for the rest of the nineteenth century and remains contested. Young took the view that the Admiralty, as the funder, had the power, and that individuals of excellence would anyway continue to cultivate scientific practice themselves. Herschel argued for a greater role for the men of science in the decisions, and state support for more robust scientific activity. Croker, with Barrow, had to negotiate his way through. But he ultimately retained control, as politicians tend to do, even if he and Barrow were guided and influenced by their expert advisers on individual decisions.

Banks, outside formal structures of government, was able to pursue his private interests, those of a landowning gentleman, through the public sphere. That would become harder over the century, as the demarcation became more explicit, and a cadre of employed scientific managers or civil servants evolved, such as George Airy at the Royal Observatory, Henry De La Beche and Roderick Murchison at the Geological Survey, Sabine at the Kew Observatory, Frederick Abel at the War Office, and Francis Beaufort at the Hydrographical Department of the Admiralty.

MICHAEL FARADAY AND THE ADMIRALTY

The Admiralty's Resident Committee of Scientific Advice, created in 1828 with Faraday, Young, and Sabine as its members, was soon down to an effective membership of one. Young died in 1829 and Sabine was posted to Ireland for seven years in 1830. Faraday had been appointed on Young's recommendation, following his work on glass and because the Admiralty wanted a chemist.[36] He effectively became the Admiralty's scientific adviser for some twenty years. For example, in the 1830s he gave advice on the quality of copper for sheathing,[37] on using lightning rods on ships, on whether oatmeal on prison transport was contaminated, and on methods of treating dry rot. This last request related to an inquiry by an Admiralty committee in 1835 to examine John Kyan's mercury-based patent wood preservative, following an apparently successful trial in the "fungus pit" at Woolwich. It was a discovery that excited much interest, and Faraday had chosen it as the topic of his inaugural lecture as Fullerian professor of chemistry at the Royal Institution in 1833. George Birkbeck had lectured on it the next year, and was one of the five members of the committee, with John Daniell, professor of chemistry at King's College London, and Alexander Hutchison, a naval surgeon. Faraday carried out experiments on the penetration of the material into wood, supplied by Marc Brunel, and the atmosphere around it. He was careful to state that he "always felt that the process must be submitted to experiment to arrive at a satisfactory result, because he felt he could not apply chemistry to prove the utility, or otherwise, of the process." But he gave a qualified opinion that the preservative would be effective, subject to a large-scale trial in a dockyard.[38] Although the commissioners had heard from Robert Smirke, George Rennie, and others that they had used the process successfully, they only gave weight in their report to the experiments by Faraday and Daniell, which established the efficacy of the process to their satisfaction. They also believed that the fumes would not endanger the health of workmen or crews. Over subsequent years some disadvantages of the treatment emerged, and it was phased out in favour of creosote.

In the 1840s Faraday's advice included naval use of the electric telegraph,[39] but it was the Crimean War, fought between 1853 and 1856, that brought Faraday into more secret work. The military strategy required the capture of the two main Russian naval bases of Cronstadt in the Baltic Sea, and of

Sebastopol in the Black Sea, to prevent forces reaching the Crimea. The elderly Admiral Thomas Cochrane proposed a bizarre plan to use sulfur-filled fireships to attack Cronstadt. Ships filled with four hundred tons of burning sulfur would, he thought, incapacitate or kill the defenders and allow forces to land and capture the port.

Faraday was consulted by a secret Admiralty committee consisting of William Parker, commander in chief of Devonport; Maurice Berkeley, a lord of the Admiralty; John Burgoyne, inspector general of fortifications; and Thomas Byam Martin, admiral of the fleet. Faraday tactfully advised that although the plan might work in theory, there were too many uncertainties, such as wind and the time available for defenders to organize themselves, for any confidence to be given. On the basis of the report and Faraday's advice, James Graham, first lord of the Admiralty, turned down the proposals. The file was not declassified and released until 1946. As ever, Faraday showed himself cautious and mindful of the need for practical testing and operational experience. He accepted no fee for his advice.[40]

Lyon Playfair played a similar role in the Crimean War, when he and Graham were asked by Viscount Palmerston to report on this plan. Palmerston accepted their advice not to pursue the idea, and it may be that neither Faraday nor Playfair knew of the approach to the other. Playfair also offered unsolicited advice, through Prince Albert, although unsuccessfully. His suggestion of creating an incendiary phosphorus shell was turned down, as was the possible use of a poison gas shell containing cacodyl cyanide. This was rejected by the military, to Playfair's disgust, on the grounds that it would be as bad as poisoning the wells of the enemy.[41]

THE RISE OF GEORGE AIRY

By the early 1830s difficulties with the project on optical glass, together with the failure of Davy's "protectors," had damaged the relationship between the Admiralty and the Royal Society. The old order changed. Wollaston and Young had both died within a few months of Davy. With the earlier demise of Banks an influential group of savants, all key figures in the Royal Society with their roots in the eighteenth century, had disappeared from the scene.

Political and administrative structures changed too. Coincident with the Reform Act 1832, Graham introduced his Admiralty bill that abolished the Navy Board and transferred its duties to a redesigned Board of Admiralty. The Lords Commissioners were jointly responsible for naval affairs, but also had individual responsibility for the work of several departments of the Admiralty, especially after 1869. A more managerial structure emerged, at the same time as the Treasury gained firmer control over budgets. The interactions of the scientific experts with the navy would now be on a different footing.

One exemplar of this managerial structure is George Airy. Airy was the son of a man who had risen from being a farm laborer to become collector of excise in Northumberland. A gifted mathematician, he matriculated at Cambridge University in 1819 and graduated as senior wrangler and first Smith's prizeman in 1823. By 1826 he held the prestigious Lucasian chair of mathematics, once held by Isaac Newton, and by 1828 he was Plumian professor of astronomy and director of the new Cambridge University observatory. In 1835, when he was thirty-three years old, Robert Peel offered him a pension of £300 per annum. Shortly thereafter he was appointed astronomer royal, in succession to Pond, a post he held until 1881.

If Banks had effectively been the government's chief scientific adviser at the turn of the century, followed in a sense by Faraday, Airy came as close as anyone to that position in mid-century. But in contrast to Banks and Faraday he was an employee of the government, and he discharged his position as a civil servant with scrupulous virtue. That included, throughout his time at the Royal Observatory, managing the exhaustive business of testing and rating the chronometers for the navy.[42]

Whereas leading figures of the previous generation—such as Banks, Young, and Gilbert—had merged public and private interests, Airy did not. For example, when approached by Tyndall to sign a memorial to Prime Minister William Gladstone about the treatment by a government minister of Joseph Hooker at Kew, Airy declined. He argued that "my position, as that of an officer under the Government, makes it very difficult for me to take any steps affecting another officer." He strongly disagreed with the manner in which Tyndall and other "outsiders" were interfering.[43]

Airy was equally clear about his view of the proper relationship between government and science. In his presidential address at the British Association for the Advancement of Science meeting in Ipswich in 1851, he said: "When any branch of science has been put in such a form that it admits of continued improvement under a continued administration, that administration should be undertaken by Government. But I trust that in all cases the initiative of science will be left to individuals or to independent associations."[44] In this respect, although not in the merging of public and private interests, he shared the views of the earlier independently wealthy gentlemen of science.

Airy was a superb administrator, introducing systematic methods that transformed the Royal Observatory and improved the quantity and quality of information for astronomy and navigation. In 1838 he created a magnetic and a meteorological department to extend the scope of the observatory, so that he could examine the effects of geomagnetic phenomena on marine compasses and explore weather patterns. His work over decades in establishing longitudes relative to Greenwich was critical to the choice in 1884 of the

Greenwich Meridian as the zero of international longitude, along with an extensive system of time signals.[45]

Much of his work was internal to the observatory and the Admiralty in general. For example, in 1838, at the request of Beaufort, hydrographer of the navy from 1829 to 1855, under whose office the Royal Observatory officially fell, Airy carried out numerous experiments and calculations on the effect of iron hulls on ships' compasses.[46] He also proposed plans for sawmills for the Admiralty, in work starting in 1842 and continuing for some years.[47]

Although Airy's power base was in the Royal Observatory and hence the Admiralty, he was soon consulted by government on a wider range of questions.[48] These are addressed elsewhere. They include Babbage's calculating engine; the railway gauge (chapter 7); tidal harbors and lighthouses (chapter 8); sewers (chapter 11), weights, measures, and coinage (chapter 14); the Atlantic telegraph cable;[49] and the Palace of Westminster and Big Ben.[50] In total, Airy served on around three dozen public inquiries.[51]

COAL FOR THE ROYAL NAVY

The nineteenth century saw several changes that had major effects on the navy. They included the gradual replacement of sail by steam power, new types of guns and explosives, and the development of armor plating. In each case there was a close relationship between civilian and military developments, and the contracting of much business to the private sector.

Advice as well as manufacturing capability could come from outside. One example is the report from a committee of the Admiralty on the use and manipulation of iron for the navy, which reported in 1846. The four members included James Nasmyth, inventor of the steam hammer. Based on visits to private sector works, the committee offered extensive advice on handling and working iron, including for the manufacture and testing of chain cables and anchors.[52]

Sometimes the Admiralty was less enthusiastic. When the British Association's parliamentary committee asked in 1861 for the publication of the results of steam trials, the Admiralty brushed them off with the response that "the ships of the Royal Navy only employed steam occasionally, and only as an auxiliary power."[53]

The Baltic fleet during the Crimean War was the first to use steam-powered warships on active service, and the supply of the most appropriate coal was an important question. The radical MP Joseph Hume raised the issue in a letter to the Admiralty in 1845, enclosing a report made for the US Navy two years earlier.[54] The Admiralty asked the Earl of Lincoln, first commissioner of Woods, if he would approach Henry De La Beche for assistance. De La Beche suggested that he could combine chemical analysis in the laboratory of the Museum of Economic Geology with research by the Geological Survey, subject to a provision of £600 from the Admiralty. That and more was forthcoming. De La Beche involved Playfair, and the work commenced in March 1846, shortly before the two of them reported on a previous commission to examine explosive and noxious gases in coal mines (see chapter 10).

Playfair was inundated with requests for help by the government at the time. Having carried out an investigation on the state of Buckingham Palace and of Eton College, he also reported on the state of graveyards for the Board of Health and analyzed town water supplies.[55] Playfair held a professorship in the College of Civil Engineers at Putney as well as his post of chemist at the Museum of Economic Geology, and experiments with boilers were made at Putney by John Wilson, later professor of agriculture in Edinburgh.

De La Beche and Playfair sent their first report in January 1848 to Lord Morpeth, now first commissioner of Woods, with a copy to the Admiralty.[56] It was an extensive analysis of the ability of different coals, in different physical and storage states, to create steam from boilers either quickly or over a sustained period. They found that "no practical result could be attained by mere laboratory research," and therefore instigated tests on a larger scale.[57] The report named the chemical analysts and engineers who had supported and worked on the experiments, including Rennie. It gave the chemical composition of the many coals analyzed, their calorific values, their ability to produce steam quickly or over a long period of time, and the mathematical formulas developed to produce the comparisons. They also pointed out other factors, such as the influence of boiler design, the effect of different storage arrangements on the physical nature and properties of the coal, the possibilities of spontaneous combustion, and the need to reduce smoke production so that a ship's position was not given away too readily. Understanding all these factors was critical to optimizing the performance of steamships, and no coal satisfied all the desirable criteria simultaneously.

The second report concentrated on the ability of the coals to raise steam, their mechanical structures and bulk storage properties, and further chemical analysis.[58] At the request of the Admiralty, different mixtures of anthracite and more bituminous coals were trialled. The third and final report was submitted to Lord Seymour, who had succeeded Lord Morpeth as first commissioner, in 1851.[59] It concentrated on burning coals in varied draughts of air and tabulating the "economic values" of the coals, in terms of the amount of steam generated per pound of coal. This work formed the basis of the selection of fuel for the navy, just a few years before the navy first deployed steam warships during the Crimean War, where they were critical to naval success.[60]

SHIPBUILDING

Technological change over the nineteenth century is nowhere more visible than in shipping. From wooden sailing ships at the beginning, to iron steamships at the end, the transformation was huge. Add new explosives, weapons, and armor, and the Admiralty had the need for the best external advice.

Significant changes had already taken place during the Napoleonic Wars. Samuel Bentham, inventor of the panopticon, developed later by his brother Jeremy as an idea for controlling prisons from a central location, had introduced scientific approaches, mechanization, and structured management systems to the dockyards as inspector general of naval works from 1797 to 1807.[61] This included the mechanised production of blocks introduced with Marc Brunel at Portsmouth from 1802. Bentham's designs influenced Seppings, surveyor of the navy from 1813 to 1832. Seppings played a critical role in applying a scientific approach to improving ship design. His designs opened up the possibilities of incorporating steam engines, propellers, and then building in iron.[62]

Of the many civilians who informed policy on ship design, two stand out: John Scott Russell and William Froude. Russell had developed a steam carriage in 1834, which ran between Glasgow and Paisley until it was withdrawn following the collapse of a wheel, which killed four people, possibly the first fatal automobile accident. His interest turned to water transport, and he presented a paper to the British Association in 1835 describing his experimental design of a ship to reduce water resistance.[63] It led him both to his "wave-line" theory, which, although erroneous, improved the speed of ships, and to a method of hull construction using longitudinal struts, widely used thereafter.

It proved difficult to find a method of calculating the power required to propel a ship at its design speed. Russell was a member of five committees of the British Association between 1838 and 1866 that sought a solution,[64] and was asked to make reports on ship construction. He was also called on to advise government—for example, giving evidence to the select committee in 1853 on mail and passenger traffic between London and Dublin, which explored "what Improvements modern Science can suggest to establish a more Speedy and Commodious Communication between the two Capitals."[65] Russell supported the concept of the wave-line ships designed by Oliver Lang and offered to build something similar.

After designing several ships in Glasgow, Russell had moved to London in 1844. He became secretary of the Society of Arts, where he was instrumental in the success of the Great Exhibition of 1851, and responsible for introducing its prime mover, Henry Cole, to Prince Albert.[66] Having worked with Isambard Kingdom Brunel to build two mail steamers, he designed the hull and paddle engines for Brunel's *Great Eastern* and built the vessel, the largest for the next fifty years. It failed commercially, causing Russell financial distress, but is remembered as the ship that laid the transatlantic cable in 1866 and 1867.

Russell soon became involved in designing and building ships for the navy, including four gunboats and twelve mortar floats, and an abortive attempt to build a submarine during the Crimean War.[67] He also suggested ideas for iron warships, and may have had some influence on the design of *Warrior*, the first armor-plated iron-hulled warship, launched in 1860.[68]

This period, in mid-century, saw the intensification of the clash between the engrained naval culture of traditional wooden ships and the new iron ships. For a time around 1850 the Admiralty forbade the construction of iron ships, on the grounds that they were less resistant than wooden hulls to shot.[69] The question became party political, as John Pakington, the Conservative first lord in the Earl of Derby's 1858 and 1866 governments, sought to attack the Liberals for failing to manage the change.[70] *Warrior* was the first ship to be fitted with William Armstrong's guns. Unlike Armstrong, who patented his guns, which enabled the government to pay to keep details secret, Russell kept his ideas out of the public eye by not taking out a patent.

The Admiralty found itself needing civilian advice as the steam-powered and iron-clad or iron warship became a reality. Russell played a role here in founding the Institute for Naval Architects in 1860 with Edward Reed and Nathaniel Barnaby, both later chief constructors of the navy. The new institute was a channel between naval and civilian architects and engineers, and he followed this initiative with the establishment of an Admiralty school of naval architecture in 1864. The influence of Russell, with the more visible Armstrong and William Fairbairn, was behind the debates that led the Admiralty to move from wooden to iron warships.[71] Nevertheless, even the technical experts within the Admiralty, the controller and the chief constructor of the navy, found themselves somewhat excluded in this period from the predominantly craft-based culture of naval shipbuilding.[72]

Froude's contributions were equally substantive as Russell's. He was a more capable theoretician than Russell, and prevailed in an argument with Russell over his development of an understanding of the rolling of ships. William Rankine, stimulated by William Thomson and James Napier, had developed a streamline theory in the early 1860s, later supplemented by the work of Thomson and George Stokes.[73] Froude's main contribution was to extend the understanding of the behavior of ships in a seaway, by showing that model tests could give reliable data. That was validated in a major Admiralty trial in 1871, and led to his involvement in Admiralty committees on ship design in 1871 and in 1877.[74]

THE SINKING OF HMS *CAPTAIN*

On September 7, 1870, the new warship HMS *Captain* sank off Cape Finisterre with the loss of about 480 lives. Only twenty-seven were rescued, and the dead included sons of Hugh Childers, first lord of the Admiralty, and Thomas Baring, undersecretary of state for war. The ship's designer, Cowper Coles, also went down with her.

This was a tragedy with a long history. Coles had gathered fame in the Crimean War by devising a raft with a turret gun, and he developed this design over subsequent years. When the Admiralty turned down his design for a ship in 1865 he orchestrated such a campaign in the press and Parliament that the Admiralty backed down, and let him build a two-turret ship. This became the *Captain*, laid down in 1867 and completed in 1870. Its production was as controversial as its initial approval. Serious concerns about the low freeboard were raised by senior naval staff, including Vice Admiral Robert Robinson, comptroller of the navy and elected a fellow of the Royal Society in 1869, but they were overruled.

In an unusual departure, the subsequent court martial called Thomson and Rankine as witnesses. The inquiry established that the ship had capsized because of its low freeboard and lack of stability for righting. Indeed, the ship had sailed before the results of inclination tests had been published. The conclusion was that the ship had been built in deference to public opinion, against the views of the comptroller of the navy.[75]

The Admiralty soon established a large committee to examine the designs of recent ships of war.[76] Chaired by Lord Dufferin and Clandeboye, and with a complement of admirals and other senior naval officers, the civilian component was a strong one. Among them were three fellows of the Royal Society in Thomson, who chaired the scientific subcommittee, Rankine, and Froude. The engineer members included George Bidder and George Rendel. The witnesses examined were mostly naval officers, while the scientific subcommittee reported on the stability and strength of various designs of ships, and Rankine added papers on the stability of mastless and sailing ships.

The report explored the tensions between speed and armor plating, and between offensive and defensive capability for different types of ships in different situations. Assuming that the navy would want to retain flexibility, the committee offered suggestions for improvement of the various types. The members did not agree with Armstrong's view that armor plating should be reduced to a minimum on all ships, given that the power of current guns and their likely development would render it of limited effectiveness.[77] Two senior admirals, George Elliott and Alfred Ryder, dissented from the report, implying that the eight members of the scientific subcommittee "who are not familiar with ships and their behaviour at sea and in action" might have been biased by the four naval members.[78] Thomson, by then Lord Kelvin, served on a subsequent committee from 1904 to 1905 on the design of the new dreadnought battleships and battle cruisers.

Rendel offers a particular example of the Admiralty's use of civilian expertise. After serving on the committee on warship design he played a major role in the 1877 design of *Inflexible*, which had four eighty-ton guns and thick armor plating. In 1882 Rendel left his position with Armstrong and became an extra-professional civil lord of the Admiralty, a new post intended to bring a "practical-man-of-science" to the Board of Admiralty. He held the post until ill health intervened in 1885.

Toward the end of the nineteenth century, the question of offensive against defensive qualities was raised in public debate at the Institute of Naval Architects when William White, director of naval construction, presented a paper defending the design of the *Royal Sovereign* as a fast ship, but with limited armor. Reed, now an MP, objected to the lack of armor. He also charged that the best scientific advice, producing high-sided seaworthy shops, was being sacrificed to demands by naval officers for unstable gun platforms such as *Captain*.[79]

But as White and his successor Philip Watts recognized, there was a need to balance the strategic demands of the navy and the operational expertise of naval officers with scientific advice on possible solutions to meet their needs. Naval architects gained their authority through their interactions with naval officers, Admiralty staff, and politicians. The best politicians and administrators were open to their ideas, while being clear about the strategic and operational priorities. This time, at the end of the century, was a period of concern that Britain's place as a leading naval power was being lost. Lord George Hamilton, first lord of the Admiralty, saw through the Naval Defence Act 1889, which massively expanded naval construction. The Armstrong and Whitworth companies merged, as the form of the military-scientific-industrial complex started to take shape.

Throughout the nineteenth century the Admiralty made great use of civilian scientific and engineering expertise, which complemented its extensive in-house capability. Those interactions generated practical and policy advice, in addition to the production of ships, weapons, engines, cables, and anchors. Over the century naval architects like Russell and Froude gained influence on policy, but had to contest their authority continuously against the judgements of naval officers, experienced shipwrights, and powerful politicians.

By the beginning of the twentieth century a balance was emerging, in which the boundaries between naval and scientific expertise and responsibility were clearer. That is illustrated by the approaches of George Hamilton in building the *Royal Sovereign* in the 1880s, and John "Jacky" Fisher as first sea lord in 1904 for the design of *Dreadnought*. Hamilton consulted with naval officers on the ship's requirements, which White then designed. Fisher went a step further by bringing together naval and civilian expertise with Watts from the outset. This was the committee to which Thomson contributed, with other ship and engine designers.[80] The design was then iterated, while supportive experiments were commissioned.[81] It was a far cry from the artisanal although still expert approach of the shipwrights of the early nineteenth century.[82]

4

WAR OFFICE, ARMY, AND ORDNANCE

THE WAR OFFICE AND THE ADMIRALTY HAD DIFFERENT CULTURES AND histories. Britain's naval forces were controlled by the Admiralty, and both civil and military aspects came under the responsibility of the first lord of the Admiralty. Responsibility for Britain's land forces was more complex. This reflected in part the political sensitivity of handling a standing army, not least at times of civil unrest when elements of the British army itself might be a potential threat, and when the Home Office was responsible for internal security.

Military control of the army lay with the commander in chief. During the nineteenth century, relatively few individuals held the post. Among notable figures, the Duke of York was commander in chief from 1795 to 1809, and from 1811 to 1827. The Duke of Wellington held the position from 1827 to 1828, and from 1842 until his death in 1852. He was no innovator. Wellington was succeeded by Viscount Hardinge. After him, the Duke of Cambridge, cousin of Queen Victoria and a champion of the conservative faction in the army, held the post from 1856 to 1895. Direct interaction of the scientific advisers with the field army was limited. Their connections were stronger with the civil administration in the War Office and with the Board of Ordnance, whose political heads generally held much shorter appointments.

Civil administration of the army at the beginning of the nineteenth century was complex, split across thirteen different departments.[1] The post of secretary of state for War, created in 1794, became that of secretary of state for War and the Colonies in 1801. This cabinet position was civil and political, with the holder responsible to Parliament for the adequacy of the establishment recommended in the annual army estimates, the plans for expenditure. Alongside this post the secretary-at-war, who ran the War Office, was responsible for financial matters. This post was occasionally a cabinet-level position, and was merged in 1854 with the re-created role of secretary of state for War, whose office was confusingly known as the War Department, becoming the War Office in 1857. The War Office and the Admiralty remained firmly separate. The first Minister of Defence was not appointed until 1946, and the integrated Ministry of Defence was not established until 1964.

Throughout the nineteenth century there was tension between civilian and military command, as Parliament took increasing control from the monarch for the direction and operation of the army and sought better financial accountability. Integration of army command and the War Office following Edward Cardwell's reorganization in 1871 did not solve the issues. For example, after the Cardwell reforms the surveyor general, in charge of supply and manufacture, became a political and civilian appointment, not a military officer as the master-general of the Ordnance had been. That led to complaints from the military side about his lack of expertise.

There was also a lack of clear agreement or specification of Britain's overall military policy. As late as 1887 the reforming military chief Viscount Wolseley could say, "There has never been any authoritative inquiry instituted as to what are the military requirements of the Empire."[2] The secretary of state for War in 1903 wrote to the prime minister, "I do not find that any definitive instruction exists as to what is the exact purpose for which the Army exists, and what duties it is supposed to perform."[3]

Underneath these broader concerns of foreign policy and high politics were technical questions of armaments, explosives, and other war materiél. While the ultimate choice of the types and numbers of guns and ships was political, optimizing their performance and designing and manufacturing new munitions required particular scientific and engineering expertise.[4] Some of that came from external civilian rather than military personnel, as already seen in relation to the Admiralty in the previous chapter.

This chapter deals primarily with the development of weapons and explosives. The health of the army was a further major concern, especially after the debacle of the Crimean War. This challenge, under the eagle eye of Florence Nightingale, may be followed elsewhere,[5] as may disputes over the siting and design of Netley and St Thomas's Hospitals, the latter bringing it into conflict with John Simon.[6] So too may the involvement of George Stokes and Frederick Abel in advising on optimum colours for army uniforms.[7]

BOARD OF ORDNANCE

The Board of Ordnance, which also supported the Royal Navy, formed a critical element of military responsibility. It dates back to 1597, and operated until 1855 when Lord Panmure, the secretary of state for War in the incoming Palmerston government, abolished it and incorporated it into the War Office in the upheavals during the Crimean War.[8] Headed by the master-general of the Ordnance, it was a huge organization that buttressed Britain's military capabilities. At the beginning of the nineteenth century, it was second only to the Treasury in power and influence. Its responsibilities encompassed gunpowder manufacture and storage, artillery and small arms manufacture, forts and fortifications, the Royal Artillery and the Royal Engineers, and the Royal Military Academy at Woolwich. The master-general of the Ordnance, a senior military figure, had a seat in the cabinet until 1828.

Surveying was important to the Board of Ordnance. That derived from building harbors and fortifications, and resulted in the creation of the Ordnance Survey in 1791. It forms an aspect of the scientific approach to measurement that underpinned several important state activities. While the Ordnance Survey mapped the land, the Royal Observatory mapped the heavens, the Hydrographic Office mapped the seas and the coasts, and later the Meteorological Office measured the atmosphere.[9] All were vital to Britain's military interests, and to colonial and domestic activities.

The Ordnance Survey, for which the War Office took responsibility on the abolition of the Board of Ordnance in 1855, became the focus of a commission established in 1858 to examine its purpose and progress.[10] Chaired by Lord Wrottesley, its members included the Earl of Rosse, George Airy, Isambard Kingdom Brunel, and Charles Vignoles. The secretary was Stokes. The commissioners recognised the importance of the survey both for the needs of the state and for individual citizens. For the state, the purposes encompassed military objectives, land taxation, and land transactions. For individuals, a general map was seen as a primary need, since the commissioners thought that its production on a commercial basis was unlikely. A general map would enable better private surveys of particular areas, for facilities such as railways, water supply, drainage, and mining.[11] The state was prepared to fund this comprehensive national survey to provide both military and civil benefits. Responsibility for the Ordnance Survey would later move to the Office of Works in 1870 and then to the Board of Agriculture in 1890.

Engineering and manufacturing capability, scientific research, and scientific education for officers were vital to the Board of Ordnance. The board had links with the Royal Society through its funding of the Royal Observatory, although that ceased in 1818. In the 1790s the Royal Society had given advice to the Board of Ordnance about the trigonometrical surveys that the board was undertaking, and on lightning conductors for magazines.[12] In 1801 the Board of Ordnance further asked the Royal Society about the best means of covering the floors of powder works to avoid the danger of fire. A committee of chemists—Henry Cavendish, Charles Hatchett, and Count Rumford, plus Joseph Banks, Edward Gray, and Charles Blagden—gave advice.[13] That was followed by a request in 1803 to advise on mothproofing the flannel used to wrap cartridges for cannon. Banks took this on himself, with the two secretaries Gray and Joseph Planta. The committee acknowledged its lack of knowledge of the habits of insects in distant climes, but recommended tight packing of cartridges in paper to exclude insects.[14] These were rare requests for external help, less frequent than from the Admiralty.

Under the Board of Ordnance were the Royal Artillery and the Royal Engineers, staffed by officers with engineering and scientific training and experience. They guarded their skills and knowledge in-house, perhaps due to a combination of professional pride and military security, but also because of the existence of the Royal Military Academy at Woolwich.

THE ROYAL MILITARY ACADEMY AT WOOLWICH

The Royal Military Academy gave the Ordnance Board access to civilian scientific knowledge. The academy had been founded in 1741 to educate artillery and engineer officers through theoretical and practical studies. The quality of recruits was enhanced from 1855 when admission to cadetships was determined by competitive examination. The Royal Military Academy was a counterpoint to the Royal Military College Sandhurst, which trained officers for the cavalry and infantry. The two were eventually merged in 1947, as the Royal Military Academy Sandhurst.

The Royal Military Academy at Woolwich, housed in a new purpose-built building from 1806, was home to teaching staff who were leaders in their fields. They were a natural channel for advice alongside their teaching duties. Given the paucity of scientific employment in the early nineteenth century, those who did not have independent means yet wanted to undertake research often had to obtain jobs as lecturers. Woolwich offered one such opportunity.

In its early days Woolwich had a first master, later called professor of fortification and gunnery, and a second master, later professor of mathematics. Leading mathematicians who served as the professor of mathematics at Woolwich in the nineteenth century included Samuel Christie from 1838 to 1854, and James Sylvester from 1855 to 1870, despite competition from Stokes. Given the demand for explosives and analysis of materials, chemical knowhow was also important. Notable professors of chemistry were Michael Faraday from 1829 to 1852, and Frederick Abel from 1852 to 1888. James Marsh was ordnance chemist at Woolwich Arsenal and assistant to Faraday at the Royal Military Academy until his death in 1846.

Faraday became professor of chemistry at Woolwich in December 1829, having started to negotiate for the post shortly after Davy's death. He managed to drive a hard bargain that enabled him to retain his position at the Royal Institution while securing £200 per annum for twenty lectures at the academy, which required him to be there a day or two each week during term time. The military was prepared to pay to attract top quality people in research and lecturing.[15] Faraday helped ensure that the quality was maintained. Abel recounted later that he was appointed in early 1849, "partly through the kind recommendation of Faraday, to instruct the senior cadets and a class of artillery officers in the Arsenal, in practical chemistry."[16]

The work at Woolwich, whether research or teaching, had to be devoted to practical ends. With the exception of the government grant scheme administered through the Royal Society from 1850,[17] the government did not finance basic research. Although the men of science believed that basic research would lead to practical applications, even if those applications could not be predicted, that view was not more widely shared.[18] Faraday's groundbreaking discovery of electromagnetic induction in 1831 was made in the time he carved out for his private researches, although he was able to use the huge battery at Woolwich for crucial experiments.[19]

Alongside his teaching, Faraday was called on from time to time for advice. For example, in 1843 he advised on the cause of an explosion at the Waltham Abbey Royal Gunpowder Factory. His report to Colonel James Cockburn, director of the Royal Laboratory, carefully set out some possible if unlikely circumstances by which, through friction or electrical action, the powder might have been caused to explode.[20] The type of granulating machine used in the building that exploded was not subsequently employed, although it is not clear if Faraday's advice was specific enough to account for that outcome.[21] In 1852 Faraday was able to provide the Committee of Artillery Officers with chemical analysis of a new type of French shell that had been captured off Gibraltar.[22]

FREDERICK ABEL

With a foot in the door, Abel succeeded Faraday as professor of chemistry in 1852, although for the same salary he was required to do more work. He was appointed ordnance chemist in 1854 and made chemist to the War Department in 1856, which became the War Office in 1857. For more than thirty years, as chemist to the War Office, Abel was the center of chemical advice to the military on explosives, alloys for guns, and related matters. He is notable among the major nineteenth-century chemists in devoting much his working life to military concerns, with a side line in work for the steel and petroleum industries. His family background was German, and he was one of the first students at the Royal College of Chemistry in 1845 under Wilhelm Hofmann. He was soon appointed an assistant there, and came to the notice of Faraday, who introduced him at Woolwich. In 1851 he also became demonstrator in chemistry at St Bartholomew's Hospital, under the chemist John Stenhouse.

Abel was behind many of the advances in explosives design during the nineteenth century. Until the appointment of Abel, the production of gunpowder in Britain was seen as a craft activity, with research on explosives carried out at the Royal Laboratory at Woolwich under William Congreve. The situation was different in France, where the chemist Antoine Lavoisier had been appointed in 1775 in a reformed gunpowder administration.

Abel's work on guncotton, which involved reducing it to a fine pulp, enabled it to be manufactured and stored safely. This research eventually led to the replacement of black powder by smokeless explosive. With Andrew Noble, a Royal Artillery officer, Abel carried out research on the nature of chemical changes on firing explosives, a dangerous undertaking. He received the Royal Medal in 1877 for his work on guncotton and explosives, and Noble received it three years later. They are two of only a handful of Royal Medals awarded in the entire nineteenth century that relate clearly to practical applications of science. Abel's experience with explosives made him a natural choice for a royal commission on examining explosions in coal mines (see chapter 10).

GUNCOTTON

Until the middle of the nineteenth century the traditional military propellant was "black powder," a mixture of saltpeter, sulphur, and charcoal. But in 1846 Christian Schoenbein discovered guncotton, a nitrated form of cellulose. It was more powerful than gunpowder weight for weight, and it burned without fouling guns or smoking, potentially helping to conceal gun emplacements. Nevertheless it took decades to develop as a military propellant for projectiles like bullets, cannon balls or shells.[23]

Abel started his research on guncotton in 1863. He achieved partial success in producing an effective smokeless propellant, which he characterised as trinitrocellulose, by 1865. The work was not considered confidential to the military, so that Abel published it in the *Philosophical Transactions of the Royal Society of London* and gave two discourses at the Royal Institution in 1864 and 1866, with a third in 1872 when he announced that no progress had been made since 1868. Indeed, guncotton was used in sporting shotguns even though it was not reliable enough at the time for use in military rifles.

The challenges were not just chemical, but related to the means of production and the physical constitution of the chemicals, the characteristics of the different rifles and cannons to be used, and safety. Production sites had a habit of blowing up, leading to a halt on manufacturing and testing this explosive until the Austrian Wilhelm Lenk von Wolfsberg produced a more stable version on which he was allowed to report to a committee of the British Association for the Advancement of Science in 1863. Abel was also permitted by the secretary of state for War to make a report. This committee of the mechanical and chemical sections had been formed in 1862 to investigate the application of guncotton to warfare, with a budget of £50, and sat until 1864.[24] When it reported, it suggested that much more experimental work was needed, and a deputation led by Sabine, president of the Royal Society and an officer in the Royal Artillery, drew its attention to the secretary of state for War.[25] The British Association committee was optimistic about substituting guncotton for gunpowder, and Abel took up his research at the request of the secretary of state.

The government appointed a Guncotton Committee in January 1864 and its reports in 1868–1869 covered the use of guncotton for military and naval purposes and for mining and quarrying.[26] Sabine chaired this committee, which included Abel, Stokes, William Allen Miller, John Gladstone, and the mining engineer Thomas Sopwith, with five military and naval officers. Abel carried out work from 1865 to 1868 on guncotton for cannon and small arms and was optimistic about the potential of pulped guncotton. Stokes then provided a favorable analysis of a series of tests on the firing of cylinders of guncotton in guns.[27]

COMMITTEES

The Guncotton Committee is one example of the centrality of committees to providing expert advice to the military during the nineteenth century. Committees may have broad or limited scope. Their membership may or may not include the most senior people, and they may involve civilians alongside serving officers. They may be reactive or proactive, intended as permanent or temporary, and advisory or decision-making. All these choices affect their mode of operation and influence. While military committees were invariably advisory, they differed greatly in other aspects. Some examples of the manner in which the War Office and the Board of Ordnance created, amended, disbanded, and reconstituted them over time is explored below. These committees shaped the context within which scientific advice was provided.

The military officers had established two part-time committees in 1765 to review technical developments from a military perspective, the Colonels' Committee and the Field Officers' Committee. These, containing no civilians, were later combined in a committee that existed until about 1830.[28] In parallel the Select Committee was formed in 1805 at Woolwich, to advise the

master-general of the Board of Ordnance on technical matters arising from
naval and military affairs. The members were invariably elderly officers, and
there was no civilian membership, so that according to one commentator, "the
Select Committee was hide-bound, steeped in traditional methods, lacking in
imagination and opposed to change."[29] With peace in Europe after the end of
the Napoleonic Wars in 1815, the committee remained unchallenged until it
was reconstituted in 1852 as the Select Committee of Artillery Officers and
confined to giving advice to the master-general on scientific and professional
matters relating to artillery.[30]

Other committees were formed on an ad hoc basis, such as a Small Arms
Committee, a Machinery Committee, and a Committee on Lancaster's Shells.
The scope and membership of the Select Committee of Artillery Officers were
broadened in 1855, under the weight of inventions and suggestions arising
from the Crimean War, when it became the Ordnance Select Committee. The
experience of the Crimean War, from October 1853 to March 1856, had challenged the entire military system. The fifth Duke of Newcastle, secretary of
state for War in the Earl of Aberdeen's coalition government, found the Select
Committee of Artillery Officers inadequate. It was slow, out of touch with
industrial advances, and unpopular. He created the Ordnance Select Committee just as he handed over to Lord Panmure.

Newcastle was explicit in his view of the government's need to obtain "other
than strictly military opinions upon questions of mixed scientific and practical character."[31] He was also concerned that the committee was purely reactive, when he thought it should have the expertise to originate ideas. Indeed,
he had to express his views quite strongly on those points to Major General
William Cator, the president of the committee, who was satisfied with the
committee's modes of accessing external advice and was resisting him.

At Newcastle's instigation two civilians, Charles Wheatstone and the civil
engineer Charles Hutton Gregory, were brought in. Gregory was known to
the military as the son of Olynthus Gregory, who had been professor of mathematics at Woolwich in succession to Charles Hutton, after whom his son was
named. Sylvester, professor of mathematics, and Abel, chemist to the War
Department, were also appointed to the new committee. They were joined by
engineer, artillery, and naval officers, and Board of Ordnance staff involved in
the manufacture of arms and explosives. That committee lasted just four years
and a much smaller Ordnance Select Committee of five full-time members, all
military, was appointed in 1859, incorporating the Small Arms Committee. It
was abolished by Secretary of State John Pakenham in 1868 on the grounds of
economy and replaced by an experimental branch.

By 1868 gunpowder had been better developed for artillery, negating the
need for research on guncotton for those weapons. Despite the recommendation

for the continuation of research on guncotton, a new Committee on Gunpowder and Explosive Substances, which was formed in 1869 and included Abel and Noble, restricted guncotton trials to small-caliber weapons.[32] The director of artillery played a key role in directing Abel's interest toward or away from guncotton. Guncotton was potentially of great value, but it was unstable to light and heat, making military deployment problematic. Abel continued with some experiments on its use in small arms, but without official encouragement or interest until the French developed Poudre B in the 1880s.

Advances in weapons technology, with breech-loading and rapid-firing rifles and machine guns, meant that a smokeless powder was highly desirable. The Admiralty also sought such a powder in the mid-1880s. But given the interests of the War Office, Abel and Noble explored gunpowder during the 1870s, as Robert Bunsen was doing in Germany, producing much slower burning forms. The committee was wound up in 1881. Abel blamed the military for not seizing the opportunity to use guncotton either for cannon or for small arms, but black powder had in any case been improved as an explosive and their priorities were elsewhere. As their employee, he had little choice.

Research continued nevertheless. The Committee on Explosives on Compressed Gun-cotton as Packed for Store, including Abel and Noble, reported on experiments on safe storage in 1871.[33] Likewise, the Committee on Guncotton and Lithofracteur, a formulation of nitrocellulose, was established in 1871. To the six military members, including Noble as secretary, were added the civilians William Odling, Hilary Bauerman of the School of Mines, and the engineer George Bidder. This committee was asked to examine the properties of lithofracteur supplied from abroad for quarrying purposes. Abel was not appointed as a member of the committee, given his personal interest in guncotton. But when he nevertheless attended experiments, the importer complained loudly both to the committee and to the home secretary about this breach of commercial propriety.[34] Doubtless to his disappointment, Vivian Dering Majendie, inspector of gunpowder works, who had attended most of the experiments, reported to the home secretary that the lithofracteur under consideration could not be safely licensed for storage and transport.[35] The work of this committee featured in a select committee on explosive substances, which reported in 1874 (see chapter 9).[36]

Cardwell, as a part of his reorganizations of the War Office that sought to increase civilian and financial control, created a new high-level Council of Ordnance in 1869. It was chaired by the parliamentary undersecretary of state for War and included more naval personnel. Again, there were no external members, and this short-lived committee, which was intended to deal with scientific and experimental questions "of more than ordinary importance,"[37] disappeared in 1870. Since the Ordnance Select Committee had been abolished, the recourse of the Board of Ordnance to deal with proposed inventions, experiments, and trials was to create ad hoc committees. Ten were still in place by 1881, when the director of artillery and stores suggested to the surveyor general that a permanent committee was needed. He gave two main reasons for establishing a permanent committee. In the first place he thought that "the public," by which he probably meant inventors, "would feel more confidence in decisions arrived at on the recommendations of a recognised Committee." He also thought that the previous involvement of civilian members had "strengthened the hands of the secretary of State," presumably by being seen as independent, although he remarked that "it appears not to have saved the Committee from being re-organized and their services dispensed with."[38]

The new Ordnance Committee was established in 1881, with the blessing of the commander in chief, provided that it stuck to advising on inventions and experiments related to artillery and explosives. The Treasury was unenthusiastic about the cost but acquiesced. With a general as president and an admiral as vice president, the committee had three civilian members. From 1881 they were the engineers William Barlow and Frederick Bramwell, with Noble, now employed at Armstrong's works. Abel was an associate member. Benjamin Baker replaced Barlow in 1888, and August Dupré joined as an associate member in 1891. Special committees operated alongside the Ordnance Committee. An example is a Small Arms Committee established in 1886, although it had no civilian members. There was also at this time an Ordnance Council that determined rewards to inventors, but it too had no civilian members.[39] These varied constitutions of committees reflected both their purposes and the views of those establishing them about the usefulness or not of having particular types of civilian expertise within the formal membership.

GUNS

The Crimean War changed the whole picture for arms and armaments. It galvanized moves to develop artillery, with guns of increased caliber and rifles with different modifications.

A discussion at the 1854 meeting of the British Association had raised the problem of the deterioration of metal used to make artillery. A committee was formed to investigate, following an inquiry instigated by a select committee at Woolwich. It reported in 1855, and made technical suggestions for better casting and manufacture.[40] This was a constructive move but it hardly touched the major challenge, which was that the development of artillery and naval guns had languished since 1815.

Hundreds of people volunteered technical improvements to military capabilities as the impact of problems in Crimea became apparent, but one man in particular transformed the design of guns, William Armstrong, from Newcastle. Armstrong's exploration of the generation of an electric charge from the escape of high-pressure steam had led to correspondence with Faraday and his election as a fellow of the Royal Society in 1846. He took an interest in water management and hydraulics, instigating a water supply system for Newcastle and designing a hydraulic crane that was soon a feature of docks around the country. Encouraged by the engineer James Rendel, who specialised in harbors and docks, Armstrong established a manufacturing site at Elswick upon Tyne in the late 1840s.

The move into armaments was facilitated by Rendel, given his involvement with Admiralty engineering projects. Isambard Kingdom Brunel had seen the shortcomings of British guns in the Crimea, and in 1854 tried to persuade the Admiralty to support his design of a floating siege gun. Despite the backing of General John Burgoyne of the Royal Engineers, the Admiralty was not persuaded. Rendel had discussed with Armstrong the idea that the lightness and strength of small arms made with wrought iron allowed the use of rifling and cylindrical bullets. Applying the idea to guns would enable a move away from cast-iron cannon and spherical shot. Armstrong thought he could do it. He developed a plan for a rifled, breech-loading gun made of wrought iron in a coiled design. It would use elongated lead projectiles. This made the weapon lighter, easier to use, more accurate, and with a longer range. Rendel was able to prevail on First Lord of the Admiralty James Graham to enable Armstrong to meet the secretary of state for War, the Duke of Newcastle. Newcastle supported the idea, as did Lord Panmure, who took over when Newcastle was forced to resign over conditions for soldiers in the Crimea. Armstrong's prototype was examined by the new Ordnance Select Committee, which suggested a larger gun. With advice from Brunel and James Nasmyth, Armstrong developed five-pounder, twelve-pounder, eighteen-pounder and thirty-two-pounder models. He also introduced a cast-iron shell with a percussion fuse.[41]

WILLIAM ARMSTRONG AND JOSEPH WHITWORTH

With these new developments by Armstrong, and by others, including Joseph Whitworth, who would become his main competitor, a systematic analysis became necessary. Up to this point, as Colonel John Lefroy, then secretary to the Ordnance Select Committee explained to a parliamentary select committee in 1862, possible innovations had been submitted to the Ordnance Select Committee. Their report was sent on to the secretary of state for War, who made the final decisions, with any additional remarks from Lefroy, who agreed that he was considered as "the scientific adviser to the Secretary of State."[42] General Jonathan Peel, brother of the former prime minister, had become secretary of state for War in the brief government of the fourteenth Earl of

Derby. He instigated a Special Committee on Rifled Cannon in August 1858, with Noble, then a captain in the Royal Artillery, as secretary, and it reported in November. In 1859 Peel described to the House of Commons that he had chosen to order the Armstrong weapons, on the basis of tests carried out at Shoeburyness. They were lighter, more destructive, and fifty-seven times as accurate as existing guns.[43] He ordered one hundred guns. Peel also explained the deal he had done with Armstrong. In exchange for funding for the Elswick works and for further development of the guns, Armstrong made his patents over to the government. He was knighted for his contributions. To enable him to exploit the patents, he was appointed engineer to the War Department for rifled ordnance, with a salary of £2,000 per annum backdated to 1856. He also became superintendent of the Royal Gun Factory, to the dismay of some military men, and was able to supervise work both at Elswick and Woolwich. This unusual arrangement lasted until 1863, when Armstrong resigned to look after the Elswick works after the government canceled some contracts over a dispute about costs. At Elswick, George Rendel, son of James, took charge of the ordnance works and Armstrong tempted Noble out of military service to manage the ammunition works.

Like Armstrong, Whitworth was not going to give up opportunities to supply the government without a fight. After a dispute about their large guns to defend coastal stations, a special committee examined the respective merits of their guns and ammunition for land and sea service over three years from 1863.[44] It was a comprehensive look at all aspects of guns, including their use against iron plates, the advantages of muzzle-loading and breech-loading, their caliber, firing characteristics, design, and manufacture. This time, Liberal Secretary of State for War George Cornewall Lewis asked for two civil engineers "of high standing" to be added "in order that the numerous mechanical questions which are likely to arise during the investigation may be carefully considered by competent authorities."[45] He chose William Pole and John Penn. Pole, professor of civil engineering at University College London, who had been an assistant to James Rendel, was recommended by Armstrong. Penn, a marine engine builder, was nominated by Whitworth. It was an adversarial process, and these representatives were replaced after their later resignations by Stuart Rendel at the suggestion of Armstrong, after John Percy had declined, and by John McDonald, manager of the *Times*, at the suggestion of Whitworth.[46] The extensive report, produced in 1865, was backed up by the publication of detailed experimental results and accounts of accidents. The latter illustrate the colonial ambitions of the time, as recent accidents were reported from places as varied as China, Japan, Fiji, and New Zealand. The highly technical final report laid out information to inform the choice of design and manufacture of guns, while leaving many questions open, such as the use of steel favored by Whitworth, for which experience at that point was limited. Apart from the mechanical questions, Abel provided some information on fuses. Edward Frankland had previously had some involvement in the question of the effect of altitude on the rate of burning of fuses, which went back to his early experiments with John Tyndall of burning a candle on the summit of Mont Blanc. With fuses provided by Abel, he was able to show the quantitative relationship between reduction of pressure and reduced rate of burning of the fuse.[47] Although Armstrong's guns were vindicated by the trials, Whitworth's held up well too.[48] Armstrong had shown himself as politically more astute than Whitworth, and more capable of providing a weapons system that encompassed both gun and ammunition. They would compete for the rest of the century.

Metallurgy was a critical science for these developments. Percy was appointed lecturer on metallurgy to the artillery officers at Woolwich around 1864, and held this post until his death in 1889. Since 1851 he had been lecturer on metallurgy and metallurgist to the Museum of Economic Geology, posts he retained until 1879. He was appointed superintendent of ventilation in the Houses of Parliament in 1865 after the death of David Reid. He was a member of several royal commissions, on coal in 1871, on the spontaneous combustion of coal in ships in 1875, and on warlike stores in 1887.

Many issues required no external help. For example, the Ordnance Select Committee, at this point entirely a military committee chaired by Brigadier General Lefroy, a fellow of the Royal Society since 1848, examined competing rifled systems for seven-inch guns in 1865 and coiled wrought-iron tubes for guns in 1867.[49] Armstrong gave evidence to the latter inquiry, although the metallurgist Percy was not called. The committee recommended an extensive conversion of existing cast-iron smooth-bored guns into rifled guns with linings of coiled iron, for secondary purposes of defense.[50]

The increased penetrating power of guns led to an arms race in the development of iron shields to withstand them. In 1867 Secretary of State for War John Pakington, in conjunction with the lords of the Admiralty, established a special committee to report on iron shields for the land defenses of Gibraltar and Malta. With a focus on the properties of iron, he appointed William Fairbairn, Pole, and Percy to the committee, which was chaired by Rear Admiral John Hay, MP.[51] Civilian witnesses included Francis Bashforth, professor of applied mathematics at Woolwich from 1864 to 1872, and Robert Mallet.[52] The report detailed modifications made to the construction of the shields after initial testing, but even then they proved too weak to resist powerful ordnance. The committee proposed stronger shields, to be assessed by further experiments.

Bashforth, who also acted as a "mathematical referee" to the War Office,[53] carried out systematic experiments on ballistics from 1864 to 1880 to study air resistance, using a chronograph that he designed. The originality and significance of his work became the study of a committee appointed by the further promoted Major General Lefroy at the instigation of Secretary of State Cardwell in 1868. According to Lefroy, Cardwell wanted to know, amongst other things, "(1) whether it is now to be considered as proved that the resistance of the air varies practically as the cube of the velocity of the shot for all the velocities in use in gunnery . . ." and "(2) whether this law of resistance is to be regarded as a new one, the discovery of which is due to Mr. Bashforth?."[54] Lefroy asked Airy, John Adams, Stokes, and Noble to report. Airy pleaded lack of time but Stokes, Adams, and Noble reported favorably.[55] Bashforth was awarded £2,000 by the government.

By the 1870s, with the development of slower-burning powder, larger guns for naval use could be built as breechloaders, avoiding the practical problems of muzzle-loading long barrels and the danger of explosion after double loading by mistake. George Rendel managed to persuade Admiral Astley Cooper Key, first naval lord of the Admiralty, of the benefits of the new developments at Elswick. He lobbied Nathanial Barnaby, director of naval construction; Admiral Richard Hamilton, director of naval ordnance; and John Simmons, inspector general of fortifications, at the same time. The Admiralty ordered several large guns in the 1880s after approval of the design by the Ordnance Committee.[56]

In time, the Armstrong factory moved into shipbuilding, making gunboats from 1868 and launching its first battleship, *Victoria*, in 1887. Armstrong became a baron in the same year, one of Queen Victoria's six "jubilee" peers. He was the first engineer, and arguably the first man of science to be raised to the peerage, although that precedence is traditionally given to William Thomson, who became Lord Kelvin in 1892. Both had made major contributions to extending Britain's power across the globe. Armstrong still supplied large guns. In 1892 the largest five guns made for the navy all came from Elswick.[57]

SMALL ARMS

If Armstrong held the balance of power over Whitworth for the production of large guns, whether for the army, navy, or coastal defense, Whitworth sought that position for small arms. Whitworth's expertise was in precision manufacture and mass production,[58] as Tyndall attested when he dedicated a Royal Institution discourse to Whitworth's achievements in 1875. Tyndall was a friend of both Whitworth and Armstrong, and volunteered to lecture on Armstrong too should he so wish, although the offer was not taken up.

The manufacture of small arms for military use underwent a revolution in 1854 as the preeminence of US systems of machine-made production became evident.[59] These lowered the cost of manufacture and made repair in the field easier. Problems with the supply and quality of small arms from contractors had led the Board of Ordnance to plan to take manufacturing in-house by constructing a small arms factory, based on a proposal by John Anderson, chief engineer in the Royal Arsenal at Woolwich. That decision was immediately examined by a parliamentary select committee.

The witnesses included many manufacturers of guns and parts, among them Lieutenant Colonel Sam Colt, who had established a factory at Vauxhall making pistols in 1853. He promoted the cost-effectiveness of making interchangeable parts. Nasmyth, impressed by Colt's factory, thought that machine manufacture would reduce the cost and improve the quality of muskets. He also argued that the policy of the Board of Ordnance to contract private companies separately for different parts was an error, as did Whitworth, who had been commissioned by the British government to visit the United States to examine its manufactories. Whitworth and others described means of rifling barrels, and Whitworth promoted his idea of manufacturing all parts accurately by machine. Lord Raglan, master-general of the Board of Ordnance, argued for a manufacturing facility under government control.

The advice from the civilian engineering manufacturers found ready purchase with the military and with the parliamentary committee. The committee recommended construction of a limited factory at Enfield along the lines of the US system, which subsequently expanded during the Crimean War.[60]

At the same time, the government asked Whitworth to improve the service rifle.[61] Whitworth's ability to make machine tools had generated huge interest at the Great Exhibition of 1851, impressing many engineers. Prince Albert was equally impressed, and urged the army to involve him. Viscount Hardinge, the commander in chief, approached him in 1853 to ask him to build a new manufacturing plant for rifles. Raglan then invited him to chair a committee for small arms manufacture, but Whitworth declined, fearing that it would limit his opportunities. Nasmyth, no ally of Whitworth at this point, was appointed instead.[62] As the Crimean War broke out, Hardinge approved a plan for Whitworth to carry out experiments, using his own shooting gallery, to improve rifle design.

Whitworth's approach took time, as he explored the use of steel, boring the barrel from solid metal, and breech-loading.[63] His weapon's barrel had hexagonal rifling and a smaller bore than existing arms. It also required a bullet that accurately fitted the bore. His rifle became the focus of the Enfield and Whitworth Committee, which was established by Secretary of State Lord Panmure in July 1857 to compare Whitworth's with the standard Enfield rifle. This was a military committee, with the addition of Whitworth himself.[64] The process was disputatious, and Whitworth withdrew from later experiments on the grounds that they did not conform to conditions originally agreed. His rifle was found to be more accurate than the existing models, but the committee was unable to come to clear conclusions about the benefits of different forms of rifling and bore sizes. The military members found it difficult to extract the information they needed from Whitworth. They were wary of the implications for the manufacturing method and costs of the different metal that Whitworth used, and of the possible consequences of any innovation for military tactics and practice.

With the government taking much small arms manufacture into its own hands, fifteen of the largest Birmingham contractors, members of the former Birmingham Small Arms Trade Association, formed the Birmingham Small Arms Company Ltd. in 1860 in order to compete. Its heyday coincided with the American Civil War from 1861 to 1865, by which time both Whitworth and Armstrong were selling field guns to both sides. It had to diversify after the war, moving into bicycles in 1880 and motorcycles in 1895. The manufacture of small arms was largely under the control of the War Office by the close of the nineteenth century.[65]

Following the report of the Enfield and Whitworth Committee, the Ordnance Select Committee continued to investigate rifling. Its 1862 report compared Whitworth's approach with three others. The papers demonstrate the huge number of considerations and specifications that the committee took into account in a complex set of tests. The verdict was in favor of the smaller bore of Whitfield's rifle as regards accuracy, and the committee noted "the relative superiority of his small-bore rifle even as a military weapon, over all the other rifles of similar calibre that have been under trial."[66] However, they recommended against issuing it to the whole army on the grounds of expense. Whitworth continued to make some rifles, but went on to develop rifled field guns on his own account to compete with Armstrong. The principles of his modifications to rifle design were widely copied and adopted. Had he been able to manufacture bullets and cartridges at scale to fit his rifled barrels, he might have had greater success.[67]

One aspect that became increasingly important, and also contested, was breech-loading. Examples of breech-loading rifles from eight manufacturers were tested by a subcommittee of the Ordnance Select Committee in 1865. As the president Brigadier General Lefroy advised the secretary of state, the clear winner was the Snider model. It was robust, had the most rapid fire, and had the advantage that its cartridge carried its own ignition.[68] Further trials in 1866 to establish the best modifications showed that the breechloaders provided twice the rapidity of fire as muzzle-loaders, with greater accuracy.[69] Major General Charles Hay, inspector general of musketry, was a firm supporter of the new rifle. It also had a smaller bore, making ammunition lighter, just as Whitworth had advised more than a decade previously. Lefroy recommended the rifle to the secretary of state, the Marquess of Hartington, as he handed over to General Peel in the incoming Conservative government. The modified Enfield-Snider rifle became the army's primary weapon until it was joined by the Martini-Henry cartridge rifle in 1871.

The choice of small arms was a military and political decision. Although the Ordnance Select Committee sought advice from Abel on explosives for cartridges, it was military performance and the implications for military deployment and operation in the field that mattered. Civilian advice was less relevant to the choice. That said, the advent of the Martini-Henry rifle in 1871 involved some civilian input. The Special Committee on Martini-Henry Breech-Loading Rifles had five military members, including Captain Vivien Majendie, then assistant superintendent of the Royal Laboratory, and three civilians. They were Charles Gregory, the MP Lord Elcho, and Edward Ross. Given some concerns about the soundness of the mechanical principles involved in the construction of the breech action, the committee invited eight people to attend to examine the question, on the recommendations of Gregory. Although Bramwell declined, on the grounds that he had already been consulted by a gunmaker, Nasmyth and Pole were among the attendees. The group had no hesitation in pronouncing the mechanism sound and superior to the existing one, and the committee unanimously recommended adoption of the rifle by both services.[70]

Problems with the Martini-Henry rifle later became apparent in the Sudan campaign of 1884 to 1885, when cartridges would often jam. The replacement of the single-shot Martini-Henry by the magazine-fed Lee-Metford again illustrates the military and political nature of the decision.[71] The possibility of rapid fire under the control of the individual soldier challenged existing systems of command and control of fire by officers, and potentially of the logistics of ammunition supplies. Choosing a technology had implications for military organization and control. It could never be a purely scientific or technical decision. The navy had been urging the change to this rifle since 1883, given its different combat requirements, and a committee recommended its adoption with smaller-bore .303 ammunition in 1887. Secretary of State for War Edward Stanhope approved the change, which brought in a rifle with a bore designed by William Metford and a bolt-action and detachable magazine invented by James Lee.

Metford had previously invented a percussion bullet, which was introduced into service by Secretary of State George Cornewall Lewis in 1863. It offers another example of the trials and tribulations of inventors. Metford was awarded £1,000 for the invention but asked for £7,000 to include his costs, comparing his reward adversely to the £16,000 provided to Whitworth to improve his system of rifling, and the sums of around £6,000 and £5,000 that Colonel Edward Boxer had been awarded for developing fuses for large shells and the diaphragm shell. This illustrates the rewards that could be available even to serving officers. Metford eventually extracted a further £1,271 for expenses.[72]

One further evolution in 1895 brought in the famous Lee-Enfield rifle, the fastest military bolt-action rifle of the day, which remained in service into the second half of the twentieth century.

In an ironic twist, a parliamentary maneuver calling attention to the inadequate reserves of small arms ammunition brought down the Liberal government in 1895.[73] The third Marquess of Salisbury's new government created the Army Board to advise the secretary of state. It was replaced in 1904 by the Army Council, which consisted of four military and three civilian members, including the secretary of state, paralleling at last the constitution of the Board of Admiralty.[74]

CORDITE

Toward the end of the century, there was a new focus on explosives. Abel had worked intensively on guncotton in the 1860s. The research was dormant for twenty years, at which point the British realized that continental developments were surpassing theirs. By the late 1880s it was clear that Britain had lost its lead in smokeless powders and that all contenders were foreign. The British government decided on the need for a better high explosive than guncotton. A significant piece of work, which Abel carried out with James Dewar, was the invention in 1889 of cordite, a smokeless explosive.

The secretary of state for War established an Explosives Committee in July 1888, separate from the Ordnance Committee established in 1881, with Abel as president and including Dewar and Dupré, under the direction of which cordite was developed. Despite the advantage of having a smokeless explosive, it was not until the mid-1880s that a smokeless high explosive started to replace the improved gunpowders. It was driven by the use of rapid firing rifles and machine guns, and the advance of continental practice. Abel came out of retirement to develop this explosive with Noble. The committee produced its massive *Final Report of the President of the Explosives Committee 1888–1891* in July 1891.[75]

Alfred Nobel had patented ballistite in 1887, a smokeless explosive based on nitrocellulose and nitroglycerin. He had patented dynamite, based on nitroglycerin, in 1867. But ballistite had one disadvantage. It used camphor as a stabilising agent, which tended to evaporate. Abel and Dewar developed cordite by substituting camphor by a different treatment of cording. Cordite was extensively tested for storage in hot and cold climates, and its potential was such, for the army and the navy, that many guns were adapted to make use of it. An improved cordite saw effective use in World War I. The British government required Abel and Dewar to patent cordite to the government for British use. The military considered that necessary as British military research practice was relatively open compared to the secrecy sought on the continent, although it allowed individuals to patent it at their cost for foreign use. Abel and Dewar designed and patented a compound of nitroglycerin and guncotton, and assigned the patent to the secretary of state for War. That led to a lengthy court case when Nobel, the inventor of dynamite, discovered that Abel and Dewar, as members of the government's explosives committee, had seen his full specifications for similar explosives. He sued for infringement of his patents but lost.[76]

The trajectory of Abel's research illustrates the constraints under which he had to operate as a civil servant. He could not pursue his research interest in guncotton substantively for many years given government priorities, such as the emphasis given to experiments with Noble on developing black powder. As a civil servant, his freedom to research was limited. It has been argued that he might have beaten the French to a nitrocellulose powder for small arms if he had been able to continue with substantive research according to his own view of the opportunities.[77] However, there were problems with implications for cartridge design to accommodate the new explosive, with its stability, and with the fact that other countries did not seem to be developing a smokeless propellant. So in addition to scientific judgments there were political and military considerations over questions such as cost and of the design of guns, rifles, shells, and cartridges to be taken into account. It was these that determined research priorities.

Although the Explosives Committee had been disbanded in 1891, the search for a better propellant than the current form of cordite led to its reestablishment in 1900 with Lord Rayleigh as president. He was joined by Noble, William Roberts-Austen, William Crookes, and Richard Haldane. This committee operated until 1906 when a new Ordnance Board was created with military, naval, and civilian members, as the politicians and administrators continued to seek to find the best and most economical means of providing technical advice within military contexts in the circumstances of the time.

It is this practical technical advice, particularly grounded in the physical sciences, mathematics, and engineering, that underlies the provision of scientific expertise to the military, on sea and land. Scientific experts, whether as serving officers or civilians, contributed to the executive arms of the military system that developed new means of waging war. These were not matters of public interest in a legislative sense, and involvements of the experts with policy matters was limited. That distinguishes the roles of experts in the military field from their roles in civil matters. The following chapters explore the contribution of scientific expertise to different areas of civil policy and administration. The context here is different, as private and public interests, social forces, and political tensions between national and local government shaped a complex environment within which experts sought to wield their influence.

PART III

FOOD

5

AGRICULTURE

A PRIMARY CONCERN OF GOVERNMENT IS TO ENSURE THAT THE POP-
ulation is adequately and reliably fed. Famines or unaffordable food prices
may lead to social and political unrest and damage economic activity. The consequences of the French Revolution and the Napoleonic Wars, and occasional
bad harvests at home, brought such a prospect in Britain into sharp focus. Not
only was famine seen as a contributory cause of the French Revolution, with
the possibility of a home-grown version alarming the British ruling class, but
Britain also found itself cut off from food supplies abroad while the country
was not self-sufficient.

The idea of improving agriculture through science had strong eighteenth-
century roots. For the landed elite, who could afford to experiment rather
than rely on rural custom, there were both utilitarian and cultural motives.
Improvement in agriculture, and participation in agricultural societies,
enhanced their material and cultural power. Their commitment to the scientific improvement of private estates could be presented as a public benefit, by increasing the nation's overall production of food. It reflects a belief
that private investment would lead to public good, even as the landowners
enclosed land, raised rents, drove off smallholders, and created a class of poor
and dependent agricultural laborers. There was faith in the benefits of science
for future progress, even if success was limited.[1] This outlook may also have
made the politically influential landowners receptive to scientific advice over
a wider range of policy.

The science of living systems is more complex than that of inanimate matter. Understanding of the physical world developed substantially during the nineteenth century, as physics and chemistry became distinct disciplines. Building on descriptive natural history, the biological, biochemical, and medical sciences were slower to emerge. If it proved difficult to transfer knowledge of the physical sciences into practical applications and policy, that was doubly the case for the biological and medical sciences. The consequence is most evident in relation to public health and disease (discussed in chapters 11 and 12), but is also relevant to agriculture and fisheries. Agricultural yields changed little from the 1770s to the 1830s, despite scientific experimentation. Although there was a rise in yields in the period to 1860, there was no further improvement until World War I. It is not even clear if the increase in mid-century was due to science, at a period when agricultural science effectively became a branch of applied chemistry. It may rather reflect the wider adoption of best practices such as crop rotation, cattle breeding, and effective drainage.[2]

Nevertheless, scientific and engineering experts were called on throughout the nineteenth century to aid agriculture. The potential benefits of science could be invoked by all political groups contesting agricultural policies. But the big questions were not scientific. They were concerned with the private interests of powerful landowners, disputes with manufacturers over trade and tariffs, the implications of land enclosures, and criticisms from commercial factions against entailed estates and the wasteful underutilization of capital. Even so, science as useful knowledge and as promise is ever present.

THE BOARD OF AGRICULTURE
AND INTERNAL IMPROVEMENT

There was no government department of agriculture at the end of the eighteenth century, and agricultural science in any modern sense was effectively nonexistent. But one man in particular had a vision of what might be done.

John Sinclair was a landowner who developed a strong interest in agriculture, aimed at improving his land. In 1786 he toured northern Europe collecting societal information, with an emphasis on systems of agriculture. He called his approach "statistical," an examination of activities relevant to the state. Back in Scotland after his tour, Sinclair organized a Scotland-wide survey of agricultural practice. It led him to the idea of establishing a Board of Agriculture to advance agriculture across the British Isles. The opportunity to realise his dream came in 1793, when France declared war on Britain and the need for increased food production became critical. Sinclair had suggested to William Pitt a means of alleviating a currency shortage. When Pitt offered him a reward for this service, Sinclair asked for his board.[3] The Board of Agriculture and Internal Improvement was created as a closed corporation incorporated by Royal Charter, with a budget granted by Parliament of £3,000 per annum.[4]

Although Pitt had re-created the Board of Trade in 1784, there was no political interest in creating a department of state for agriculture. Nor did Sinclair establish his board with any political agenda. The Board of Agriculture, which had no political control or executive government powers, was effectively a private undertaking with public funding. The thirty ordinary members who formed the board were mostly landed gentry and MPs. Of these, the fourteen "official members" included the prime minister, home secretary, and foreign secretary. They also included Joseph Banks as president of the Royal Society. Given the political roles of most of these official members, the Board of Agriculture may be viewed as a sort of hybrid between the Board of Trade and a voluntary association.[5]

Not only was the Board of Agriculture the personal creation of Sinclair, but as president he ran it as his fiefdom. Arthur Young, who had gained experience of agriculture on his own farm, and developed a reputation as a writer on agricultural improvement, was appointed secretary. He had been chairman of the Committee of Agriculture of the Society for Arts, which had encouraged and supported agricultural improvements for many years through a system of prizes, variously in the form of cash or medals.[6]

These two leading figures, Sinclair and Young, had a wide knowledge of agricultural practice and a desire to spread that knowledge. The aim of the Board of Agriculture was to provide farmers with information on the best methods and to "excite a spirit of industry and experiment."[7] Sinclair believed in the power of his statistical surveys of types of agriculture and production, and commissioned county surveys of varying quality across the country over many years. There were large tracts of common and unproductive land that Sinclair sought to bring into cultivation to increase national food production. Over many years he attempted to introduce a General Enclosure Bill to reduce the cost of enclosing such land. He was opposed by the legal profession, unwilling to give up the lucrative fees associated with the need for each enclosure to have separate parliamentary approval, and by the Church of England, concerned about the impact on tithes. Parliament did pass a limited Enclosure Consolidation Act in 1801, at a time of food shortage, but a wide-ranging General Enclosure Act had to wait until 1845.

Despite the Board of Agriculture's quasi-public role, the government took no notice of it in policy terms. Perhaps it was a convenient political fig leaf for Pitt to be able to say that he was doing something about agriculture, for which there was no national policy.[8] This lack of policy, together with the minimal administrative structure of the time, may be responsible for the fact that advice of a scientific nature was neither sought nor taken when it was offered. The government could have used the board for advice, but it did not employ it for anything except administration of funds for drainage and potato growing.[9] Despite major problems in 1795 and 1800 in the supply and price of wheat, the government never asked the Board of Agriculture for advice, even though it made forecasts. Although it did not have political impact, the board did improve farmer and landowner confidence at a crucial time, and in this respect provided a major service.[10]

Banks took a hand in 1802. In 1799 he had hosted a meeting that included seven members of the Board of Agriculture, to establish the Royal Institution of Great Britain. This new organization bought premises in Albemarle Street, near the Sackville Street office of the Board of Agriculture. The Royal Institution suffered a financial crisis from 1802, and Banks suggested that it should organize lectures from Humphry Davy to the board on agricultural chemistry.[11] Davy, now made professor of chemistry at the Royal Institution, gave six successful lectures in 1803. The Board of Agriculture asked him to repeat them annually, which he did until 1811, after which Thomas Brande gave a series of lectures before the board's demise.

Davy was appointed professor to the Board of Agriculture, mirroring his appointment at the Royal Institution, and was required to carry out some soil and chemical analysis, which he undertook in the Royal Institution's laboratory. His book *Elements of Agricultural Chemistry*, published in 1813 and based on his lectures to the board, may be seen as a transition between the semiscientific approaches of the old Board of Agriculture and the new Justus Liebig–generated interest in the 1840s of using chemical methods to improve agricultural practice.

The end of the Napoleonic Wars in 1815 brought a new politics. It changed the economic situation and the emphasis of agriculture. The Board of Agriculture took political sides, aligning itself with the interests of farmers and landowners. By 1820 it was in financial trouble and tried to subsist on voluntary income as a membership society, but gave up in July 1821. It may not have had substantive policy impact, but it had provided farmers with pertinent information and helped "excite a spirit of industry and experiment." With agricultural science in its infancy, a situation recognized by the board, the practical support and advice it offered was important.

Parliament occasionally made a direct intervention on specific issues relating to agriculture. An example is the select committee inquiry in 1817 on machinery for manufacturing flax, after petitions had been submitted by inventors of improved methods of preparing flax and hemp for spinning. In addition to laborers and people who used the machines, the committee took evidence from John Millington, professor of mechanics at the Royal

Institution. They came to the conclusion, based largely on Millington's evidence, that the new methods would afford "an increase of employment to many Thousands, and an augmentation of the national wealth to the amount of many Millions," as well as generating cattle feed. Millington had also asked Brande to undertake a chemical analysis for nutritious matter, which was otherwise wasted in the traditional manner of harvesting flax.[12]

Sinclair continued to try to construct a channel of advice about agriculture into government. In 1823 he attempted to persuade Lord Liverpool to provide him with an office in the Board of Trade, and in 1827 to persuade William Huskisson to merge the Board of Trade into a Board of Agricultures, Manufactures, and Commerce. Both attempts were unsuccessful. Nevertheless, there was still a need for a body that could encourage and support the development of agriculture. The founding of the Royal Agricultural Society of England in 1838 as a more open membership body addressed that gap for a national agricultural institution. But its founder, the third Earl Spencer, was clear that it would "scrupulously avoid politics,"[13] by which he meant party politics.

THE CORN LAWS

The trade in "corn," meaning grains including wheat, oats, and barley, had long been regulated. Broadly, the policy from the mid-eighteenth century was to encourage import and restrain export, yet prices had risen and concerns had grown about the risk of dependency on foreign produce.[14] From 1773, restrictions dependent on the price of grain were placed on both imports and exports, to protect the home market. Changes that further favored landowners and farmers were introduced in 1804, after lobbying in response to a fall in grain prices. That depression of grain prices continued for several years, as a result of abundant harvests and of increased land brought into tillage, perhaps due to some extent to the exertions of the Board of Agriculture. But prices then increased substantially, raising concerns about social unrest when people were unable to afford grain.

With the landed interest powerful in Parliament and petitions flooding in from across the country, both the House of Commons and the House of Lords established select committees to examine the issues in 1813 and 1814.[15] The aim stated by the House of Commons committee was to increase self-sufficiency and reduce the price of grain.[16] A notable feature of these inquiries, and of all such parliamentary inquiries up to the abolition of the Corn Laws, is their focus on commercial matters of trade restrictions and tariffs. Improvements to agricultural productivity were regarded as a matter for private enterprise. So scientific advice was almost irrelevant to the political arguments. In the reams of evidence provided to both committees, two short submissions were made from the Board of Agriculture. The House of Commons committee asked the board five questions about the potential for increasing tillage. In a document of less than a page, Sinclair was encouraging, commenting that "there has been a great increase of tillage during the last ten years; that the land now in tillage is capable of being made much more productive by the extension of the improved system of cultivation, and that much land now in grass is fit to be converted into tillage."[17] For the House of Lords, Young gave evidence on the costs of producing different crops, which was relevant to establishing appropriate levels of tariffs.[18]

Following these reports, the Tory government under Liverpool legislated in 1815, passing a corn law that effectively prevented imports and kept the price of grain, and therefore bread, high. Riots resulted. However, prices subsequently fluctuated hugely. A series of changes tinkered with the fine details of the regulations in subsequent years, against the background of a growing movement toward free trade and the abolition of all tariffs.

The House of Commons appointed another select committee in 1833, which produced a mammoth report.[19] It noted that problems of low prices remained, while the cost to landowners had increased, and that the country was still not ordinarily self-sufficient in grain. It recognised the impact of farming improvements: "The spread of the Drill System of Husbandry, a better rotation of Cropping, a more judicious use of Manures, especially of Bones, extensive Draining, improvement in the Breed both of Cattle and of Sheep, have all contributed to counterbalance the fall of Price, and to sustain that surplus-Profit in the culture of the Soil on which Rent depends."[20] The report reinforced the importance of protection given to corn as an insurance against overseas dependence and potential famine. But it otherwise imagined that the market should be left to itself.

The political heat increased with the founding of the Anti-Corn Law League in 1838. The free traders, led by Richard Cobden and John Bright, argued that free trade would reduce the price of bread, increase real wages for working people and stabilize employment, and increase the efficiency of agriculture itself by raising the demand for its products. Robert Peel picked up the challenge. Peel had first been prime minster in the short-lived Conservative government of 1834 to 1835. When he became prime minister for the second time, in 1841, the country was in dire straits. Britain was at war with China, there were major disputes with France and the United States, there was depression in trade, distress in manufacture, and widespread poverty, rioting, and rick-burning. Peel moved to an increasingly free trade stance, which his budgets in 1842 and 1845 advanced by removing a number of duties and barriers to agricultural trade and raising income tax.[21] The 1845 budget removed duties on several hundred products, to the extent that protection for corn stood as an anomaly. Peel had been persuaded by the argument that free trade would increase commercial prosperity, advantage working people, and reduce poverty. Yet the Corn Laws remained, a symbol of the power of the landed interests against the growing mercantile classes. Captured within this saga is a remarkable story of Peel's patronage of a rising young star of science, Lyon Playfair.

LYON PLAYFAIR AND ROBERT PEEL

Playfair grew up in St. Andrews, Scotland. Though destined by his parents for a mercantile career, he enrolled as a medical student at Anderson's University in Glasgow and then at the University of Edinburgh. But he had always been interested in chemistry, which he studied under Thomas Graham in Glasgow. After abandoning his medical degree he went to London in 1838 as a laboratory assistant to Graham, who had moved to University College London as professor of chemistry. In 1839 Graham suggested that Playfair should study with Justus Liebig at Giessen University. Liebig, Europe's premier organic chemist, was developing his chemical approach to agriculture, based on this new science.[22] Playfair impressed him, discovering new chemicals in nutmeg and cloves in his laboratory.

When Liebig published his major work, *Organic Chemistry in Its Applications to Agriculture and Physiology*, in 1840, he invited Playfair to translate it into English, and to present it at the British Association meeting in Glasgow that year. For several years Liebig had an enthusiastic following in Britain, with his vision of chemical knowledge transforming both agriculture and medicine, until concerns about his chemical fertilizers from the mid-1840s led to a decline in his reputation.[23]

Playfair now came to the notice of significant people in British science. In Glasgow he met William Buckland, the geologist and theologian, and Henry De La Beche, director of the government-funded Geological Survey of Great Britain, which had been established in 1835 following a report by Buckland, Adam Sedgwick, and Charles Lyell. Playfair had earned his doctorate at Giessen in 1841, held employment for a short period at a dye works in Clitheroe, then accepted a position as professor of chemistry from 1842 to 1845 at the Royal Institution in Manchester, with Robert Angus Smith as his assistant.

Peel had long been interested in agriculture, and in scientific developments. Of all the prime ministers of the nineteenth century it is Peel who arguably had the strongest attentiveness to scientific matters. He was invited three times to be president of the Royal Society, but declined on each occasion. Peel argued consistently that the president should be a man of science, and declined the position of president of the British Association on the same grounds.

Liebig made a tour of Britain in 1842, accompanied by Buckland and Playfair, although contrary to some accounts it did not include a visit to Peel at Drayton Manor.[24] Nevertheless, Playfair had already been drawn to Peel's attention. At the instigation of De La Beche, Buckland, who was a close friend of Peel, had written to Peel urging an inquiry into the use of organic chemistry to improve agriculture. He had suggested that Playfair would be the ideal person to carry out experiments.[25] Peel was not disposed to create a school of agricultural science, nor to introduce state funding of agricultural research, on the grounds that if one area of science were given support, all would clamor for it. That was left to private enterprise, such as the activities of the Royal Agricultural Society. Playfair attended the meeting of the Royal Agricultural Society that July and was appointed consultant chemist.[26] He then received a letter from Michael Faraday, suggesting that he take up the post of professor of chemistry in Toronto.[27] He was inclined to accept, seeing no prospect of immediate employment in Britain. But De La Beche asked Buckland to intercede again with Peel, to try to keep Playfair in Britain. To his astonishment, Playfair, still only twenty-four years old, received an invitation from Peel to visit him at Drayton Manor. He found himself, with Buckland, in the company of the Earl of Lincoln, later the fifth Duke of Newcastle, and others keenly involved in agriculture. Here, in what for Peel was an unprecedented act, he committed to finding Playfair a position in Britain if he renounced the idea of going to Toronto.[28] Playfair agreed, but it proved not to be straightforward. De La Beche tried to employ him at the Museum of Economic Geology, but failed initially as Richard Phillips would not resign his post, and Peel was not prepared to force him to do so.

Although Buckland acted in some respects as Peel's unofficial scientific adviser, it was to Playfair to whom Peel now turned for a range of tasks. His invitation to Playfair to act on a royal commission on the state of large towns and populous districts in 1843 is described in chapter 11. But it was Playfair's involvement in agriculture that was significant as the political debate over the Corn Laws came to a head. In December 1844 Playfair was with Peel at Drayton Manor, with others including Buckland; the agriculturalists James Smith of Deanston and Philip Pusey, one of the founders of the Royal Agricultural Society; along with Richard Owen and George Stephenson. He spoke at a meeting Peel had arranged for his tenants on the latest information about vegetation, manure, feeding animals, and drainage.[29]

By the autumn of 1845, the failure of the potato crop in Ireland was becoming disturbing. Playfair was invited again to Drayton Manor, where Peel asked him to go to Ireland as a commission of inquiry, with the botanist John Lindley. Their speedy report to Peel recommended a means of storing potatoes that led to a lessening of the rapidity by which they were attacked by disease, but they could do little except observe and report on the emerging dire circumstances. Their formal report to James Graham at the Home Office was equally pessimistic,[30] and Peel wrote to a colleague that the accounts from Playfair and Lindley, whom he called "the first chemist and first botanist," were "very alarming."[31] They reported again in January 1846, stating that none of the thirty-two counties in Ireland had escaped the failure of the potato crop.[32]

The issue was not just the failure of the potato crop but the high price of grain, which together would lead to famine for impoverished Irish peasants. Peel realised that he would have to remove the duties on corn, since charitable funding by the state was not politically feasible, and that once removed they could not be reinstated. This was a policy supported privately by Prince Albert and Queen Victoria,[33] but Peel was unable to persuade all of his party to back him. With the support of Lord John Russell and the Whigs, the Corn Laws were repealed in 1846. That final step to abolition of the Corn Laws was a momentous one for Britain, as it split the Conservatives and ushered in a period of several decades of predominantly Liberal administrations.

After the repeal of the Corn Laws there was increased pressure on landowners to improve productivity, along with a gradual increase in livestock farming and away from grain. New approaches in chemistry, initially through the chemical agriculture of Liebig and his fertilizers, offered one approach. Picking up on these possibilities, the Royal Agricultural Society established an unsalaried post for an analyst, and the Royal Agricultural College in Cirencester opened in 1845 as a teaching institution to include agricultural chemistry, but it proved hard to convert chemical knowledge from research into profitable results. The systematic and significant influence of science on agriculture had to await, on the one hand, the Haber process for fixing nitrogen, developed in the first decade of the twentieth century, and on the other the development of sciences such as bacteriology, virology, and genetics. Only at the end of the nineteenth century did understanding of insects, animals, and plants start to make a significant difference.

SEWAGE

The use of sewage was intimately connected with agriculture. Prior to the development of water closets and flushed systems of drainage, human excrement was collected from cesspits to be taken outside city or town boundaries and used as a fertilizer. As part of his master plan to provide London with a unified system of water and drainage (see chapter 11), Edwin Chadwick imagined that sewage from the metropolis could be piped out and sprayed over the surrounded countryside. That would both increase food supply and generate revenue to help fund the huge engineering works required.

Apart from economics, there were good chemical and theological reasons to advocate for this recycling. Liebig's ideas of continuous chemical change or "metamorphosis" in living systems required the recycling of matter, with putrefactive processes breaking down the large organic molecules to smaller ones that could be taken up by plants. On theological grounds, those such as James Johnston argued that these chemical cycles were ordained and maintained by God.[34] In both cases, working with the grain of natural law was the appropriate response. This approach led also to the idea that "filth" was matter that had ended up in the wrong place. By being spread on the land, and filtered by the soil and living matter as it passed through, it could in principle be purified before any remaining water returned to the rivers.[35]

When a bill was introduced in 1846 to enable the Metropolitan Sewage Manure Company to build a system, the government turned to De La Beche, George Stephenson, and Playfair. Their report was critical both of the lack of proper engineering plans submitted with the bill and of the assumption that no unification of water and sewage responsibilities would be undertaken in the metropolis. They reported that they did not find that the undertaking would give "sufficient protection from injury to the public health, or from nuisance and inconvenience to the public."[36]

A select committee then examined several schemes, from the relatively small scale proposed in the bill to one by Thomas Wicksteed to carry off the entire sewage of London in a single tunnel. Chadwick gave evidence, along with engineers, including James Smith of Deanston and Thomas Hawksley, and the chemists William Allen Miller, John Cooper, and Brande. Chadwick, who had met Playfair when staying with James Thomson in Clitheroe in 1842, revealed that he had suggested that the liquid waste from Thomson's calico works could be piped onto the land as manure. Smith had overseen the drainage, and Chadwick believed it was the first application of industrial waste as a fertilizer in the country. Miller reported an analysis of the dissolved and solid matter in sewage, which he claimed showed the useful nitrogen content in the liquid. The questioning by members of the committee went into extensive detail both on the engineering proposals and on the chemistry of the sewage and the feasibility of particular chemical treatments.[37]

The committee decided that the benefits of liquid manure and sewage water as fertilizers had been demonstrated, and that the system of pipes and hoses had been shown to be practicable. They recommended that the bill pass. Nevertheless, the company failed to put the plan into practice.

The General Board of Health (the creation of which in 1848 is described in chapter 11), advocated forcefully for the wider use of sewage as manure. In 1852 the superintending inspector, William Lee, produced a long report for local boards of health. He documented the many places where a system had been implemented but also called for powers for local boards to be able to extend systems beyond their immediate boundaries. He came to the conclusion that the only method consistent with sanitary principles was house drainage for collection and transmission through pipes, and the use of hose and jet for application. He claimed that sewer water was far cheaper than solid manure and that its use quadrupled the value of land.[38]

THE SEWAGE UTILIZATION ACT 1865

The possible use of sewage as fertilizer remained live for decades. It brought together a complex set of factors. The landed interest saw the potential for agricultural improvement to increase food production and provide economic benefit, while public health champions envisaged a solution for the swift removal of town sewage. Against these were concerns about water and river pollution.

The problem was handed to a royal commission on distributing the sewage of towns in 1857. Its characterisation as a problem of engineering, public health, and chemistry is illustrated by the choice of commissioners. The sixth Earl of Essex, an agriculturalist, chaired the commission, which included the sanitary engineers Robert Rawlinson and Henry Austin, the medical men Thomas Southwood Smith and John Simon, and the agricultural chemists John Way and John Lawes. Simon's call for sewage to be diverted onto the land for public health reasons, to remove the fecal-contaminated water that might lead to epidemic diseases, may have played a role in the creation of the commission.[39]

The commissioners took the view that they should look not only at the benefits of applying sewage for agriculture but also at the position of town populations for disposal of their sewage. They attached a plan for dealing with the urgent problem in London. This involved building embankments beyond the existing mudbanks, leaving reservoirs in which solid matter could settle and liquid be deodorised before release into the River Thames.[40] It was an ambitious scheme, offering in addition new recreation, transport, and navigation opportunities, but was never realised in that form.

After a lengthy exposition of the public health challenges, and the promise of further research, the commissioners turned to the utilisation of sewage. Southwood Smith, Way, and Austin visited Milan in 1857 to examine the irrigation system there. The commissioners compared irrigation systems in Britain, using channels flowing under gravity, with those using pipes and hoses. They thought that either could be used, and that any public health and nuisance concerns could be managed. They pointed to the need to match the amount of land to the supply of sewage, and showed that this treatment was best for grassland, which could stand regular watering, unlike many crops. They also explored the challenge of precipitating the solid elements using lime, as a means of treating sewage that could not be used for agriculture, following analysis by Wilhelm Hofmann and Henry Witt.[41] Again they decided that these processes could be carried out without damage to public health or creating nuisances. Overall, they declared the pollution of rivers by sewage and industrial waste to be an "evil of national importance."[42] The Treasury authorized an examination of river quality and experiments on deodorising sewage following the report.

The commissioners' second report, published in 1861, summarised the new research. This concentrated on documenting the dire conditions in rivers but also outlined tests of different methods of treating sewage. Methods included precipitation by iron salts, which Hofmann and Edward Frankland had reported to the Metropolitan Board of Works in 1859, although the chemists William Odling and Henry Letheby disagreed with them.[43] The commissioners still saw the use of liquid sewage as a benefit for agriculture, but could show little progress. The commissioners concluded again that river pollution was a national evil, and that it was getting worse, affecting water supply, canals, fishing, and even navigation. They put the problem down to the lack of any general jurisdiction over the country's waters, and the lack of protection given to the many interests connected with rivers. The commissioners had just one recommendation. That was for the need for the creation of "responsible conservancies throughout the country," with powers vested in independent local authorities such as commissioners of sewers to undertake improvements.[44]

Six months after this report was officially completed, a select committee took up the issue of utilizing sewage. Its first report was hardly ambitious, perhaps reflecting the fact that publication of the royal commission report had been delayed by the deaths of Austin, who was putting it together, and Southwood Smith. The committee opined, after taking evidence from experts, including Lawes, Hofmann, Way, Frankland, Angus Smith, and Augustus Voelcker, consulting chemist to the Royal Agricultural Society, that more careful and exact experiments were needed. It noted that some were being carried out at Rugby, and recommended that these should be continued.[45] The second report concluded that sewage could be of great agricultural value and used for profit. The committee thought that there was now enough information for towns to act according to their own circumstances.[46]

Two years later, still with no legislative action, another select committee chaired by Lord Robert Montagu looked at plans for dealing with the sewage of the metropolis and other large towns with a view to its utilization for agricultural purposes. This committee restated the view that conveying sewage by pipes to the country would be beneficial to agriculture and benefit the ratepayers in towns providing the sewage. They took the Metropolitan Board of Works to task for not making enough effort, effectively a criticism of the chief engineer Joseph Bazalgette, designer of London's major sewage system.[47] Lawes again described the agricultural benefit.

Much of the evidence related to river pollution. The problem was that the removal of cesspits and the widespread introduction of flushing toilets, to remove filth quickly from houses, had resulted in sewage being washed straight into rivers. The assumption by the Commissioners of Sewers and the Board of Health was that this was preferable on health grounds, and that steps would then be taken—for example, under the nuisance laws, to prevent or remove any pollution. That had not happened to any significant degree, given the expense of bringing prosecutions and the difficulty of showing the specific cause of any damage.

Cause and effect were hard to prove. For example, Robert Rawlinson, now principal inspector in the Local Government Act Office and one of the sewage commissioners, stated that there was no evidence to show that mortality was increased by the foul atmosphere of polluted rivers. But Way and other witnesses gave evidence that water contaminated by sewage could not be purified for drinking by any known means. William Ffennell, the chief inspector of fisheries, was clear that a considerable increase in food, and revenue to owners of rivers, would accrue from removing pollution. But as Henry Acland, Regius professor of medicine at Oxford University, and others pointed out, rivers could only be freed effectually from pollution if the Local Government Act applied to entire watersheds. The committee wanted powers to compel local boards to make sewage innocuous by applying it to the land.

At last, in March 1865 and after eight years, the sewage commission made its final report. There was nothing new to say. Just a reiteration, based on continued experiments, that the right way to dispose of town sewage was to apply it continuously to the land, by methods depending on local circumstances. The evidence was that this could be profitable, and even where not, that the demand on the rates would be small. The commissioners wanted towns to be prohibited from discharging sewage into rivers, and to have powers to take land for sewage application if necessary.[48] Even Liebig weighed in, writing to the Lord Mayor of London about the opportunities and the imperative to recycle sewage, stimulating acrimonious debates over different approaches.[49] Liebig had argued that it would remove the need to import guano, which was widely used in mid-century, although Lawes disagreed.

One consequence of all this effort was the Sewage Utilization Act 1865, the first general act passed for the more useful application of sewage. The act imposed a new layer of sewer authorities, although affairs were still complicated in practice by the diversity of local government arrangements. These authorities were given powers to ensure supplies of water and effective drainage, which could be exercised beyond their district and included applying

sewage to land for agricultural purposes, along the lines the select committee
and commissioners had recommended. They were not to put sewage matter
into watercourses. The system was strengthened by the Sanitary Act 1866 (see
chapter 12) and by a further Sewage Utilization Act 1867. Both were amended
in 1870, with consolidating measures in the Public Health Act 1875. Many
engineering and scientific organizations contributed to discussions of how
best to utilize sewage. For example, the British Association had a standing
committee on sewage from 1869 to 1876. Sewage farming could sometimes be
made to pay, at least on grassland, but it was a complex business dependent
on local conditions and agricultural knowledge.[50] In economic terms, the use
of imported guano and the new superphosphate and other fertilizers being
developed by Lawes would generally prove superior.[51]

By now, despite some useful local initiatives, the disastrous levels of pollution in waterways were all too evident. Whether it was to avoid the difficult
challenge of legislating, a charge made by the select committee chairman Lord
Montagu,[52] or to gather evidence to tackle this more general problem, the
government immediately instituted a further royal commission to explore the
best means of preventing the pollution of rivers. Its tortuous deliberations
may be followed in chapter 11.

POTATO BLIGHT

The potato blight had not disappeared in the 1840s. In 1845, when Lindley
and Playfair had reported on the devastating crop failure in Ireland, the available biological and chemical knowledge was limited. There were several lesser
potato crop failures in the 1860s and then in 1879. This time, when a select
committee in 1880 considered how to reduce the frequency and extent of crop
failures, they could call on greater but still uncertain knowledge. In addition
to potato growers, five scientific witnesses gave their views: Voelcker; William
Thiselton-Dyer, assistant director of the Royal Botanic Gardens at Kew;[53] William Carruthers, head of the Botanical Department at the British Museum;
Thomas Baldwin, superintendent of the Agricultural Department of the
National Board, Ireland; and George Worthington Smith. He was credited
in 1875—incorrectly as it turned out—with discovering the cause of potato
blight.

All agreed on the nature of the disease, the fungus *Peronospora infestans*,
and on its spread by spores. But they differed on whether it could be carried
by the wind and on how it survived over the winter. The scientific experts
did not think that any antifungal or caustic treatment to seed or soil would
halt the disease. Only Baldwin thought the disease could be stamped out—
for example, by cutting and pulling up the stalks. The Champion potato was
praised for its resistance, and all witnesses thought that new varieties which

might be resistant or early ripening should be sought. The committee recommended that the government support this work, either directly or through agricultural societies, as potato growers themselves had not done so.[54]

This was a challenge on which Charles Darwin spent much effort in his final years, in support of James Torbitt of Belfast, who was trying to produce a blight-resistant variant by selecting plants that survived the disease, in line with Darwin's principle of natural selection.[55] Darwin took Torbitt's side, supported by the botanist Joseph Hooker, in a dispute with Carruthers over whether his method of growing potatoes from their true seed could work. He spoke to Thomas Farrer at the Board of Trade, whose brother-in-law was Stafford Northcote, chancellor of the exchequer, a possible source of funding. Farrer instead approached the politician James Caird, a keen agriculturalist, but no government support was forthcoming. Instead, Darwin set up a private subscription, from which Torbitt received £410 between 1878 and 1881. Despite this support, Torbitt's ideas gained little traction. In the event, it was the Champion variety that triumphed.

Potato blight remained a concern throughout the century. After the establishment of the Board of Agriculture, Patrick Craigie, director of the Intelligence Department, which was concerned with statistics and agricultural education, produced a report on the effectiveness of a copper treatment.[56] His report, including a summary of the disease by Charles Whitehead, technical adviser to the board, and reports of trials by the Royal Agricultural Society, was supportive of the potential but considered that more experiments were needed.[57] The verdict after a further year was positive. The report recommended the treatment to be used,[58] and copper treatment had subsequently proved helpful.

ESTABLISHMENT OF THE BOARD OF AGRICULTURE

With the growing involvement of government in agricultural affairs, the Committee of Council for Agriculture was formed in 1883, and the veterinarian George Brown made his first report for this new Agricultural Department of the Privy Council Office for that year.[59] Brown's role concerned animals only, which were the primary focus of all his subsequent annual reports.[60] In 1888 and 1889, his reports extended to include research summaries. These were provided by Edgar Crookshank, professor of comparative pathology and bacteriology at King's College London. Crookshank, originally a physician who had worked with Joseph Lister, had visited Robert Koch's laboratory in Germany, and then established one of the first bacteriological laboratories for human and veterinary pathology. His seven reports in 1888 covered anthrax in pigs and horses, tuberculosis in cattle, and actinomycosis in cattle and humans,[61] followed by reports on cowpox and horsepox in 1889.[62]

Some issues were important enough to require extensive investigation. People had linked the development of scarlatina, or scarlet fever, to eruptive diseases of the teats and udders of cows, which was causing public unrest and tension with farmers. Brown, in communication with George Buchanan, principal medical officer of the Local Government Board, instigated a dozen reports from Crookshank and others, which he divided into "practical" and "scientific." The latter were based on observations and experiments by Crookshank, John McFadyean, who became director of the Royal Veterinary College in 1894, and the bacteriologist and outspoken proponent of vivisection Emanual Klein. Klein had inoculated cows with "diseased products from human patients affected with scarlatina," and McFadyean with the blood of scarlatina patients, without effect. Brown was sceptical of the idea that scarlatina might be passed from the cow to humans through milk, but did comment that the insanitary conditions in many dairies could well lead to the transmission of other diseases both among cattle and to humans.[63]

In order to bring together different government responsibilities for agriculture and for other aspects of land into a single body, the Board of Agriculture was formed in 1889. The landowner Henry Chaplin was its first president, with a seat in the cabinet. Chaplin had been a member of the royal commission on agricultural interests from 1879 to 1882, which had recommended the appointment of a minister for agriculture. He was appointed by the third Marquess of Salisbury in an effort to placate the agricultural interest.[64] Many of the board's functions were concerned with the management of land through enclosures, drainage, and improvement, but it also took over from the Privy Council Office the responsibility for dealing with contagious diseases of animals and combating destructive insects. It was required to collect agricultural statistics and to make or support inquiries, experiments, and research to promote agriculture or forestry. Unlike the previous Board, it was a public body with the president serving in government. It took responsibility for the Ordnance Survey in 1890 and for the Royal Botanic Gardens at Kew in 1903.

INSECTS AND CROP DISEASES

The Veterinary Department of the Privy Council Office had been established as a consequence of the cattle plague in the 1860s,[65] and its focus was on animal diseases. Whereas potato blight had been the major concern for crops, many other plant diseases now came under scrutiny. But the initial impetus was the voluntary work of Eleanor Ormerod, a wealthy independent woman, who started publishing annual reports of observations on injurious insects in 1877.[66] She is one of the few women in the nineteenth century who had any close involvement with scientific advice to government. Although the Destructive Insects Act 1877 gave the Privy Council Office powers to prevent the introduction into this country of the Colorado beetle, the Privy Council gave no systematic attention to pests or diseases of plants until 1885.[67] The Royal Agricultural Society had established the Entomological Department in 1882, and appointed Ormerod as honorary consulting entomologist. Whitehead, who had started out as a fruit and arable farmer in Kent, was instrumental in establishing this committee and in appointing Ormerod. He had previously persuaded the society to create a Botanical Department in 1872, with Carruthers as the consulting botanist. Like Voelcker's work in analyzing fertilizers to detect fraud, Carruthers's role included detecting the adulteration of seeds.

Whitehead, who had also given evidence on hop and fruit growing to the royal commission of 1879, thought that the Agricultural Department of the Privy Council Office should issue reports directly, and offered to prepare some.[68] His first report of 1885 was concerned with insects injurious to hop plants, and was swiftly followed by other reports on corn, grass, pea, bean, clover, and fruit crops.[69] These summarized the nature of each problem and suggested remedies. No recommendations were made for government intervention. Reports followed on subjects such as the Hessian fly and insect attacks on root crops, in a similar pattern of descriptive life histories and treatments, and with extensive help from Ormerod.[70] The spread of infestation of corn by Hessian fly in 1886 and 1887 led to Whitehead and Charles Gray, a farmer and Conservative MP, being appointed as commissioners to report on it.[71] Again this was a scientific report, with suggestions for prevention and for remedies. Fortunately, a wet summer removed the danger.

After these efforts Whitehead was appointed agricultural adviser to the Committee of Council for Agriculture, and made his first annual report in 1887.[72] It was effectively a continuation of his previous separate reports. By the time of his third report in 1889, Whitehead was technical advisor to the newly formed Board of Agriculture, which his representations may have helped to create. From 1892 his reports were produced in color, making identification of insects easier. When he retired after ten years in 1899, the board decided to change how it obtained technical advice on questions relating to agricultural botany and economic zoology. That advice in future would be sought from the Royal Botanic Gardens at Kew and the Natural History Department at South Kensington, respectively.[73] The days of the semivoluntary individual were over. The Royal Society retained an involvement too. When damage by insect pests to stored grain and flour became an issue during World War I, the society's Grain Pest Committee advised on the problem.[74]

SCIENTIFIC RESEARCH AND AGRICULTURE

Agriculture in Britain suffered a long depression from the mid-1870s, in conjunction with a trend away from grain and toward more livestock farming, to which farmers slowly accommodated. Indeed, a key rationale for fertilizing grassland with sewage was to increase meat and milk yields.

A severe drought in 1893 had exacerbated the situation and the problem was again given to a royal commission. The royal commission on agriculture, which sat from 1893 to 1897, identified the fall in prices of produce as the cause, but could see no responses which would be more than palliative.[75] Protection, even in the unlikely event of being politically possible, did not appear to be a solution.

In a supplementary report most of the commissioners, including Chaplin and Whitehead, saw the problem as one of international monetary policy, which they urged the government to address.[76] Science, it seemed, had nothing to offer in responding to the depression. In the event, the depression reached its nadir around 1896. There was no suggestion, even from the Board of Agriculture's technical adviser, that scientific advances could make British agriculture more competitive. The same was the case for the earlier royal commission on agricultural interests from 1879 to 1882. This put the main cause of the growing depression down to successive years of bad weather and to foreign competition.[77] It did suggest the need for more scientific and technical education, but not for research to improve productivity. Voelcker gave evidence on the effective purchase and use of fertilizers, but his work was mostly on characterizing fertilizers to avoid fraud, rather than research to improve agriculture.[78]

It was many years before the state supported agricultural research in any systematic way. As in so many areas of science, research in relation to civil applications was mostly left to private enterprise in Britain throughout the nineteenth century. Much of that private enterprise, given the wealth of the landed interest, was devoted to agricultural improvement and can be traced back to the original Board of Agriculture and the formation of organizations like the Royal Agricultural Society. Specific initiatives complemented these, such as the establishment of the Rothamsted Experimental Station in 1843 by John Lawes and Joseph Gilbert. It is now a public sector research establishment.

Lawes, despite his enormous contributions with Gilbert to the business of agricultural improvement since the 1840s, was sceptical in 1881 that science had offered much to agriculture, although he thought it might do more in future.[79] That did not happen until around the turn of the century. In those years agricultural departments in universities were becoming established and many research institutes were created. An example is the emphasis on agriculture at University College Reading in the early 1890s, given its importance to the local industries of biscuits and seeds. Thomas Middleton, professor of agriculture at Cambridge University, moved to the Board of Agriculture in 1906 to take charge of education and research, bringing that expertise and influence to the board. He became one of the commissioners for the Development Fund established in 1910, which supported research.[80]

In 1930 the government created a single authority to supervise agricultural research, the Committee of the Privy Council for Agricultural Research. Three years later that became the Agricultural Research Council. In turn it was incorporated into the current research councils, now brought together under UK Research and Innovation.[81] The private enterprise of the early Board of Agriculture and Internal Improvement had been replaced by a government Department of Agriculture and state-funded research.

Throughout the nineteenth century, government showed itself loathe to intervene in agricultural affairs. Agricultural improvement was seen primarily as a matter for private landowners. Only when public interests became significant, as in the potential agricultural use of sewage, or the pollution of waterways, was Parliament and government stirred to action. The special case of fisheries, in the next chapter, provides particular examples of clashes between private and public interests, in local contexts, under conditions of limited scientific knowledge.

6

FISHERIES

As the Labour politician Aneurin Bevan said in 1945, "This island is made mainly of coal and surrounded by fish. Only an organizing genius could produce a shortage of coal and fish at the same time."[1] Fishing, like coal mining, was of national importance. It was bound up with questions of class, economics, and national security. Salmon fishing, for example, was an upper-class sport, leading to conflict with poachers. In economic terms, sea-fishing generated employment and revenue for coastal communities. Considerations of national security encompassed both food security for an island nation and the development of a pipeline of skilled seamen suitable for the Royal Navy. The practices and consequences of fishing were therefore contentious in both national and local politics. Questions of trade, tariffs, and quotas set the parameters for debates on economics, while conflicts over rights between landowners and fishing communities, and between those who fished in the sea, estuaries, and rivers, raised social questions of rights and responsibilities. It was the worsening of water pollution through industrialization, and the need for knowledge of the natural history of different fish species, that brought the scientific experts into discussions on policy options.

HERRING

Parliament grappled with the herring industry throughout the nineteenth century.[2] It was important for trade with Europe and of significance to remote coastal communities, including in Scotland and Ireland, where it was a source of food and employment for many people ejected from the land by enclosures and clearances. As part of its mercantile trade policy, the British government passed several acts in the early nineteenth century to regulate imports and the bounty paid to encourage exports. These regulations had to respond to the extent of the catch and to demand from the West Indies, where salted herring was a staple diet for enslaved people.[3] The government also tried to prevent the deterrent effects of fishermen being pressed into the navy, although encouragement of the fishing industry was seen as a means of training men who could be called upon in times of war.

In 1808 Parliament passed an act to regulate the British white herring industry. Focusing on Scotland, the main center of the herring industry, this established six commissioners, who appointed two general inspectors to cover the west and east coasts of Scotland. These inspectors were required to ensure appropriate standards of fish for export, to brand barrels that passed inspection, and to support the conservation of fishing stock. The last they did by seeking to implement regulations on the size of the mesh for nets, so that small fry would not be caught. The commissioners worked in conjunction with the navy, which both policed disputes between boats on the seas and sought to keep foreign boats out of British waters.

While the focus of legislation and inspection was on the quality of herring and on dispute management by the inspectors or by the navy, scientific knowledge was relevant to conservation. It was important to distinguish between sprats and small herring, and research in marine biology around 1840 showed that sprats and herring kept to separate shoals and could be distinguished.[4] But with marine science in its infancy there were many questions that could not be resolved to general agreement, including the impact of new technologies of netting.

The possible impact of new trawling methods was one problem, which led to the establishment of a royal commission in 1863. This proved to be the first commission on which Thomas Huxley served. Huxley was just emerging as "Darwin's bulldog," following the publication of Charles Darwin's On the Origin of Species in 1859. Although his public activities were devoted more toward government support for science education and research, and to debates over the relationship between science and religion, he was drawn into policies on fishing over two decades.

Huxley's fellow commissioners, who spent considerable time touring the coast of Scotland, included Lyon Playfair, who had co-opted him.[5] Their report is notable for its exposition on natural factors affecting herring populations and on the natural history of the herring, including experiments carried out by George Allman, professor of natural history at the University of Edinburgh.[6] The commissioners argued that changes in herring populations were affected far more by natural factors than by particular types of fishing, and viewed some of the repressive measures as no more than the consequence of "class interests," so they advised the relaxation of some of them. For example, the prohibition against ring-netting, which had been banned in 1851 on the grounds that it led to overcatching and destroyed spawn and fry, was repealed in 1867.

SALMON

The case of salmon provides a linked but contrasting example to that of herring. Herring live only in salt water, but salmon swim the rivers too. Compared to herring fishing, many more conflicting interests were affected by fishing for salmon. They encompassed the landowners upstream and downstream holding fishing rights, tidal fishermen, mill owners whose dams blocked the streams, and the growing number of factory owners spewing their manufacturing effluent into the nearest river.

An act that aimed to protect salmon in English rivers had been passed in 1818. It banned spearing, although poachers continued to use it illegally and, like previous legislation, the act was largely unenforced and ineffectual.[7] Following a number of petitions, and a noticeable decline in the number of fish in the rivers, the Scottish MP Thomas Kennedy secured a select committee to examine the issues in 1824, with the emphasis on Scottish rivers. He argued that the natural history and habits of the salmon were little known but could now be investigated, that different modes of fishing should be properly examined for their effects, and that the conflict between private and public rights needed clarification.[8]

The committee made an initial report, suggesting further work would be necessary in the next parliamentary session.[9] The evidence published is notable for a two-page treatise by Humphry Davy, as president of the Royal Society and a keen salmon fisher.[10] He described the salmon's breeding habits, and urged the abolition of stake fishing, which cut salmon off from the rivers. He advocated the availability of free passages for salmon upriver, the use of nets of limited mesh size, and a longer close season when fishing was forbidden.

In its eventual report in 1825, based also on evidence from English rivers, the select committee recommended legislation to extend the close season. It added technical measures to require gratings to prevent fish being caught up in mills, to ban the use of lights, and to regulate the size of meshes for nets. For enforcement, the committee wanted the regulations monitored by conservators or water bailiffs paid by the owners or lessees of fishing rights. The committee noted, without suggesting any solution, that it was "indispensable to guard against the admission into all rivers, streams, estuaries and lakes, in which Salmon exist, of any matter proceeding from manufactories of any description which is known or deemed to be poisonous or deleterious to Fish."[11]

The second report concentrated on obstructions to the movement of salmon upriver and on pollution from manufactories. Here, the committee members found the conflicting rights and economic forces too difficult to resolve. While bewailing the "injurious tendency of all obstructions extending across Rivers," they could only urge that "persons interested should endeavour to accommodate differences."[12] For the industrialized areas, they simply gave up, arguing that "in those Rivers on which large commercial cities are situated, and on which the interests of manufactures have led to the expenditure of vast capital, it is not to be looked for that the Salmon Fishery should flourish; and while it may be from those causes nearly extinct, it would be chimerical to expect that it should ever be restored." Most of the evidence was given by men with a practical knowledge of salmon fishing, although Thomas Telford was called to offer evidence on the design of fenders or gratings for mills, which might be required in a future act. His advice covered the size and orientation of the bars and of the whole grating, and its impact on water flow.[13]

An Act for the Preservation of Salmon Fisheries in Scotland was eventually passed in 1828. It instigated a uniform close season for Scotland. A further select committee examined the operation of this act in 1836.[14] This committee took the view that fish numbers had declined since the passing of the 1828 act, although the members noted that "in some instances the Returns of the produce of River Fishings have been withheld from them."[15] Evidence indicated that salmon ran at different times in different rivers, so the committee recommended that the close season ought to match local conditions, and a weekly close time be enforced. Mills, weirs, and other barriers were still seen as major obstacles to the movement of salmon. The committee took evidence from James Smith of Deanston, who explained his "salmon stair," an invention to enable salmon to surmount dams, supported by the engineer Robert Thom.[16] Smith further advised that harmful matter from the water of gasworks, which was becoming an increasing nuisance, could be readily removed before it was released into a river.[17] But no legislation resulted.

Further acts tinkered with the situation over the next twenty years, although legislation suffered from a lack of detailed scientific knowledge, making it "vexatious and futile."[18] Then, following petitions about the decline of salmon and disputes over proposed local acts for particular rivers, the House of Lords established another select committee in 1860. Evidence presented to this committee on the decline, or not, of salmon over the years was incomplete, although it tended to suggest a decline. Having been offered much conflicting evidence, the committee recommended the prohibition of all fixed nets and devices, the enforcement of powers to require free passage for salmon past dams and mills, and the appointment of a central board or commission to regulate the crucial industry in Scotland.[19]

THE ROYAL COMMISSION ON SALMON FISHERIES

Given the complexities revealed by the select committee, the government decided in time-honored fashion to establish a royal commission "with the view of increasing the supply of a valuable article of food for the benefit of the public."[20] Its focus was restricted to England and Wales, and the commissioners referred to Irish practice and to a recent report on Scotland, to which their recommendations were similar.

The three commissioners were William Jardine, natural historian and geologist; William Ffennell, the Irish fishery inspector; and George Rickards, counsel to the speaker of the House of Commons, who had recently been Drummond professor of political economy at Oxford University. The commissioners toured the rivers of England and Wales to make their inquiries, finding "many erroneous opinions respecting both the nature and habits of the salmon, and the true interests of those who possess of claim rights in the fisheries. Indeed, it is, beyond doubt," they argued, "to such false notions and short-sighted views that the decline and partial ruin of the English salmon rivers is mainly to be ascribed."[21] This exhaustive inquiry took evidence from several hundred witnesses in its travels, many of whom gave information on the movement of salmon from their practical experience.

The commissioners were convinced by the evidence that fish stocks had substantially declined. They pointed out that the clash of rights over fishing different parts of the river system, which were not or could not be resolved by cooperation, led to decline for all, as no individual had an incentive to invest in methods of conservation. It was therefore necessary to legislate against private rights to meet the wider public interest of food supply. The litany of problems included obstructions to the passage of fish; defective regulation of close times; pollution from mines, factories, and gasworks; illegal fishing; legal uncertainties; and the lack of an organized system of management of the rivers.

The commissioners grappled again with the extent to which salmon fishing should have precedence over industrial activity in particular rivers. It was an almost insoluble conundrum of the time, and the context within which any scientific advice would be considered. Nevertheless, the commissioners were optimistic that legislation could remedy the problems to a great extent, and drew on the example of Ireland, which had already established an inspection system, in making their recommendations.[22] They justified the cost of establishing a central board or commission on the grounds of developing salmon fishing as a public benefit for the provision of food.

The Liberal home secretary, George Cornewall Lewis, brought forward a bill to implement the recommendations. It met opposition from private land-owners and industrialists, sufficient to prevent the formation of an interfering board. But the resultant Salmon Act 1861 did abolish fixed nets and forbade the construction of mill dams and weirs without fish passes. It established two inspectors of salmon fisheries for England and Wales.[23] One was Ffennell, and the other was Frederick Eden, also a fishery commissioner for Ireland and who had been the royal commission's secretary.

Although the inspectors were initially optimistic about the operation of the act, problems soon emerged in relation to conflicting property rights, lack of enforcement, and the lack of statistical information and scientific evidence to support particular regulations.[24] The Salmon Act 1865, introduced by Home Office minister Thomas Baring against similar opposition, introduced further powers. It set up locally elected boards with powers of licensing and inspection, and gave three commissioners, initially including Eden himself, further powers, although not to make bylaws. But in 1866 Eden became severely ill and had to retire, and Ffennell died in March 1867.

FRANK BUCKLAND

Eden and Ffennell were replaced by Francis Buckland, known as Frank, and by Spencer Walpole. Buckland, a friend of Ffennell,[25] was the eldest son of William Buckland, friend and scientific adviser to Robert Peel.[26] He had studied chemistry with Justus Liebig in Giessen and was an enthusiastic natural historian, committed to advancing the nation's pisciculture, who had established a collection of fishery displays that he moved into the Science and Art Department buildings in South Kensington. Walpole was the youngest son of the home secretary, also called Spencer Walpole, which resulted in attacks on the grounds of patronage at a time when appointments in the civil service were increasingly on merit following the Northcote-Trevelyan report. Buckland was not known as a leading researcher, but he brought a commitment to science to his work as an inspector, complemented by Walpole's quiet and diplomatic efficiency.

Buckland voiced the belief that fishing needed to operate in accordance with natural law. He saw the need for a better understanding of salmon migration and reproduction, for ways of improving fish passes and ladders, and for better measurement of pollution. At South Kensington he initiated experiments ranging from fish-pass technology to artificial breeding. Ever solicitous for the fish, while awaiting the construction of one fish pass he stuck a notice on the weir that read, "No road at present over this weir. Go downstream, take the first turn to the right, and you will find good travelling water upstream, and no jumping required."[27] He consulted Edward Frankland on the measurement of pollution in rivers, and instigated statistical studies of salmon markets, licenses, and prosecutions. Like many inspectors, Buckland and Walpole offered policy advice for tightening up the weaknesses in the 1861 and 1865 Salmon Acts, such as penalties for discharging pollution and the formation of local fishery boards that could police and implement regulations. Joseph Dodds, MP for Stockton-on-Tees and with fishing interests, instigated a select committee to address the problems raised.

The Select Committee on Salmon Fisheries convened in 1869 and again in the following parliamentary session in 1870 to complete its inquiry. Its meetings in 1869 included evidence from the Scottish engineer William Rankine, who gave a long exposition on hydraulics in relation to weirs, fish passes, and mills, although he was not prepared to comment on anything "dependent on the habits of the fish,"[28] such as the actual design of fish ladders or passes. He was clear that the practical effects on mill operations of design changes to river installations could be determined with great accuracy. That could enable the calculation of compensation for mill owners if the flow to their mills was reduced by changes to protect salmon. Buckland, in his evidence, highlighted the problems of river pollution, which were being addressed at the same time by the royal commission on the pollution of rivers (see chapter 11).

The committee's second report in 1870 reiterated the conclusions of the royal commission of 1861.[29] It proposed a central authority to supervise the salmon industry and local elected boards with the ability to pass bylaws to regulate matters such as close times, net sizes, gratings, licensing, and conservation. The committee also suggested transferring responsibility for fisheries from the Home Office to the Board of Trade until the industry was sufficiently developed to justify a fisheries department, although that did not happen until 1886. The driving force was the desire to increase food production and sales. The eventual Salmon Fisheries Act 1873 established local boards of conservators able to frame bylaws, a further move toward the protection of salmon and the development of the industry.

THE ROYAL COMMISSION ON SEA FISHERIES

From the 1860s on, questions beyond the individual interests of herring or salmon fisheries became more apparent. Attention started to turn both to fishing as a whole, whether freshwater or marine, and to questions of pollution arising from the growth of mining, manufacturing, and gasworks discharging their waste into the rivers alongside human and animal excrement. Two royal commissions examined these issues with different emphases in the 1860s and early 1870s: the royal commission on sea fisheries from 1863 to 1865, and the troubled royal commission on the pollution of rivers from 1865 to 1874.

The royal commission on sea fisheries brought in Huxley as a commissioner, extending his work on investigating trawling for herring. His fellow commissioners were James Caird, MP, who chaired the commission, and George Shaw Lefevre, who would become a lord of the Admiralty in 1866 and serve in many ministerial positions. Huxley, not being a member of Parliament, was paid £3 per day for 104 days work.

The motivation for setting up this commission was again the desire to increase food production. The commissioners were asked to examine whether fish supplies were increasing or not, whether methods of catching were wasteful, and whether any legislative restrictions were harmful to fisheries. They visited and took evidence from eighty-six places around the coasts of England, Wales, Scotland, and Ireland. They found that the supply of fish was increasing, even in the English Channel, where in 1833 a select committee had found problems.[30] The exceptions were a depression in oyster supply, and in Ireland, where there were more widespread problems. The latter they attributed to the impact of the famine of the 1840s and its effect on fishing communities. Many fishermen had emigrated as people could no longer afford to buy fish, a stark reminder of the long shadow that the famine cast over Ireland. The commissioners concluded that trawling in the open sea or in bays was not wasteful or damaging, and that any restrictions would result in a decline in fish production. Limits to production, not concerns about conservation, remained the prevailing concern right through to the turn of the century.

The commissioners recommended, based on their optimistic position on the natural supply of fish, that fishing in the open sea should be unrestricted. But recognizing the lack of robust evidence about fishery statistics, which led to disputes between competing interests based on minimal evidence, they called for the systematic collection of such information.[31] Proper attempts at the conservation and management of fish stocks could not be made until both statistical evidence and understanding of fish populations were much improved.

THE INSPECTORS INSPECT

Following the passing of the Salmon Fisheries Act 1873, the salmon inspectors took it upon themselves to examine other freshwater fisheries, and some coastal and sea fisheries. They were given responsibility for oyster fisheries, previously the responsibility of the Board of Trade, since these showed similar characteristics to the salmon fisheries. In 1875 they examined the Norfolk crab, lobster, and herring fisheries after complaints by local anglers.[32] Their inquiries resulted in local legislation for a close season to protect spawning fish, and a minimum net size.[33] Then in 1877 they looked into crab and lobster fisheries across the United Kingdom and at the herring fisheries of Scotland.[34] Even dynamite came under their purview. They proposed a ban given the indiscriminate effects on marine life of its increasing use for stunning fish, which had started in Cornwall. Parliament passed the Fisheries (Dynamite) Act 1877 that year to prohibit the practice.[35] These studies, extending to an examination of the destruction of fish generally, led to the Freshwater Fisheries Act 1878, which enabled conservators to establish fishery boards for char and trout rivers, and extended close season regulations to all freshwater fish. Although salmon and trout fishing was primarily an aristocratic sport, the formation of the boards also satisfied demands of coarse fishermen who had been lobbying for some time for protection. Angling was seen as a popular sport, a healthy and appropriate activity for working-class recreation that would keep people from drinking and lawbreaking. Coarse anglers carried political weight given the extension of the franchise, which they used to protect their sport both for leisure purposes and to an extent for food.[36] Further consolidation of the law was obtained in 1880.[37]

THE RETURN OF THOMAS HUXLEY

At the end of 1880, Buckland died after a long period of ill health. His enthusiasm, allied to Walpole's steadiness and the ability of both of them to engage with the diverse fishing interests, had resulted in a large increase in the value of salmon fisheries and steered a course in which there was acceptance of some state control and of penalties for breaking regulations.

The local power to establish bylaws under the Salmon Fisheries Act 1873 needed the permission of the home secretary, making the inspectors' advisory role important. When Buckland died, William Harcourt, home secretary in the recent William Gladstone administration, approached Huxley to fill his place. The invitation reveals Harcourt's view of the post, and the limited number of such public positions available even so late in the century for men of science. Harcourt sold it to Huxley as a "pleasant occupation," a "holiday task" in "picturesque localities" that would "not require much labour or time." It was not a "grand place," worth "only £700 per annum."[38] Huxley took the post while insisting that he could keep all his other positions, and looked forward to "jamming common sense down the throats of fools."[39] His appointment was objected to by those who would have preferred a practical man rather than a professional man of science. While Huxley did make one significant scientific contribution, by investigating a fungal disease in salmon,[40] and although he got on well with Walpole, his impatience, indeed arrogance toward the fishing community, did not make him effective in the broader inspectorial role. Stressed as he was by other aspects of his life at the time, this required a diplomacy which he did not demonstrate. When Walpole left in early 1882 and was not replaced by Charles Fryer until 1883, conservators longed for the return of the practical approach of Buckland.

In addition to his inspection duties, Huxley found himself on another commission in 1883, chaired by the thirteenth Earl of Dalhousie, to investigate complaints of line and drift-net fisherman about the use of the trawl net and beam trawl.[41] William McIntosh, professor of civil and natural history at the University of St Andrews, was employed to conduct observations. When the recommendations of the commission incensed some of the fishermen, they marched to his house and burnt an effigy outside it.

Huxley soldiered on until late 1884 when, depressed and seeking to recover his health, he left for Italy. On his return in 1885 he gave extensive evidence to the select committee on salmon fisheries in Ireland.[42] But he was still depressed, failed to sign the commissioners' report on trawling, and resigned not only the fisheries post but also his professorship and the presidency of the Royal Society. Arthur Berrington, respected in the fishing community, replaced him, and sought to continue Buckland's approach. But in 1886 responsibility was transferred from the Home Office, where there had been light supervision, to the Board of Trade, with its heavy weight of bureaucracy. Requests for better statistics and consolidating legislation were rebuffed and Buckland's Museum of Fish Culture was neglected.[43]

TOWARD A RESEARCH AGENDA

The royal commission on sea fisheries had come to the conclusion in 1865 that the Board of Fisheries in Scotland served no useful purpose. It argued for example that quality control through branding barrels of herring should be left to market forces rather than being a responsibility of government.[44] But because the board had gathered so many functions that could not easily be managed elsewhere, a reconstituted Fishery Board for Scotland was created in 1882. Answerable to the home secretary, and including responsibility for herring and salmon fisheries, its remit included oversight of research in marine and fisheries biology. In its first report it set out a research program concentrating on the most important food fishes such as herring, cod, ling, haddock, mackerel, sole, plaice, and flounder. The list of research objectives was broad, reflecting the lack of detailed scientific knowledge. They encompassed the food, life history, distribution, migration, and spawning of individual species. Added to those were research into the protection of fish, predators, the use of new species, artificial cultivation, and the effects of atmospheric and water conditions.[45]

Britain hosted the International Fisheries Exhibition in London in 1883, at which Huxley restated in his opening address his position from the commission in 1866 that fish stocks could never be exhausted by existing methods of fishing.[46] Others differed, including Edwin Lankester, who took a broader ecological approach. A committee of fourteen biologists—among them John

Lubbock, Philip Sclater, Michael Foster, John Burdon Sanderson, and George Allman—was formed to establish the British Marine Zoological Laboratory. Huxley did not join the committee. The British Association supported the proposed laboratory, which was envisaged as similar to those established in France, Trieste, and Naples. That led to the formation of the Marine Biological Association in 1884, with the recalcitrant Huxley tempted back as its first president and Lankester as secretary. Huxley then resigned in 1889 when the association proposed that its annual grant from the government should be increased to allow it to investigate the North Sea fisheries, and especially the effects of trawling on immature fish.[47] The association opened a laboratory in Plymouth in 1888, with funding of £1,000 per annum by 1892 from the Treasury, although the Scottish Fishery Board had a larger budget.[48]

As the nineteenth century drew to a close, marine biology and its potential to influence public policy started to become systematic. But many problems persisted. Regulations on close times, the free ascent of salmon, and the prevention of pollution were still widely disregarded.[49] The question of pollution remained a public health issue, and little progress was made in the context of fisheries.

The subject of fisheries encapsulates so many of the factors that constrained the influence of scientific advice on policy. They include changing trade policies, conflicting rights, tensions between national and local government, and limitations of scientific knowledge. The best of the inspectors had taken a pragmatic approach to methods of improvement in the practice of fishing, for negotiating between people with conflicting rights and priorities in the rivers and on the seas, for developing appropriate local and central powers, for improving statistics, and for better understanding of the natural history of fish species. But without state support for better statistical information and for sound environmental research, they were hamstrung by a lack of knowledge to inform policy.[50] In addition, the inspectors came up against a bureaucratic approach at the Board of Trade at the end of the century that confined them largely to London, while the local boards of conservators dealt with fishing in practice, often clashing with sea fisheries committees established by the Sea Fisheries Regulation Act 1888. Fishery work was transferred to the Board of Agriculture in 1903, as political and administrative integration continued. The department became the Ministry of Agriculture, Fisheries and Food in 1955.

PART IV

INFRASTRUCTURE
AND TRANSPORT

7

TRANSPORT INFRASTRUCTURE AND ENGINEERING

DURING THE LATTER PART OF THE EIGHTEENTH CENTURY, A SMALL group of men emerged from the artisanal traditions of millwrights, masons, carpenters, and other practical trades, to undertake large-scale engineering works and make steam-powered engines.[1] Land drainage for agriculture, and canal and roadbuilding for transport and trade, formed the focus for many of these men, alongside bridge building and harbor improvements. These engineers developed a role as consultants. They soon established close connections with Parliament, because they often needed to present their plans for approval so that their client landowners could secure private acts to enable them to implement local schemes. In time, this demand for expertise extended to public buildings and related works, such as the rebuilding of the Houses of Parliament, the Crystal Palace housing the Great Exhibition of 1851, the National Gallery and South Kensington Museum, and even the Serpentine.[2]

These initial parliamentary relationships were of a different nature to those of the scientific and medical experts, whose advisory roles were concerned more with informing national legislation and its implementation. But throughout the nineteenth century, engineers were called upon to inform legislation too, in areas ranging through water and sewage systems, railways, gas and chemical works, ship construction, boiler safety, agricultural drainage, and even fisheries. They also took up many of the increasing number of inspectoral roles, not least on the railways. Many of these engineers came from humble origins, without the benefits of property and family connections. During the Victorian era, several would rise to become household names in the heroic tradition promoted by Samuel Smiles in his book *Lives of the Engineers*, people such as Thomas Telford and George Stephenson.

These engineers were busy men, often based in the midlands and the north, the centers of industrialization, and few became members of Parliament. The most notable engineers who became parliamentarians were Robert Stephenson and his colleague Joseph Locke in the early part of the century, and Daniel Gooch later on. More visible in Parliament were civil engineering contractors such as William Cubitt, Samuel Moreton Peto, Thomas Brassey, John Laird, William Jackson,[3] and the "Railway King" George Hudson.

In 1771 John Smeaton, one of the leading engineers of his day, had founded the Society for Civil Engineers. The term *civil engineer*, which Smeaton used, differentiated this cadre of people from the military personnel in the Royal Engineers. He sought to create a society that would bring engineers together in professional exchanges, but in the gentlemanly mold of the Royal Society dining club.[4] By 1800 there were around forty members, about a dozen of whom were fellows of the Royal Society. Joseph Banks was elected as a gentleman member in 1793, as were Humphry Davy in 1808 and Davies Gilbert in 1811.[5]

It was the elite nature of this society at the beginning of the nineteenth century that saw it superseded by the creation of the Institution of Civil Engineers in 1818. Several engineers saw the need for an organization that could offer more opportunities for contact between fellow spirits and encourage engineering training and education. So, the Institution of Civil Engineers may be seen as the first professional engineering institution in Britain. Well before the reform of the Royal Society, which was never a professional institution, the Institution of Civil Engineers was careful that there should be no election of "promiscuous" members.[6] Quality and commitment to developing the profession were paramount.

The early years of the Institution of Civil Engineers saw the transformation of Britain by railways. As canal building petered out, engineers piled in to benefit from the railway developments of the 1830s and 1840s. While the Institution of Civil Engineers was the initial professional home for many of these engineers, it did not adapt itself well to the new railway specialists, who were also concerned with the development of steam engines and other mechanisms. So, several railway engineers formed the Institution of Mechanical Engineers in Birmingham in 1847, with George Stephenson as the first president, succeeded by his son Robert in 1849.[7] Although at first a nonmetropolitan institution, like so many springing up at the time, it soon developed a national role and eventually moved to London in 1877.

Paralleling the moves in science and in medicine, many other more specialist engineering institutions were formed during the nineteenth century, such as the Institution of Naval Architects in 1860, the Iron and Steel Institute in 1869, the Society of Telegraph Engineers in 1871, which became the Institution of Electrical Engineers in 1888, and the Institution of Municipal Engineers in 1873.

Although engineering education was seen in the early part of the nineteenth century as best provided by practical experience and personal mentoring, deriving from the tradition of apprenticeships, engineering later developed as a university discipline. Its linkage to the sciences is exemplified by the formation of a section of the British Association in 1832 initially called "arts." In the nineteenth-century sense, that meant inventions and technological improvements, which could include the applications of science. In 1836 it was named the mechanical science section, with Gilbert as president. Listed last in the roll call of sections, it could not claim the privileged scientific status of mathematics and physics.

John Millington was arguably the first professor of engineering in Britain, as professor of engineering and the application of mechanical philosophy to the arts at the new London University in 1827, although he resigned shortly after appointment. A chair of civil engineering was then established in 1841, held by the railway engineer Charles Vignoles until 1845, but for the most part engineering courses were sparse in universities in the first half of the century. People like William Rankine, professor of engineering at the University of Glasgow from 1855 to 1872, and a leading advocate of the value of engineering science, and Fleeming Jenkin, professor of engineering at the University of Edinburgh from 1868, were thin on the ground. Even well into the nineteenth century, there was a widespread suspicion among engineers of the value of theoretical studies. In time, the position changed, and engineering became a recognized university discipline. The university professors, including Rankine, would be called on alongside the practicing engineers to offer advice to government and Parliament.

ROADS

If there had been a chief engineering adviser to government in the early nineteenth century, the post would doubtless have been held by Thomas Telford. As Britain's preeminent civil engineer in the first three decades, Telford's constructions encompassed canals, roads, bridges, harbors, and railways. He became central to the development and delivery of government-supported public works, frequently reporting to Parliament on options and costs.[8] Telford epitomises the artisanal approach of the early leading engineers. He came from a poor background in Scotland, became a stonemason, and worked his way up through practical experience and by reading the books of leading practitioners.

From 1726 the government had started investing substantially in roads, initially for military purposes in Scotland, but then as a means of mitigating famine and distress, and encouraging trade, particularly from England to Scotland and Ireland. This is the first substantial example of central government funding for national infrastructure, and it resulted in engineers becoming an early element of an expanding state administration that introduced regulation of road structures based on engineering expertise.[9] By the first two decades of the nineteenth century, engineers were already becoming integral to state activity, by convincing government that they could offer rigorous principles for roadbuilding and quantifiable metrics for delivery.[10]

Telford became involved in government surveys of the Scottish Highlands in 1801 to 1802 and in the establishment of commissions for the Caledonian Canal and for roads and bridges in Scotland. After effective lobbying by Irish and Scottish interests, the state financed 1,700 miles of roads radiating from London between 1803 and 1835, primarily to strengthen links with Dublin and Edinburgh for commerce, the mail, and defense purposes. Telford was appointed engineer to the highland roads commissioners from 1803, which resulted in the building of 1,100 bridges, 1,200 miles of roads, and more than forty harbors across Scotland. In 1815 he became engineer to the Holyhead roads commissioners, and masterminded the road from London to Holyhead, so important for communications and trade with Ireland, and developed many other schemes, including a total of more than one hundred harbors, docks, and piers. In 1817 he was appointed as adviser on all works involving engineering for the Exchequer Bill Loan Commission, which oversaw public works designed to help counter the effects of economic depression following the Battle of Waterloo. His advice supported expenditure of £1 million in the first ten years, including canals, bridges, and harbors. His approach of centralized, planned engineering inspired Edwin Chadwick's later vision of a coherent system of waterworks, drainage, and sewerage.[11]

REGULATION OF THE ROADS

The dreadful state of the roads at the beginning of the nineteenth century exercised both the public and Parliament. An early inquiry demonstrates the parliamentary interest, which included both technical and safety matters.

In 1806 a select committee investigated the best shape of broad wheels for ease of travel and for minimizing damage to the roads as well as the effects of overcrowding on stagecoaches. By law, no more than six people could be carried on the outside of a coach, but in practice there could be twenty or more. The result was fatal accidents and potential damage to roads caused by the extra weight exerted through the wheels, as the committee first reported.[12]

The committee's extensive second report offers a fascinating treatise on the practical experimentation that people involved in coaching had undertaken. Many factors were considered for reducing wear on the roads and for increasing safety, including carriage design, the shape, height, and breadth of wheels, their spokes and rims, and the positions of their axles.[13] The most technical advice was given by the engineer William Jessop, on conical and cylindrical wheels.[14]

This committee was chaired by John Sinclair, who had already taken an interest in the shocking state of the roads in his role as president of the Board of Agriculture. As he would later explain, he considered that agricultural as well as commercial and manufacturing activities would be improved by better roads. Such improvements would lead to reduced prices of goods and the need for fewer horses, freeing up agricultural land for "the production of food for Man" instead of for horses.[15]

The committee's first report offered dozens of recommendations on the maintenance and management of roads, with technical suggestions on the design of carriages and wheels.[16] The second sought again to reduce the number of passengers and weight carried, on the grounds of accident and cruelty to horses, recognizing that coaches were frequently driven by "intoxicated or unskilful Coachmen."[17] The third report, which included reference to work by engineers Jessop and Telford, and by inventor John Boswell, suggested paving key roads out of the metropolis.

Boswell's letter to Sinclair is a powerful statement of his understanding of the relationship of scientific theory to practice. He wrote that "science is decried under the invidious appellation of Theory, and the scanty portion of information which an individual, however illiterate, can obtain from casual observation, is preferred and extolled under the fostering name of Practice." But for Boswell, theory was an "Index to Science," the result of the accumulation of knowledge. It was clear to him, following Francis Bacon, that "all Science was practical in the end."[18] The reports of Sinclair's committee illustrate this practical science well. The committee's report ended by stating that it considered that no system of roads and carriages would be complete "unless some form of Public Institution or Department of Government" was entrusted with enforcing the laws on roads and turnpikes.[19]

While no such department was created, regulations were first introduced in 1810, on weights, footpaths, and the use of cylindrical rather than conical wheels, and were gradually extended until 1835. The system of roadbuilding itself was specified. John McAdam initially won the debate on principles of construction over Telford, by promoting uniform standardization rather than methods that might vary locally, even if the latter might be cheaper, based on scientific advice, although Telford's approach later prevailed.

However, this centralization and standardization soon came up against the desire for local control. A further select committee in 1820 wanted a more uniform approach, including reports of the expenditure of public money by turnpike trusts and the merging of trusts under commissions that could deal with longer stretches of roads. But it was emphatic that management by central government would be "utterly destructive of the objects they have in view." Instead, it thought that "general information and scientific plans, modified by accidental circumstances, combined with local management, more or less extensive, as particular cases may require, will, in the opinion of Your Committee, form the most perfect system for road-making that has yet appeared in any part of the world."[20]

That tension between central control and local accountability became evident, as Parliament sought to balance the encouragement of local responsibility with some common standards and regulations in the public interest. Expert advice tended to imply the need for national standards, which cut against vested local interests and a political culture that instinctively favored local autonomy. No new roads were built at parliamentary expense after 1836, and the massive central funding and standardization ended in mid-century as local rather than national interests triumphed.

When a further select committee examined turnpikes in 1836, it called on scientific evidence from Dionysius Lardner, who had been the first professor of natural philosophy and astronomy at University College London. Lardner had form. He had disagreed with Isambard Kingdom Brunel over the design of the Box Tunnel, losing the argument when Brunel pointed out that he had ignored air resistance and friction when calculating the acceleration of a train downhill. Lardner argued that the road surface should be hard and inelastic for efficient travel, which would also help durability and drainage. He claimed that the best scientific and engineering knowledge had not been applied to roads, unlike the case with railways, and thought that the best civil engineers should be put in charge of any new road projects.[21] The committee made no recommendations in this vein.

By the 1830s, there were more than one thousand turnpikes responsible for about twenty-two thousand miles of road in England and Wales. In the years following Telford's death in 1834 it became apparent that the roads were suffering serious competition from railways, leading to financial problems for the turnpike trusts from decreasing tolls and for the creditors who had lent money to them.[22] A royal commission into the state of the roads, established in 1839, recommended consolidation of trusts, which legislation supported over the following years.[23] Toll roads were gradually handed over to local administration by highway boards. But the consequence was a steady deterioration and regional inequality as tolls remained predominantly in poor areas. The Local Government Act 1888 gave responsibility for maintaining main roads to county and borough councils, and all tolls were finally abolished in 1896. The wheel turned part circle in 1936, when the trunk roads were nationalized.

STEAM CARRIAGES

One technology that failed to take off was the introduction of steam carriages on the roads. Telford's involvement with railways was limited, as his career was coming to an end just as the railways arrived. But he recognized that railways, as opposed to roads and canals, created monopolies for traffic. As a solution, he promoted the idea of steam engines on roads.

A select committee in 1831 gave him the opportunity to put his views before Parliament. This committee on steam carriages, which included Gilbert, was set up to examine possible tolls and prospects for this mode of transport.[24]

Telford advocated using self-propelled steam carriages on roads because they would be safe and would obviate the potential impact of the monopoly of transport on the railways, where the company owned both the track and the vehicles running over it. Noise, fumes, potential explosions, and wear on the roads were significant issues, so the committee examined "several very scientific engineers," including Telford and Richard Trevithick. Gilbert gave evidence too, offering technical advice on the advantage of speed of steam-driven over horse-drawn vehicles.

The committee was supportive of replacing animal by steam power but thought that it should be a gradual process, dependent on experiment. It left those issues to private enterprise, and rejected concerns about noise, smoke, and safety. But the committee did appreciate the high tolls being levied in some places, despite evidence that steam engines would not damage roads as much as horse-drawn carriages. Overall, it concluded that steam carriages would become a cheaper mode of conveyance than horse-drawn coaches. In the end, high tolls, mechanical problems, and vested interests stood in the way of their widespread introduction, especially when they lost their visible advocate after Telford's death.[25]

HARBORS

As an island and maritime nation, harbors were vital to British interests. The Royal Navy, commercial trade, and fishing all relied on safe havens. The new breed of large-scale civil engineers was soon advising governments and building and improving harbors. Throughout the nineteenth century, the increasing size of ships, the replacement of sail by steamships, and the changing nature of cargos that needed specialized facilities, all led to the need for continuous improvement in harbor and dock facilities. At no point did governments attempt to introduce a national strategy for harbors and ports, so development was piecemeal and relied on private acts of Parliament. Ministers were loath to spend money on local works that they believed should be raised locally. They were also unwilling to identify particular sites for fear of causing dissension from the owners of sites not selected, or who had already funded works themselves. Competition was rife—for example, between the ports of Liverpool and Birkenhead, or between Hull and Grimsby.

In the early years of the century, many improvements were carried out as part of extensive public infrastructure works undertaken after the Napoleonic Wars to support economic development. In these early civilian projects, Telford and his contemporaries acted separately from the military engineers. There seems to have been little technology transfer between the engineers of the Royal Dockyards and those on the first civilian projects.[26]

Parliamentary committees frequently reported evidence from engineers on costs and practicalities of works, which would then inform the drawing up of bills to authorize expenditure on specific harbors. Examples include the elder John Rennie's estimates for work over many years on Holyhead and Howth Harbours for the major transport link between London and Dublin.[27] Likewise, Rennie gave evidence with others, including William Cubitt and Vignoles, on the important Edinburgh harbors of Leith and Newhaven, and Cubitt on Dover Harbour.[28] This involvement of engineers continued throughout the century. For example, in 1884 a select committee on harbor accommodation would declare itself unanimous in agreeing with the opinion of the eminent engineers it examined, that piers and breakwaters were best built in solid monolithic walls of concrete rather than on rubble foundations.[29] George Airy had suggested as early as 1845 on scientific grounds that vertical solid walls would be the best shape to build a breakwater.[30]

HARBORS OF REFUGE

Throughout the nineteenth century there was concern about the lack of "harbors of refuge" around the coasts, as places where ships could find safety in storms and that naval vessels could use in emergency in wartime. Worried particularly about a lack of safe harbors off the northeast and northwest coasts of England, a select committee was established in 1836. This committee found that there were no safe harbors between the River Forth and the River Thames, as all were tidal.[31] It suggested an urgent need, on commercial grounds and for public safety, for building harbors of refuge, but was not prepared to recommend government expenditure.

The Select Committee on the Causes of Shipwrecks, established later in

1836, also identified a lack of harbors of refuge as one cause of significant loss of life.[32] It did not recommend specific action, although a harbor commission in 1840, including James Walker and Cubitt, examined some sites on the south coast.[33]

The subsequent Select Committee on Shipwrecks in 1843 went further in recommending annual national expenditure on creating harbors of refuge but abstained from recommending particular sites "from a conviction that these points will be best decided on by a body composed of scientific and competent persons."[34] Prime Minister Robert Peel took this forward in 1844 by appointing ten commissioners to advise on whether, and if so where, a harbor of refuge should be built in the English Channel. Chaired by Admiral Byam Martin, who had been comptroller of the navy, the only nonmilitary commissioner was Walker, president of the Institution of Civil Engineers. Walker had taken over plans for developing Dover Harbour after Telford's death. Cubitt, Rennie, and many other engineers gave evidence about possible developments and their costs. Surveys were made of all candidate harbors. Henry De La Beche, as director of the Ordnance Geological Survey, advised the commissioners on the movement of shingle and sand,[35] and Richard Phillips at the Museum of Economic Geology was employed to examine water at different states of the tide in Dover to measure silting.[36] The commissioners recommended the construction of breakwaters at Dover, Portland, and Seaford.[37]

Ten commissioners, consisting of eight from the original commission with the addition of Francis Beaufort, hydrographer to the Admiralty, and Captain Henry Brandreth, director of Admiralty Works, reported on the plans for the harbor at Dover.[38] The interpenetration of the Royal Society and military personnel at this time is illustrated by the fact that seven of the ten commissioners, including Beaufort and Brandreth, were fellows. Two commissioners, Lieutenant General Howard Douglas and Captain William Symonds, objected to the findings. Symonds, surveyor of the navy, thought the planned harbor was too large, would be difficult to access, and would silt up. He also attacked the idea of an upright wall, which, while favored by "the majority of theorists," was "not borne out by a single practical experiment in deep water."[39] Douglas agreed, and set out a detailed analysis of all the scientific and engineering evidence presented to the main commission.[40] Airy, for example, had argued for a sloping wall in the deeper water, and upright only above it.[41]

Approval was given in 1847 for work on four harbors of refuge—at Dover, Portland, Holyhead, and Alderney—although progress was slow. Despite the continued pressures to build harbors of refuge around all the coasts, governments were not generally prepared to spend significant sums of money from national revenue that might benefit one local community over another. For example, despite the lack of a harbor of refuge between the Thames and the

Forth, and a favorable report by the engineer David Stevenson, a bill to enable construction of the Victoria Harbour and docks at Redcar did not proceed.[42]

As the issue did not go away, a select committee established in 1857 examined possible harbors of refuge for the whole British Isles. Many military and civil engineers again gave evidence.[43] The committee urged the expenditure of substantial sums of public money on the grounds that not to do so would seriously disadvantage trade and cost lives and money at a time when marine trade was increasing. While they identified the coastal locations on which they proposed that harbors of refuge should be built, they recommended a royal commission composed in part of "professional persons" to determine the actual sites.

The government agreed, and a commission of seven was established in 1859. Chaired by Rear Admiral James Hope, its mostly military members included John Washington, who had succeeded Beaufort as hydrographer to the Admiralty in 1855, and Major General John Frederick Smith. John Coode, then responsible for harbor works at Portland, was the only civil engineer. The commissioners maintained that harbors of refuge were national undertakings and so should be built in whole or in part at public expense, without any passing toll being levied, and placed under direct government control.[44] They recommended specific sites on the coasts of England, Scotland, Ireland, and the Isle of Man, at a public cost of some £2.4 million with £1.6 million provided locally.

Again, government did not take up the challenge. Smith, one of the commissioners of 1859, tried unsuccessfully in the House of Commons in 1865 to pass a motion demanding that government implement the recommendations. Thomas Milner-Gibson, president of the Board of Trade, had previously indicated that the government was planning to introduce some measures.[45] But he now responded that since shipping companies were unwilling to pay, the matter could not be considered a priority and the government should not step in.[46]

Others tried from time to time, also without success. A select committee on harbors accommodation reported in 1884 that nothing had been done beyond a loan to the Tyne commissioners.[47] The committee expressed "in the strongest terms" the need for the harbors of refuge, both for military purposes given advances in warfare and for the safety of mariners.[48] But when the matter was raised in Parliament in 1886, both Joseph Chamberlain, president of the Local Government Board, and Anthony Mundella, president of the Board of Trade, gave the familiar explanation that the government was averse to spending general revenue on local purposes unless there was a significant local contribution, since it was difficult to allocate public money fairly between competing locations.[49] This was despite an average annual loss of 3,000–3,500 ships off the coasts, an annual loss of property of over £2 million, and a loss of life of 800–900 persons. Not until 1897 was a contract let for the completion of the harbor of refuge at Dover.

TIDAL HARBORS

Despite much harbor building, there were growing concerns in mid-century that tidal harbors were silting up, due to landowners enclosing land and preventing the scouring of rivers. That was detrimental to trade and to the availability of ports in case of war. Joseph Hume raised the issue in the House of Commons in 1844 and called for a royal commission on the viability of tidal harbors and the extent to which the Admiralty had or had not used its powers to prevent encroachments. The commissioners of harbors of refuge had not reported at this point, but Peel acknowledged the problem, although he thought that natural factors were also significant. The motion was agreed, and the Tidal Harbours Commission appointed.[50]

The ten commissioners were chaired by Rear Admiral William Bowles, Conservative MP and brother-in-law of Viscount Palmerston. There was a strong naval contingent, including Beaufort and Washington, and further scientific expertise in the form of Airy. Airy had made a particular study of tides. In 1842, at the request of Colonel Thomas Colby, director of the trigonometrical survey, Airy had advised him on setting a reference height based on the mean height of the tide.

The commissioners' first report in 1845, based on a few harbors and the evidence of a dozen civil engineers, demonstrated the sheer scale of the problem. The engineers knew their business. Cubitt declared that he had examined about fifty tidal rivers and harbors in his career, and Walker more than sixty.[51] Airy submitted a report on Wexford Harbour, despite never having visited it. Based on previous surveys and charts, he offered a theoretical analysis that he thought "introduced important principles."[52] The commissioners pointed to the "great want of accessible harbors along the whole of the coasts," especially with the growing use of steamships. They expressed the need for "detailed inquiry and local examination into the state and condition of every port and navigable river."[53] Responsibility was mixed between commissioners for individual places and the Admiralty, so the report proposed a single Board of Conservancy, in connection with the Admiralty, to oversee all tidal harbors and navigable rivers.

The second report contained a litany of criticisms about poorly built and maintained structures, and financial inefficiencies or improprieties.[54] It found several harbors "governed by numerous, self-elected, irresponsible commissioners."[55] Around 1,600 pages of surveys and reports were appended, giving a detailed insight into the extent of civil engineering involvement in developing harbors and ports around the coasts. From as far back as Smeaton's on

Wells Harbour in 1767, reports featured Telford, John Rennie, his son John, Cubitt, Walker, James Rendel, and many more. The commissioners reiterated their call for a Board of Conservancy to oversee improvements and the proper expenditure of public funds.

As with many other proposals for central boards, that recommendation was not acted upon. It was not until 1863 that responsibility for harbors was transferred from the Admiralty to the Board of Trade's Marine Department. The Harbour Department of the Board of Trade was created in 1864. When it became separate from the Marine Department in 1867 it further took responsibility for sea fisheries and for railway schemes that might affect navigation from the Admiralty. In 1874 it took over matters such as wreck, salvage, quarantine, and registry of ships. Having taken over salmon and freshwater fisheries from the Home Office in 1886, the Board of Trade created a Fisheries Department, which merged with the Harbour Department in 1898 to form the Fisheries and Harbour Department. Bureaucratic responsibilities were becoming integrated into more managerial structures and systems for accommodating scientific advice.

RAILWAYS

The concept of a railway was not new to the nineteenth century. Tracks, originally made of wooden rails, had been used since the early seventeenth century to carry coal short distances in wagons. The first public railway, the Surrey Iron Railway, opened in 1803 for carrying goods, and the Swansea and Oystermouth Railway first carried passengers in 1807. Both were horse-drawn. But the passenger railway age is generally traced to the Liverpool and Manchester Railway in 1830, which ran its own steam engines and stock with a regular timetable.[56] Within ten years a profusion of local lines had been built, with some long-distance connections between major cities in England.

Throughout the century, railways remained in the private sector. Each railway required an act of Parliament to establish its line, but such independent private initiatives could not take account of the public or national interest. It soon became apparent that the nature of a railway led to companies operating as monopolies. That tended to result in higher prices for passengers, unreasonable levels of dividends for investors, and the potential neglect of safety issues. While the economic aspects of railways were one major focus for policy and legislation, another was passenger safety and the political implications of frequent accidents. Both led to often-grudging intervention by government in these private businesses. It was accidents in particular, and the implications of new technologies, that brought in the engineers and scientific experts to inform policy and legislation.

Following a petition to Parliament in 1839 complaining of monopolistic practices, Charles Poulett Thomson, president of the Board of Trade, moved for a select committee "to inquire into the state of communication by railways."[57] The importance of the subject is indicated by the calibre of members of the committee, which included Peel, James Graham, Lord Stanley, and Lord Seymour.

The committee was persuaded that there was real conflict between private and public benefit, affecting the carriage of goods, fares for the poorer class of passengers, and raising the danger of accidents. They asked, given the complexity, for further examination in the next parliamentary session, although they suggested the need for a supervising authority, which might be "a Board to be annexed to the Board of Trade, of which the President and Vice-President should be members, together with one or two engineer officers of rank and experience."[58] Stephenson, one of the railway engineers called to give evidence, agreed with this proposal on safety grounds.[59] This was an approach seeking to balance private and public interests, analogous to approaches in poor law and health legislation, which sought to balance the powers of local and central government.[60]

Henry Labouchere, who had replaced Poulett Thomson at the Board of Trade, secured the reappointment of the select committee in 1840.[61] Its first report reiterated the recommendation of a supervisory board, since railway companies seemed unable to resolve some issues affecting safety themselves.[62] In its third report, recognising the need for control by a single railway company of its line for safety purposes, the committee set out recommendations for supervision and inspection by a central authority.[63] The fourth report addressed the potential ability to move troops quickly round the country under the Mutiny Act in case of emergency. That raised the question of the best means of rapid long-distance communication.

The committee had questioned Charles Wheatstone, professor of experimental philosophy at King's College London, and inventor with William Cooke of the electric telegraph, which was already being trialed between London and Drayton on the Great Western Railway. The committee was impressed by the "great ingenuity in the various modes in which Mr. Wheatstone has applied the power of electricity to alphabetical communications."[64] The members could see the potential for preventing accidents and reducing delays but also the risk of leaving such a means of communication in private hands when it could also be of use to the government. They were "anxious to fix the attention of The House and of the Public on a discovery which is no less susceptible of useful than of dangerous application."[65] Technology, they knew, could be used for good or ill. It also offered the government a powerful means of control.

The Railway Regulation Act 1840, which implemented most of the recommendations of the select committee, was taken through Parliament by the Whig Lord Seymour, with government support. It resulted in the Board of Trade creating a railway department under George Porter, the head of the statistical department, as it was envisaged that collection of statistical data on matters such as passenger numbers, fares, and accidents would be a prime responsibility.

Inspectors with engineering experience were appointed to ensure that railway companies complied with their acts. They were also to examine the companies' lines, buildings, engines, and carriages, thus bringing together the combination of central supervision with professional expertise that character-ised so much regulatory legislation in the nineteenth century. This approach offered posts of influence to the engineers and the men of science and medi-cine. Inspection soon focused on more than legal compliance, as safety issues became significant. Thirty-five accidents involving personal injury, including twenty-eight fatalities, were reported to the board in the first five months alone.[66]

A late addition to the act specified that inspectors should have no con-nection with railway companies. That requirement made it almost impossible to appoint inspectors with practical railway experience. The Board of Trade turned to the Royal Engineers, not least since cost savings could be achieved by appointing officers already paid from the public purse. For the rest of the century and beyond, that remained the position, even though the ban was repealed in 1844.

The first inspector general, Lieutenant Colonel John Frederick Smith, was soon at work. In addition to his inspection duties he acted in 1841 with Peter Barlow, professor of mathematics at the Royal Military Academy at Wool-wich, to carry out a survey for the Treasury of whether an east coast or west coast route between Scotland and England was preferable.[67] Smith was suc-ceeded from 1841 to 1846 by Major General Charles Pasley, a fellow of the Royal Society since 1816. Smith himself was elected a fellow shortly before he left his post in 1841, with Porter, Barlow, and Pasley among the signatories. He gave evidence to another sitting of the select committee in 1841, which was set up to examine whether the Board of Trade should have discretion-ary powers to issue regulations to prevent accidents. Opinions were split, not least because in the limited state of development of railways much remained unknown about the potential causes of accidents. Isambard Kingdom Brunel, engineer to the Great Western Railway, was thoroughly opposed, accusing the inspectors of being without "any sufficient knowledge of the practical work-ing of the system with which they propose to interfere."[68] George Stephenson, by contrast, was more relaxed about possible powers of the Board of Trade over regulations. Smith did not believe that it was possible to clearly define the powers that would be required, given the current state of knowledge. The committee came to the view that supervision should be exercised "in the way of suggestion rather than that of positive regulation," a system that seemed to have worked well to date.[69]

The Whig government fell in August 1841, and Peel returned as prime minister with William Gladstone as vice president of the Board of Trade. That political change did not signal any significant change in emphasis toward the railways. An act of 1842, brought in by Gladstone, tightened up some safety provisions by empowering inspectors to postpone the opening of new railways on safety grounds, and required the notification of accidents involving personal injury. Two years later Gladstone, now president of the Board of Trade, chaired a select committee to examine his own proposals for intervention in the public interest.[70] This included the option of nationalization, first systematically advocated by the solicitor William Galt from 1843. In the end it was World War I that forced the resolution of this issue. Under the Regulation of the Forces Act 1871, passed during the Franco-Prussian War, which had highlighted the vital military role of railways, the government eventually nationalised the network.

Following the select committee, the government passed the Railway Regulation Act 1844, which established the direction of railway development for many years. It ensured that companies ran "parliamentary trains" that were affordable to poorer people, and it set parameters for regulating fares and for possible nationalization.[71] The basic pattern of price and quantity regulation that the act laid out was not abolished until the Transport Act 1960. Companies were also obliged to carry mail, troops, and police officers, and to make their telegraph systems available for public use.

The passing of the act coincided with the creation of a Railway Board within the Board of Trade, chaired by the tenth Earl of Dalhousie, the vice president, as Gladstone sought to distance himself from the avalanche of railway business. Porter became a member of the board and would become joint secretary of the Board of Trade in 1847.[72] The next two years saw the height of "railway mania" as hundreds of proposals came to Parliament for approval through private acts, putting huge pressure on the railway department and on Parliament.[73]

Gladstone resigned from the Conservative government in 1845 over the Maynooth Grant, and perhaps due partly to his family interest in railways,[74] leaving Dalhousie as president. He lasted until the Whigs came to power again in 1846, when the government created the Commissioners of Railways, independent of the Board of Trade so that they carried more weight than the short-lived Railway Board, given the importance of railway business. John Russell, the incoming prime minster, asked Dalhousie if he would preside, and undertake not to oppose the government. That is an indication of the relative lack of party politicization of railway matters, although Dalhousie declined, wishing to retain his independence.

By 1848 the four commissioners included Labouchere, already president of the Board of Trade, and Earl Granville, who was then vice president. That made the commission effectively part of the Board of Trade, which took back formal responsibility in 1851. Several new inspectors had been appointed in 1847. They soon became more involved in activities to promote safety in addition to their original remit to ensure that lines were constructed in accordance with the relevant act.

THE RAILWAY GAUGE COMMISSION

The railways had developed in the absence of any national strategy. One consequence was the existence of two different gauges. Most lines were laid at a narrow gauge of 4 feet 8½ inches, or 5 feet 3 inches in Ireland, while some, led by Brunel at the Great Western Railway, were at the broad gauge of 7 feet.

In June 1845 the Radical free trader Richard Cobden raised in Parliament his concerns about the situation, arguing that "there were two questions involved in this subject. First, the prevention of injury to the passengers and traffic from the want of uniformity in the gauge; and, second, the possibility of bringing the existing lines of railway into one uniform system."[75] His motion for a commission to examine the issue was approved and Peel named the three commissioners a fortnight later. They were Smith, previously the inspector general; Barlow; and Airy. Airy had been approached by Dalhousie, who had also passed on Smith's application.[76] Airy added this task to his work on the Tidal Harbour Commission, which had started a few months earlier. The commissioners, all public employees, were unpaid. They examined forty-eight witnesses over thirty days,[77] and observed a competition between trains on the two gauges.

The broad gauge appeared to offer marginally faster travel, and possibly better safety for passengers, but the commissioners considered that outweighed by other considerations and by the needs of general traffic. Having ruled out technical solutions such as telescopic axles and dual gauges, they insisted on a single gauge. The compelling arguments to them were the many public and potentially military inconveniences of switching between gauges and the desire for national uniformity.[78] Although they recommended the narrow gauge, the Gauge Regulation Act 1846 did not entirely compel it, limiting it only to particular districts, and both gauges ran in Britain until later in the century.[79] The different gauge in Ireland is still retained. Airy's "nerves were shaken" by the vigorous combat between the disputing and buccaneering engineers—who included Brunel, Bidder, Gooch, John Hawkshaw,

Vignoles, and Robert Stephenson—and retired to the waters of Wiesbaden to calm himself.[80]

Buccaneering engineers were not Airy's only problems with the railway companies. On several occasions, around 1835, 1845, and 1863, the Royal Observatory found itself threatened by plans to build a tunnel under Greenwich Park, driven by the desire of the railway companies involved to find the cheapest route.[81] It was feared that vibrations from passing trains would make accurate astronomical observations impossible, an ironic situation since the railways relied on the observatory to provide them with an accurate time. Indeed, it was demand from the railways that led to the introduction of a common time across Britain in the 1840s, measured by the Royal Observatory and replacing the different local definitions, although the government did not legally adopt a standard time until 1880.[82] On each occasion that the observatory was threatened, Airy and his fellow astronomers responded, seeing off the schemes. The power of railway shareholders in Parliament was substantial, but consistent lobbying by the astronomers, allied with Admiralty concerns, overcame the commercial interests of the railways.

THE DEE BRIDGE COLLAPSE

On May 24, 1847, a train crossing the Dee Bridge on the Chester and Holyhead line fell through the girders into the river, killing five people and injuring nine others. The inspector sent to investigate, Captain and later Field Marshal John Lintorn Simmons, reported that the problem was a failure of the cast-iron girders, and that the wrought iron trusses fixed to them in a design by Robert Stephenson did not reinforce them. The accident horrified public and Parliament, and Stephenson was fortunate not to be charged with manslaughter.

Edward Strutt, president of the Commissioners of Railways, told the House of Commons that he had recommended that the government appoint a commission "constituted partly of gentlemen of eminent scientific experience, and partly of practical engineers" to carry out experiments.[83] They were to inquire "into the conditions to be observed by Engineers in the application of Iron in Structures exposed to violent concussions and vibration" and to "illustrate by theory and experiment the action which takes place under varying circumstances in Iron Railway Bridges which have been constructed."[84] Airy was asked to preside, but was relieved to discover that the Earl of Auckland, his ultimate boss as first lord of the Admiralty, objected.[85] So Lord Wrottesley took the position, with Robert Willis, Jacksonian professor of natural philosophy at Cambridge University; Captain Henry James, who later became director general of the Ordnance Survey; the engineers George Rennie, William Cubitt, and Eaton Hodgkinson, professor of the mechanical principles of engineering at University College London. The secretary was Lieutenant

Douglas Galton, who would soon play more substantial roles. All were, or soon became, fellows of the Royal Society. In the event, Airy was dragged in too, and the commissioners thanked him in their report for observing some of the experiments and for making many "valuable suggestions."[86]

The commissioners soon became aware that calculations of the required size and shape of cast-iron girders had been made only on the basis of the stationary load they could carry. They did not take into account the effect of the massive vibrations set up by passing heavy trains, nor had any scientific experiments to that end been carried out. So a huge range of experiments was ordered, a large proportion of which was instigated by Hodgkinson. Others were carried out by Willis, James, and Galton, many at Portsmouth Dockyard with the support of the Admiralty. Several engineers—including Joseph Glynn, Robert Stephenson, Hawkshaw, Fairbairn, and Gooch—wrote in with information. On the theoretical side, Willis found the calculations so complex that he had to call on George Stokes to help him with the mathematical analysis.[87]

The commissioners reported in July 1849. Given the state of knowledge, they were not able to be definitive, declaring that "on the whole, the art of railway bridge-building cannot be said to be in that settled state which would enable an engineer to apply principles with confidence."[88] Theoretical understanding of the effect of a weight traversing a bar was also lacking, with the commissioners noting that "unfortunately the problem in question is so intricate that its complete mathematical solution appears to be beyond the present powers of analysis except in the simplest and most elementary case."[89]

Because of this lack of precise practical and theoretical knowledge, the commissioners did not recommend legislation that might constrain research and development. Instead, they offered some guidance on the appropriate load-bearing characteristics of girders to try to ensure safer designs.[90] Stephenson subsequently used wrought iron instead of cast iron for the Britannia Bridge across the Menai Strait.[91] Brunel, who had argued that a commission would stop further improvements by regulating, and that it was up to engineers to solve the problems, likewise moved to using wrought iron.[92]

INVESTIGATING ACCIDENTS

While collapsing bridges were a dramatic illustration of the potential dangers of traveling by rail, crashes and derailments, often with fatal consequences, were frequent. Although since 1840 companies had a duty to report accidents, they did not always do so, with the result that the railway department also scanned the newspapers for intelligence.[93] Nevertheless, companies invariably offered every facility to investigating inspectors, even though the inspectors had no legal power to enquire.

Over the years, the inspectors worked extensively by persuasion and consent, and publication of their findings in concert with media and public interest doubtless encouraged the companies to cooperate. It was frequently argued that greater regulation would reduce the companies' own sense of responsibility, so that measures to increase public safety would in fact decrease it.[94] That included resistance to the introduction of some safety devices, which it was argued would reduce the need for railway staff to feel responsible. Blame could therefore be placed on individuals rather than on the management or technical characteristics of the system as a whole.

From 1852, the inspectors' reports on accidents were published separately from the annual report, with suggestions on causes and possible remedies. At the same time, inspectors, including George Wynne and William Yolland, challenged the idea that individual staff such as drivers or signalmen were necessarily at fault, rather than their management or the safety technologies deployed.

Two reports from Wynne are notable for introducing John Tyndall to government service. Tyndall had left school in Ireland for employment as a lowly surveyors' assistant on the Irish Ordnance Survey in 1839. Lieutenant Wynne, also born in Ireland, was the officer commanding his division.[95] Despite their different social backgrounds, the two developed great mutual respect. Tyndall, having studied for a doctorate at the University of Marburg, soon come to the notice of influential men of science, after his initial work on magnetism. With Edward Sabine's patronage he became a fellow of the Royal Society in 1852, and was appointed professor of natural philosophy at the Royal Institution in 1853, where he worked alongside Michael Faraday. The following year Wynne asked him to help to examine Professor Gluckman's invention of an electric bell system to enable guards to communicate with engine drivers. Tyndall "expressed himself very favourably with regard to the action of the instrument, and its non-liability to derangement."[96] The report was raised in the House of Lords after the attempted murder of a policeman on a train, following a suggestion that such a means of communication should be made compulsory.[97]

A year later, now a Lieutenant Colonel, Wynne called on Tyndall to examine the explosion of the boiler of the *Actæon* at Gloucester Station. Tyndall realized that the cause was electrolytic corrosion, and offered possible remedies, having checked his conclusions with Faraday.[98] He suggested the use of different metals, of sacrificial anodes, or a different design to reduce the possibility of corrosion. This science-based solution was not practicable for the railway companies, who continued to employ the same materials. What these inquiries did do was to establish the need for more frequent examinations of boiler interiors to detect serious thinning before it led to failure.[99]

The gauge commissioners had found in 1846 that accidents were caused by "collisions, obstructions on the road, joints wrongly placed, slips in cuttings, subsidence of embankments, a defective state of the permanent way, loss of gauge, broken or loose chairs, fractures of wheels or axles, &c.; and, lastly, from engines running off the line through some other cause."[100] By contrast, a select committee in 1858, to which the inspectors Wynne, Yolland, and Henry Tyler gave evidence, found the main causes to be inattentive staff, defective works, or rolling stock and excessive speed. The committee was not prepared to recommend "direct legislative interference" since that might relieve railway companies of their responsibility.[101] They urged better punctuality of trains to avoid the incentive to speed to catch up with the timetable, means of communication between guards and drivers, and use of the telegraph. Although inventors suggested many technological solutions to safety problems, the railway department rarely investigated their merits, leaving it to the companies. There were some exceptions, as in 1859 when Fairbairn asked for government funding toward experiments on strengthening wrought iron bridges. It was granted, although it took him five years to report the results.[102]

Braking systems, communication devices to the driver, block working, and signaling proved to be the technical areas on which safety efforts were focused. Communication between guards and drivers was a frequent topic of interest. Some companies installed a rope attached to a bell, and by 1868 it became compulsory for passengers to be able to use it as well as guards. Block working, in which a train was not allowed to enter a section until a telegraph signal showed that the previous train had left it, was introduced. In 1853 Prince Albert had urged it on Edward Cardwell, then president of the Board of Trade, who replied that he had no power to enforce it but had long exhorted the companies to adopt it.[103] Signaling improvements, such as linking systems together for easier and safer operation, were also introduced.

Advice on matters such as these was included in the requirements for opening new lines issued by the railway department from 1858. That gave the department some control over the companies and led to gradual improvements as the requirements were revised over time, although accidents continued to occur with relentless regularity. In 1858, 276 deaths and 556 injuries were reported to the department.[104] That compares with 216 deaths and 486 injuries reported in 1852. The length of line open had increased in those six years from 7,000 to 9,500 miles and passenger numbers from 90 million to 145 million,[105] which implies a small improvement in safety per number of passengers.

Accidents remained a perennial concern. Although the Royal Commission on Railways, established in 1865, was focused on economic matters, including emphatically rejecting state purchase of the railways,[106] the commissioners did comment on accidents. They noted a marked diminution since the 1850s, and explained that safety relied on the common law and Lord Campbell's act, which gave a right of compensation for death or injury. They thought this better than giving the Board of Trade additional powers, although they did recommend that the board should be able to compel the attendance of railway staff at hearings and the production of books and documents, and to make their reports public.[107]

THE ROYAL COMMISSION ON RAILWAY ACCIDENTS, 1874–1877

Between 1844 and 1868, the railway department's influence grew, although without significant additional legal powers. Those gradually increased thereafter, in relation to safety, fares, and the settlement of labor disputes, as more stable two-party politics and wider public pressure following the Reform Act 1867 led the government to legislate for more detailed regulation. The Regulation of Railways Act 1868 required a means by which passengers could communicate with the driver or guard in an emergency. A select committee in 1870 focused mostly on compensation for accidents but suggested no safety legislation, although it urged the use of block and interlocking systems and continuous brakes.[108] The subsequent Regulation of Railways Act 1871 made formal provision for the investigation of accidents. It permitted the Board of Trade to appoint a tribunal to investigate accidents, although the power was rarely used, an exception being the Tay Bridge disaster.

In April 1874, shortly after the arrival of Prime Minister Benjamin Disraeli's new government, the seventh Earl De La Warr rose in the House of Lords to propose a royal commission into the prevention of railway accidents.[109] He had tried to bring in a bill to require use of the telegraphic block system and interlocking points and signals, which the select committee had rejected on the grounds that the railway companies were already doing enough. Earl De La Warr argued that accidents had recently increased, with 1,100 killed and more than 3,000 injured in 1872. The sixth Duke of Richmond, president of the Privy Council, intimated that the government would support a royal commission which Charles Adderley, president of the Board of Trade, confirmed in the House of Commons.[110] There were some objections, including from those such as Lord Houghton, who wanted no further intervention in the railways and from others such as George Bentinck, who called for more government intervention. Bentick argued that if the government allowed "the question to be dealt with by Committees and Commissions it neglected those duties which it was imperatively called upon to perform, and was responsible for these accidents."[111]

The Royal Commission on Railway Accidents was chaired by the third Duke of Buckingham. Among other members were Thomas Harrison,

president of the Institution of Civil Engineers, providing engineering expertise, and Lieutenant General Simmons, bringing his deep experience of the
railway department. He had been appointed a railway inspector in 1847 and
secretary of the railway department in 1851, and had been the inspector on
the Dee Bridge collapse. The formal recommendations suggested giving more
discretionary powers to the Board of Trade, subject to judicial review by a tribunal, so that safety might be improved. The commissioners further suggested
that trains should have particular braking capabilities, that companies should
provide increased compensation to delayed passengers in order to encourage
punctuality, and that the civil liability of companies and criminal liability
of employees for actions endangering life should be increased. No legislation
followed.

The royal commission's deliberations had at least instigated trials of continuous brakes, but it was not until an accident due to inadequate brakes in
Armagh, Ireland, in 1889, when 80 people were killed and 260 injured in a
Sunday school excursion train, the worst rail disaster of the century, that the
Railway Regulation Act 1889 finally compelled companies to introduce automatic brakes and to complete systems of block operation and interlocking of
signals.

THE TAY BRIDGE DISASTER

The interaction between legislation, regulation through inspection, and scientific and engineering expertise on the railways reveals the multiple tensions
and interests that influenced legislators and the executive during the nineteenth century. It also reveals the influence that the scientific and engineering
experts were or were not able to bring to bear. Most of the issues that directly
affected legislation were concerned with widely held beliefs that private capital and a competitive private sector would offer the best outcomes for passengers and for traders in goods. Added to that was the principle that companies
and individuals should take responsibility for their actions.

It was frequently argued that enforcing safety measures would lead to
worse outcomes, if companies and individuals had some of that responsibility
removed. But even inspectors differed on this. For example, giving evidence to
a select committee in 1870, Yolland urged powers for the Board of Trade over
the safety of passengers, as was already the case for shipping.[112] By contrast
Tyler, chief inspector in 1871, was emphatic in his annual report on accidents
for 1870 that any interference that would relieve companies of their responsibility "would have a mischievous rather than a beneficial tendency."[113]

Legislation attempted to strike a balance between the interests of the companies and the public, while seeking to maintain the sense of responsibility of
companies and individuals, by either incentivising it or punishing the lack of

it through the courts. Legislation to protect workers lagged far behind, as it did in other industries. The Railway Employment (Prevention of Accidents) Act was not passed until 1900, when inspectors were given powers to investigate accidents to railway staff.

Nevertheless, behind the scenes, the persuasive interactions of the inspectors with the railway companies led to steady improvements over the years. Technological issues were always in the background, but it often took a major accident to bring them to public and parliamentary attention. The Tay Bridge disaster offers an example.

During a violent storm on the evening of December 28, 1879, Tay Bridge, spectacular marvel of British engineering, admired—among others—by President Ulysses S. Grant, the emperor of Brazil, and Queen Victoria herself, collapsed with a train on it. Everyone aboard, about seventy people, was killed as it crashed into the river below.

The tribunal to examine the causes of the disaster was assembled within a few days. It consisted of Henry Rothery, wreck commissioner; Yolland, now chief inspector of railways; and William Barlow, president of the Institution of Civil Engineers. The engineer Henry Law was contracted to make a full structural report. The effect of wind was one significant factor explored by the inquiry. In 1873 Airy had been consulted by Barlow and Thomas Bouch, the engineer of the bridge, about the effect of wind pressure, and his report was much referred to in the inquiry.[114] In addition, Airy, Stokes, and Robert Scott were asked to examine the effects of wind in detail. Scott was the secretary to the Meteorological Council, which had replaced the Meteorological Committee in 1877.

The design and construction of Tay Bridge illustrates the overconfidence of the time and the seemingly small factors which added up to disaster.[115] Bouch had a reputation for building cheaply and for lax supervision. The tribunal found, in the words of Rothery, that the bridge was "badly designed, badly constructed, and badly maintained."[116] Rothery placed the chief blame on Bouch for faulty design and supervision, and concluded that the disaster was caused by the effect of wind acting on a structure badly built and maintained. Yolland and Barlow believed that they had been asked to report the causes, not to apportion blame. They wrote a separate report,[117] although Rothery stated that they did not differ overall. Yolland and Barlow suggested that the Board of Trade should establish rules for wind pressure, which Rothery rejected on the grounds that it was for the engineering profession not the Board of Trade to formulate them. As so often, the question of whether or not to regulate or legislate came up against uncertainty of knowledge and the desire not to constrain further research and development by setting rules too soon.

Later studies have broadened blame to include poor quality control, the overconfidence of directors and shareholders of the North British Railroad, and the lax inspection by Major General Charles Hutchinson, the Board of Trade's inspector.[118] Indeed, all these are touched on or implied in Rothery's report. Hutchinson would later investigate the Armagh rail disaster and serve as chief inspector of railways from 1892 to 1895.

While Tay Bridge was made of cast and wrought iron, the new Forth Bridge would be made of steel, and overdesigned in compensation. Given the lack of knowledge of the effect of wind pressure on railway structures, in 1880 the president of the Board of Trade, Joseph Chamberlain appointed a committee of five, all fellows of the Royal Society— William Armstrong, Barlow, Hawkshaw, Stokes, and Yolland—to investigate and report. Their report, using extensive observations of wind speeds, suggested five rules, based on assuming a maximum wind pressure on bridges of fifty-six pounds per square foot, higher than that used for Tay Bridge's design calculations.[119]

NEW TECHNOLOGIES IN THE METROPOLIS

Toward the end of the nineteenth century, new technologies for railways and unanticipated effects of existing railways called for further inquiries. Two examples give a flavor of the challenges. They are the potential for the use of electrification and cable systems, and the problems of ventilation.

When the question of electrification appeared, along with the possibility of cable-drawn railways, a joint committee of five members of the House of Lords and three of the House of Commons was established to provide advice for bills under consideration.[120] It included two with strong scientific backgrounds in the recently ennobled William Thomson as Lord Kelvin and Nevil Story Maskelyne, a Liberal Unionist MP. The City and South London Railway had used an electrified line since 1890, although the first such railway in Britain was Volk's Electric Railway in Brighton, opened as a pleasure attraction in 1883.

Major General Hutchinson, inspector of railways, declared that electricity would suffice as motive power. The two high-profile technical witnesses, in addition to several consulting engineers, were William Preece and Alexander Siemens, and it was Lord Kelvin and Story Maskelyne who did most of their questioning. Preece, electrician to the General Post Office, had visited the United States to observe the development of electric railways there. He outlined their advantages in terms of economy, public perception, and the ability to climb steeper gradients. Siemens was sure that the proposals were technically feasible. The committee reported that it was satisfied by the evidence on the capability and potential of electrification and of cable traction, especially for steep gradients.[121]

In 1897 public concerns about the ventilation of the Metropolitan Railway, which used steam locomotives, led to the appointment of a committee of the Board of Trade. Members included Douglas Galton and John Scott Haldane. The problem was that no act required ventilation, and no remedial measures of making openings or using fans had been found to work. Henry Armstrong and Haldane carried out tests that demonstrated high levels of carbon dioxide, but they were not able to report figures on the poisonous carbon monoxide. The committee assumed that more dangerous impurities would be proportional to carbon dioxide, and recommended a limit on carbon dioxide slightly higher than that suggested by Haldane.[122] Its summary conclusions were that electrification would be preferable, as the cost of fans would be prohibitive. The central London lines were electrified by 1907.

These examples illustrate, as the twentieth century opened, the particular use of expert committees to advise on the practicalities of new technologies, as in so many other areas of potential application. Taking the nineteenth century as a whole, policy toward transport infrastructure illustrates the power of the private sector, the belief in personal responsibility, and the weight given to local determination, against the desire for national standards advised and monitored by technical experts. These factors, in this particular area of civil policy and practice, embody the four overarching principles and beliefs within which I argue that politics and expertise were contested overall: the sanctity of private property, the laissez-faire approach to capitalist private industry, the emphasis on individual freedom and responsibility, and the importance of local government.

8

SHIPS, LIGHTHOUSES, AND THE BOARD OF TRADE

PICTURE A BRITISH CITIZEN'S TRANSPORT OPTIONS AT THE BEGINNING of the nineteenth century. Roads were uneven, dusty, or muddy. You could walk, ride a horse, or travel in a stagecoach. Sailing ships plied the coast, while canals took inland waterborne traffic. There were no railways, excepting short tracks for mining and quarrying, and almost no mobile steam-powered vehicles. But in just a few decades, there was a complete transformation. Railways crisscrossed the land, altering social patterns and revolutionizing opportunities for travel and trade. Steamships changed forever the reliability of seaborne transport.

Trade and transport are intimately connected. It is therefore understandable that the Board of Trade became the government department most concerned with transport. Its antecedents are committees and commissioners established from the early seventeenth century to advise the Privy Council on colonial questions of settlements and trade. After falling into abeyance, it was reconstituted in 1784, because of the need to regulate trade between Britain, its colonies, and the now-independent United States of America following the Revolutionary War.[1]

The new Board of Trade soon found itself grappling with the trade implications of rapid industrialization, and with technological developments in the means of transport that facilitated that trade. Campaigns for free trade and against protectionist duties and tariffs became irresistible, as British industrialists became convinced that they could compete abroad and grow their profits in a free trade environment. Duties were gradually lowered and simplified.

By the mid-1840s, in concert with the abolition of the Corn Laws, duties were designed to provide revenue to the Exchequer more than to control trade. But free trade required regulation to ensure a level playing field in the public interest, to ensure fair competition between capitalist enterprises, and to avoid deleterious effects of monopolies.

Nowhere is this more apparent than in the growing responsibilities of the Board of Trade in regulating transport systems, particularly those of shipping and the railways. Even ardent proponents of market forces recognized that the public interest was not best served by the existence of monopolies that would tend to raise prices and reduce the level of attention to the safety and welfare of passengers and workers. The engineers and scientific experts became essential agents in the development and regulation of these transport technologies.

SHIPPING

British shipping had little competition in the early nineteenth century. The economic policy of mercantilism, which aimed to keep the benefits of trade within Britain and its colonies by minimizing the loss of profits to foreigners, resulted in a range of protectionist measures. The Acts of Trade and Navigation, dating to 1651, set the tone. British ships had a monopoly of trade between ports in Britain and its colonies, and around the British coast. The consequence was complacency, leading to large numbers of unseaworthy ships, unqualified masters, and ill-treated seamen. Under pressure from free traders, some of the provisions of the acts were relaxed in the 1820s. Competition from foreign shipping rose, but many British shipowners cut costs rather than invest in improvements, exacerbating the problems. The navigation acts were finally repealed in 1849. In the meantime, the systemic problems of the shipping industry had become so apparent that Parliament repeatedly stepped in.[2]

Up to 1815 the problems had been hidden, as losses could be attributed to the misfortunes of the Napoleonic Wars. But by the 1830s the drowning of some two thousand seamen annually and the foundering of hundreds of ships could no longer be ignored. Strange as it may seem, one factor underlying the disasters was the definition of a ship's tonnage.

TONNAGE

A ship's tonnage measures its carrying capacity. That in turn set levels of taxes and fees—for example, for passing a lighthouse, using a harbor, or taking on a pilot. It mattered to shipowners and the Exchequer. The rules for measurement changed over the centuries, so that by the 1800s the definition of 1776 applied. This required measuring the length and breadth of the ship in a particular way, subtracting three fifths of the breadth from the length, multiplying the result by the breadth, multiplying that result by half the breadth, and dividing the product by ninety-four.[3] It was wide open to abuse. Not only could certain construction devices reduce the measured length and breadth, but the lack of reference to depth meant that ships could be made deeper without changing the tonnage. That could make them slower and less stable, as well as depriving the Exchequer of revenue.

In 1820 the Board of Trade asked the Admiralty to examine the issue. The Admiralty approached the Royal Society, and in 1821 the Board of Longitude established a Tonnage Committee, which included Davies Gilbert, Thomas Young, Henry Kater, Robert Seppings, and Nicholas Robilliard, the surveyor for the Acts of Trade and Navigation. The involvement of scientific expertise illustrates again the challenges of converting scientific and mathematical analysis into practical policy. The committee toyed with the idea of working from the difference between the waterlines when a ship was empty or loaded but found the idea impractical. Instead, they offered a mathematical formula, more complicated than the existing one but that included the depth. It was ignored.

The problem was revisited in the 1830s when William Parsons, later the third Earl of Rosse, was asked by the Admiralty with the Board of Trade's endorsement to explore new means of calculation. That led to a committee that reported in 1834.[4] It was chaired by Gilbert and also included Robilliard; Francis Beaufort; Edward Riddle, a mathematician; and Henry Raper, an authority on navigation.

This committee likewise rejected the idea of using displacement as a measure. It developed rules for determining the internal capacity, which were introduced by Parliament in 1835. These rules ran alongside the existing system, as owners of older ships refused to accept retrospective changes. This lack of clarity led to a further committee being appointed in 1849, chaired by Lord John Hay, one of the lords of the Admiralty, and including Parsons, now Lord Rosse, and George Moorsom, who became the surveyor general for tonnage.[5] This committee recommended a system of external measurement, but the bill to introduce it was withdrawn after opposition. Moorsom persisted, and his subsequent method, based once more on internal measurement, was finally introduced in the extensive Merchant Shipping Act 1854.

STEAMSHIPS

Although some experiments were made with steam-powered vessels in the eighteenth century, the first commercial service was operated by *Comet* on the River Clyde in 1812. In the early days, steamships used paddle wheels for propulsion, with propellers becoming efficient by the middle of the century, at which point the Royal Navy started taking them seriously.[6] The weight and bulk of the engine and fuel made them uncompetitive with sail for many purposes, so they were first used for short high-value trips carrying mail or passengers, or as tugs.

The business was competitive, leading owners to cut corners on safety, such as boiler design and maintenance. These boilers had a tendency to explode, dramatically enough in 1817 for a select committee to be established.[7] This was at a time when loss of life at sea was an accepted fact, not viewed as a responsibility of government, and the appointment of a select committee was not the regular occurrence it became later in the century.[8] But the novelty and visibility of the technology, concerns of local authorities, the media publicity, and the fact that members of the public had been injured made some response imperative.

The impetus for the select committee of 1817 was the tragic explosion on Good Friday of the Norwich steam packet *Telegraph*, the first such disaster in Britain. Nine people were killed, including women and a four-year-old child.[9] The committee, chaired by Charles Harvey, the MP for Norwich, called about twenty men, almost all engineers, to give evidence. The roll call included figures such as Bryan Donkin, Timothy Bramah, and Henry Maudslay. This is perhaps the first example of a select committee grappling with the problems of new technology introduced by the Industrial Revolution.

It became clear that the boiler had been improperly constructed and the safety valve overloaded. The committee recognized that it had to balance the public interest with the desire to let private enterprise flourish without interference. It called for the registration of steamboats and the inspection of boilers and safety valves, which, if introduced, would have preceded the inspection of factories by many years. But chance events prevented it. A bill to introduce regulations fell when Parliament was prorogued soon after the committee reported. Harvey reintroduced it in the next session, but it fell again when Parliament was dissolved in 1818. Harvey then left the House of Commons and no one took up the baton.

The topic was revisited in 1822 when a select committee explored the possibility of steamboats carrying mail as part of the London to Dublin link. This committee examined the form and strength of vessels; the engines, boilers, and machinery; and the use of sails. Engineers including Donkin, Bramah, and Marc Brunel gave evidence. So did Michael Faraday, who explained the mechanism by which sea water could cause corrosion of boilers. He also suggested a means of avoiding it, and of preventing the deposition of salts in the boilers.[10]

Despite remarking that "nothing could be more baneful than the interference of the Legislature with this new branch of science," the committee recommended that the General Post Office should build at least four boats to a particular design by Seppings, using wrought iron for the engine, and should install a fire extinguisher and two lifeboats. It also suggested that reducing the fares, which had been increased following the introduction of steam, would increase revenue.[11]

It was disasters such as the wrecking off the coast of north Wales of *Rothsay Castle* in 1831 with the loss of 130 lives that stimulated a further select committee. This was established by the arch-Tory Colonel Charles Sibthorp, that scourge of new technological contraptions such as the railways and steam engines. Much of the evidence provided by some forty-five men related to rules for the operation of ships, but several engineers and shipbuilders were also invited, including Maudslay and Seppings. This committee, again seeking to balance private and public interest, recommended the registration of steamboats, their regular survey, limitations on passenger numbers, speed restrictions on the Thames, and the requirement to carry lifeboats and lights.[12] As with the previous committee, the recommendations of this one fell into abeyance when Parliament was gripped by the political convulsions over the Reform Act.

It was not just steamships that foundered. Overall shipping losses were such that a select committee on the causes of shipwrecks was established in 1836,[13] unfortunately at the hurried end of a parliamentary session and without government support. The committee explored a whole range of operational and design matters. Most of those giving evidence were directly involved in shipping, but scientific evidence was brought to bear on accidents caused by compass readings being affected by iron on the ship, which could lead to ships veering off course and being wrecked. Peter Barlow, professor of mathematics at the Royal Military Academy at Woolwich, described means of minimizing the problem using a compensatory iron plate.

The committee identified ten causes of shipwreck, ranging from modes of construction to drunken masters and crew. Its recommendations were comprehensive, including the establishment of a Mercantile Marine Board to regulate the industry. There was no time for action, and when a bill was brought forward in the next session, it was swiftly defeated as being against the country's shipping interests.[14] But although this initiative failed, many of the committee's recommendations became law in a piecemeal manner from 1850 onward.

Steam-vessel accidents remained in the news. The Board of Trade commissioned a report in 1839 from John Pringle and the engineer Josiah Parkes. Pringle was a veteran Royal Engineer and founding superintendent of the Geological Survey of Ireland in 1826.[15] Parkes was a drainage engineer, but with extensive experience of steam engines. Their report classified ninety-two accidents according to four causes: wreck (forty), explosion (twenty-three), fire (seventeen), and collision (twelve).[16] They concluded that there was a need for a registry, for periodical surveys, and for the licensing of steam vessels.

The committee proposed regulations that covered those three principal aims and included provisions for "rules of the road" and carrying lights. Like the 1836 committee, it proposed a board to oversee the arrangements. But in 1839 a weak Whig government, which was in financial difficulties, did not prioritize adding new administrative activities and costs. Nor was the technical evidence always compelling. The engineers did not agree among themselves about the safest designs of boilers and valves. Pringle and Parkes commented with respect to the response they obtained from engineers that it was "difficult to classify the opinions of these gentlemen."[17] That made it easier to avoid legislation. Indeed, introducing regulations might have restricted technological development, by setting standards before different methods had been tried. Although the Board of Trade accepted the proposals and Henry Labouchere as president introduced a bill in the next parliamentary session, he met strong opposition and had no qualms about dropping the measure.

It was not until 1846 that legislative action was taken, following the example of other countries and of sometimes contradictory initiatives by local authorities. This followed another select committee inquiry on shipwrecks in 1843, after further petitions. It was a high-powered committee that included William Gladstone, vice president and soon to be president of the Board of Trade, Robert Fitzroy, Admiral James Dundas, and Admiral Charles Napier. Like previous committees, it was concerned with all factors relating to safety, including the construction of ships, the competency of masters and mates, pilots, harbors, lighthouses, and charts and compasses. It called a preponderance of expert naval witnesses, although the civil engineers William Cubitt and John Rennie were examined about harbor design.

In addition to general recommendations for safety at sea, such as watertight compartments and the carrying of lifeboats, this committee advocated the superintendence of all passenger steamers.[18] Now, with reduced opposition from the shipping industry, the vice president of the Board of Trade, George Clerk, saw the Steam Navigation Act 1846 through Parliament. It required the inspection of passenger steamboats and set rules for safe practice. As a result of this act, the Board of Trade established a Steamboat Department. It appointed the naval officer Henry Denham, a fellow of the Royal Society who had carried out extensive coastal surveys, as the first inspector of steamboat accidents. His role included the appointment of local inspectors and the investigation of specific accidents. Denham argued for surveyors appointed by the Admiralty and chosen for their expertise, rather than being appointed by shipowners. He also wanted each surveyor to ensure that boilers were fitted with tamper-proof safety valves to reduce the likelihood of explosions. These requirements were embodied in an act of 1851, and incorporated into the Merchant Shipping Act 1854.

Safe means of securing ships using cables and anchors also exercised government over many years. That was especially important from mid-century, as iron ships became so much larger and heavier,[19] and the developments over several decades can be followed elsewhere.[20]

EXPLOSIONS AND SPONTANEOUS COMBUSTION

Despite tightened regulations, ships continued to catch fire,[21] and boilers to explode. A select committee in 1870 concentrated on steam engine boilers on land, but called Robert Robinson, surveyor general of steam vessels to the Board of Trade, and Robert Galloway, the chief surveyor of steam-vessels, to give evidence.[22] The committee was chaired by John Hick, an inventor, industrialist, and Conservative MP, and included Lyon Playfair. Hick also served on a select committee of 1874 on the efficiency of the testing of chain cables and anchors.

In addition to Robinson and Galloway, the committee questioned many engineers, including James Nasmyth and William Fairbairn, who had proposed and now represented the Manchester Steam Users' Association, formerly the Manchester Association for the Prevention of Steam Boiler Explosions, founded in 1854.

Galloway described the inspection of steam boilers for passenger ships, and claimed that there had only been seven explosions in twenty years, at a much lower rate than land-based boilers. He put this down partly to inspection and partly to properly qualified operatives. Fairbairn suggested that the government should compel inspection, although not by government inspectors.

The committee's discussion centred on the questions of compulsion, insurance, and of who would certify and appoint inspectors, the government, or local and voluntary associations. Fairbairn's view—a common one—was that government inspection would remove responsibility from private businesses, leading to inattention. The same applied to insurance for public liability, which was seen for much of the nineteenth century as leading to complacency by the responsible individuals in mines, factories, and railways.[23] These were not technical matters, and the views of the engineers differed. Robinson thought inspection should be compulsory, with inspectors appointed locally but approved by the Board of Trade, similar to the manner in which boards of guardians appointed medical officers, subject to the Poor Law Board.

The select committee returned to the subject in 1871 calling, among others, the engineers Charles Vignoles, Frederick Bramwell, and Thomas Hawksley. Vignoles proposed a central board, to include "a scientific member" and one or two engineers appointed by the Board of Trade. When asked if people would serve, he commented that he thought it was "almost the duty of a scientific man to do anything of that kind which is likely to tend to the public safety."[24] Bramwell, unlike Vignoles and Hawksley, was against compulsory inspection, and had dissented from a proposal for compulsory inspection made by a British Association committee on steam boiler explosions in 1870. He thought it removed the responsibility from individuals and would constrain engineering progress.

The committee agreed with Bramwell. It came down against compulsory inspection on the grounds that there were only about fifty explosions and seventy deaths per annum, and recommended that the burden should be on the owner to report and prove the quality of the boiler and its operation to a coroner, should there be an explosion. But coroners only investigated deaths, and often had difficulty in calling or paying for expert witnesses.

In 1871 and 1872 Hick introduced a bill "to provide a more efficient remedy to persons injured and property damaged by the explosion of steam boilers through negligence." The attempts failed, and although explosions and loss of life continued, and trades unions argued for government certification of those operating boilers, nothing was done until the Boiler Explosions Act 1882. This provided for explosions to be notified to the Board of Trade, which could order an inquiry. The Employers Liability Act 1880 had by this time made employers liable for injuries caused by their carelessness or preventable causes.

ROYAL COMMISSION ON THE SPONTANEOUS COMBUSTION OF COAL IN SHIPS, 1875–1876

It was not only boilers that exploded. The problem of coal-bearing ships bursting into flames became so acute in the 1870s, with about seventy ships sailing for overseas ports suffering damage and casualties in 1874, often resulting in loss of the ship, that a royal commission was established to identify preventive measures. One stimulus to the inquiry was that courts examining such accidents invariably exonerated the ships' officers and recommended better ventilation, which was then disregarded by shipowners and underwriters. The Salvage Association thought the courts had little influence as they were not expert, and called for a commission of men "of high scientific attainments."[25]

The eleven commissioners were chaired by Hugh Childers, who had been first lord of the Admiralty until 1871 in Prime Minister William Gladstone's government. Frederick Abel and John Percy were the two men "of high scientific attainments." Their paper formed a significant part of the report,[26] and John Ferguson, professor of chemistry at the University of Glasgow, submitted a chemical analysis of different types of coal.[27] Abel and Percy put the causes down to the presence of iron pyrites and to some hydrocarbons in coal, both of which could combust when exposed to oxygen in the atmosphere. That caused the coal itself to burn. They observed that porous coal, which broke down into smaller pieces with a greater surface area to absorb oxygen, could also lead to spontaneous combustion. In addition, they noted that fresh coal could continue to release marsh gas, which could explode when mixed with air and ignited by a spark. They advised that ventilation throughout the hold would in fact be more likely to lead to combustion by any of these routes, contrary to recommendations from previous courts of inquiry, but that some surface ventilation might help to remove combustible gases. They did not recommend legislation. They expected shipowners to take note, and suggested that inspectors of mines should hold inquiries if coal from their areas spontaneously combusted.[28]

The navy also had problems with explosions in bunkers. After the loss of the *Dotorel* in 1881, an Admiralty committee including Abel and Warington Smyth reported on the incident. Unlike the Board of Trade, the Admiralty swiftly issued new regulations for the storage and use of inflammable materials.[29]

REGULATION OF SHIPPING IN THE NINETEENTH CENTURY

Were it not for the contingencies of parliamentary timetables, steam shipping might have had a regulatory and inspection system as early as 1818, driven by the potential danger to the traveling public. It was the danger to the public, seen as unable to protect themselves, rather than to the seamen that made the issue salient. Specific protection for the latter had to wait until the 1860s. This parallels developments in factory inspection, which initially protected vulnerable pauper apprentices before being gradually extended to children, young people, and women. Adult men had to wait for most of the century for direct protection.

The regulation of tonnage and of steamships were just two aspects of shipping frequently occupying Parliament and government in the early 1800s. A third, which demanded extensive engineering and scientific advice, was that of lighthouses, addressed below. Many other issues, some requiring scientific advice, were also tackled. They included the use of pilots, the rules of the road for passing ships, registration, stowing cargo on deck, training shipmasters and crew, the welfare of seamen, and the conditions of emigrants as passengers and of coal-whippers who unloaded coal at the docks.

Until 1854, most of these concerns could only be addressed in isolation, as powerful interests on a single issue could torpedo a bill attempting to cover many. But at that point Edward Cardwell, as president of the Board of Trade, managed to consolidate these strands in the vast Merchant Shipping Act 1854, with its 548 clauses. It set the parameters for the rest of the century and was not substantively revised until the Merchant Shipping Act 1894. Some further additions were made. For example, the Merchant Shipping Act 1867 established a medical inspectorate, as John Simon had urged, which soon eradicated the scourge of scurvy.[30]

Whereas many different government departments had previously had responsibility for some aspect of shipping, those responsibilities were now focused within the Board of Trade and its new Marine Department, created in 1850 as the Naval Department and renamed in 1853. The department was headed by Thomas Farrer, who would become the permanent secretary of the Board of Trade in 1867. The Board of Trade had become an executive department of government rather than an advisory committee to the Privy Council.

Safety at sea remained a preoccupation. The Liberal MP Samuel Plimsoll had attempted to pass a bill for a safe load line on ships in 1870 and 1871, even if his technical evidence for it was limited, but was defeated by shipping interests.[31] His continued pressure led to a royal commission on unseaworthy ships, which reported in 1873 and 1874.[32] Charles Merrifield, principal of the Royal School of Naval Architecture and Marine Engineering, along with the contractor Thomas Brassey, brought technical expertise as commissioners. The witnesses included shipowners, builders, repairers, insurance agents, and civil servants. This commission stated the familiar position that it thought the Board of Trade had ample discretionary power to detain unseaworthy ships, and advised that "the wisest policy will be for the Board to interfere only where there is ground for suspecting some gross mismanagement."[33] The commissioners did not recommend a compulsory load line. The responsibility, they thought, should lie with the shipowner.

To address some of the continuing challenges, the government introduced a bill in 1875. It proved problematic to shipping interests, and when Benjamin Disraeli announced that it would be dropped, Plimsoll called objecting members of the House "villains," creating an uproar.[34] But after pressure from popular feeling, the Merchant Shipping Act 1875 was passed, adding some safety regulations, although no load line. It remained a live issue, and in December 1883 a departmental committee including the naval architect Francis Elgar was appointed by the Board of Trade to review the question of the load line. The committee reported in 1885, and tables for establishing buoyancy were applied voluntarily by owners.[35]

But still, the losses mounted. Between 1874 and 1883 there had been a marked increase in losses of both ships and lives, prompting another royal commission.[36] This one came to the sobering conclusion in 1887 that "the hopes and intentions of Parliament to reduce the loss of life at sea by the various provisions of the acts referred to and through the general powers entrusted to the Board of Trade, have not been successful; and that all the measures adopted have hitherto had little or no effect."[37] This was despite the fact that it was now a criminal offence to send a dangerously unseaworthy ship to sea. The possible remedies still relied, in the commission's view, on placing responsibility on shipowners, insurers, and crew.

Scientific experts are notable by their absence as witnesses to this commission, except for the provision of statistical information. Scientific and engineering advice could offer little more, when legal complexities and political judgements so largely determined the question of safety at sea. At last, in 1890, the Merchant Shipping (Load Lines) Act made the marking of load lines compulsory.[38] These Plimsoll lines recognize Plimsoll's earlier exertions.

LIGHTHOUSES

Today we take for granted the powerful illumination emanating from lighthouses guarding treacherous coasts and the booming sound of foghorns when the mist comes down. At the beginning of the nineteenth century there were only thirty lighthouses around British coasts and no foghorns. New construction and lighting technologies were just emerging, so that simple wood or coal fires had been replaced by candles and parabolic mirrors and then by the oil-burning Argand lamp.

Responsibility for these lights was dispersed. In Scotland and Ireland it was vested in the Commissioners of Northern Lights and the Commissioners of the Ballast Board in Dublin, from which emerged the Commissioners of Irish Lights. Both were established in 1786. For England and Wales, Trinity House was responsible. Trinity House, a seamen's guild that had been incorporated in 1514 by King Henry VIII, acquired powers to regulate pilotage and to provide navigational aids on land and at sea. Although it had the authority to build lighthouses, it largely left this to private bodies and collected rents for income. Its membership consisted mostly of former shipowners and merchants, many of whom had ties to central government. The members had no technical expertise. The same was true in Scotland and Ireland. The Commissioners of Northern Lights were all lawyers, and the Dublin Ballast Board commissioners were merchants, bankers, magistrates, and railway directors.

Most lighthouses had been built by private individuals. For example, the totemic Eddystone Lighthouse, leased by Trinity House, built by John Smeaton and operating from 1759, was funded privately by charging dues on ships entering Plymouth Harbour. When Trinity House took it back in 1807, it replaced the candles with lamps and reflectors. That improved the brightness sixteenfold and made the light visible from Plymouth for the first time. This example encapsulates the two factors that would shape lighthouse policy throughout the nineteenth century—development of new technologies and the conflict between private and public ownership. They were closely interlinked.

Parliamentary interest in lighthouses at the beginning of the century was driven by economics. The first select committee to investigate them was one of a series on trade policies established to tackle challenging economic conditions after 1815. It was looking both to boost trade and cut costs, and it set out to examine complaints about excessive dues for passing lights. By 1822, when the committee assembled, there were twenty-two lighthouses in private hands, and thirty-three operated by Trinity House.

The committee took evidence from forty-two witnesses, mostly from those with interests in shipping and in the financial business of lighthouses. There was minimal discussion of the quality of lights, although none of it critical of them. Just two witnesses offered technical evidence, which was marginal to the committee's inquiry. Robert Stevenson, engineer to the Commissioners of Northern Lights, outlined the technologies used for building and operating lighthouses, but was questioned mostly on costs and income.[39] Rennie provided information on the cost of works in Ramsgate Harbour.[40] The committee latched on to what it considered excessive profits of £55,000 in three years on £90,000 of dues, and declared that a reduction in charges was indispensable to expanding foreign commerce.[41] It recommended placing the control of all lights under Trinity House.[42]

THE FRESNEL LENS

Just as the committee of 1822 made its deliberations, lighthouse technologies were taking a vast step forward with the invention of new lenses by the Frenchman Augustin-Jean Fresnel. Using oil lamps and Fresnel lenses, the brightness of lighthouses could be increased a thousandfold. That made coastal lighthouses far more effective and transformed the prospects for safe coastal navigation. David Brewster had tried to interest British lighthouse authorities in his polyzonal lenses in 1812, but when private companies failed to take up these new technologies in the 1820s and 1830s, he railed against what he saw as private interests preventing public benefit, and accused the lighthouse boards of lacking scientific involvement.[43]

Part of the problem was the nature of these new technologies for coastal lights. Although it was straightforward to collect dues for harbor lights, it was not so easy for coastal lights, without government intervention to specify and operate means of collection. The risk associated with the expense of installing new technologies, or of building new coastal lighthouses, was not attractive to private enterprise. France had solved this problem in 1792 by abolishing the right to private lighthouse ownership and giving the responsibility to the navy. Fresnel was employed by the state, which funded the necessary research. At this time Britain had neither a public lighthouse system nor a publicly funded research program.

JOSEPH HUME STEPS IN

Joseph Hume chaired a select committee in 1834 to examine the entire business. A radical MP and son of a shipmaster, Hume was committed to reducing unnecessary expenditure. As in 1822, the focus of the 1834 committee was on costs and the collection of dues, although more engineers gave evidence. The report started by recognizing the need for the best lighthouses "which the state of science can afford," and stressed their importance to the safety of merchandise and of people traveling by sea.[44] The problem it identified was the lack of any common management, resulting in the system of lighthouses "heavily taxing the Trade of the country, for the benefit of a few private individuals."[45]

The report proposed reducing dues to the minimum required to build and maintain lights, with a central public board to manage all the coastal lights. This would reduce costs, with the advantage that "the introduction everywhere of the very best Reflectors, Lenses, Lamps, &c.; and the results of the Experiments which will be made on every new Proposition, will tend to keep up the Lights with the state of Science."[46] Robert Peel had earlier suggested that "it would be better to put the whole of the lights under some public department."[47]

The majority of the committee believed that Trinity House should constitute the managing board, with additional naval and scientific membership. Hume favored the proposal made by Thomas Drummond for "a Board consisting of Four Persons; one to be a Seaman, and the Hydrographer of the Admiralty, another to be a scientific Chemist; a third a Member of the Royal Society and an Optician; and the fourth, the President or Vice-President of the Board of Trade."[48] Drummond suggested that the "Optician" should be the astronomer royal George Airy. He regarded an engineer as unnecessary to the board, which he thought should be concerned with scientific developments, employing an engineer when necessary to build new structures. But he suggested the need for a role for "scientific persons recommended by the Royal Society" to visit and report on the state of lighthouses and plans for their improvement.[49]

The 1834 report did review the latest scientific and engineering position, although the committee, perhaps complacently, was satisfied with the quality of lights used. It did not examine in any systematic manner different designs of lights and their lenses, fuels, or burners.[50] Evidence was given by the engineers Robert Stevenson, his son Alan, James Jardine, and Drummond. Both Brewster's polyzonal lens and Drummond's limelight were discussed. Jardine emphasized the importance of Brewster's lens,[51] which Peter Barlow had certified as twenty times better than reflectors.[52]

In 1835 Hume sought permission to bring in a bill to implement the recommendations, including putting all the lighthouses under one board. He emphasised the loss of life "for want of adequate lighthouses on the coasts of Great Britain and Ireland" and abuses in management.[53] The bill failed, partly because the government was benefiting from the dues from leased land for lighthouses.

By 1836, fourteen lighthouses were still in private hands. That year, stimulated again by Hume, the Whig president of the Board of Trade, Charles Poulett Thomson, introduced a bill for the government. This bill gave responsibility for all English and Welsh coastal lighthouses to Trinity House, leaving those in Ireland to the Commissioners of Irish Lights and those in Scotland to the Commissioners of Northern Lights. Harbor lights would remain local responsibilities, with oversight by Trinity House.

Hume explained the choice of Trinity House, rather than the Board of Trade or the Admiralty, on the basis of wanting "hard-working practical men" in charge.[54] Peel supported the bill, and in doing so revealed his attitude toward the proper place of the men of science. In response to Drummond's suggestion to the committee that there should be a small board of scientific men, he said that he "did not wish to underrate the value of the services of scientific men in matters of this kind, but he was of opinion that it would be found a much more efficient plan to offer rewards to them for the discovery of such improvements as might from time to time be wanted, than to place them permanently upon a Board of this kind."[55] He would have agreed with a later Conservative prime minister, Winston Churchill, who is reported to have said that "scientists should be on tap, but not on top."[56]

Trinity House acquired the power to purchase and maintain all private lights, a process completed in 1842. It did not, as recommended, admit men of science, but it did employ a scientific adviser in the form of Faraday to guide future development.

MICHAEL FARADAY AS SCIENTIFIC ADVISER

With Trinity House becoming in effect an arm of the state by taking responsibility for public provision of the lighthouse service, it needed a means of assessing scientific proposals and carrying out scientific tasks. Faraday had worked on optical glass and was already a renowned chemist. In 1836 Trinity House appointed him "scientific adviser to the Corporation on experiments in Light."[57] Faraday's position with Trinity House was as a salaried adviser.

Faraday's work was determined by requests made by Trinity House. The suggestions and recommendations in his reports were generally taken up, a tribute to his communication skills in detailed reports to civil servants and military men.[58] His tasks included chemical analysis of oils, water, and white lead; experiments to examine technical improvements constructing a photometer; comparing different types of lamps; and improving the optical adjustment of lenses.[59] He would also be consulted by James Walker, Trinity House's contracted engineer.

In the early 1840s, Faraday invented a chimney for lighthouses that prevented combustion products settling on the glass, and spent several years working to improve this method of ventilation.[60] It was installed in all Trinity House lighthouses as well as in the Athenaeum and Buckingham Palace.

A significant part of his work involved investigating the merits of technical inventions, which often required visits to lighthouses on land or sea. In the 1850s, the possibility of electric lights drew his attention. An arc light developed by Joseph Walker did not find favor, despite its brilliance, given its dependence on the maintenance of a large battery.[61]

In the later 1850s, Frederick Holmes invented an arc light powered by a steam-driven generator. Faraday organized trials at South Foreland near Dover from 1858.[62] Impressed by its brightness and reliability, he worked with the glass manufacturer James Chance to improve its adjustment.[63] Introduction of electric lights was slow. Just five lighthouses in Britain were operating with electric lights in 1879,[64] and by this time electric lighting was maturing as a technology for commercial and industrial applications. An extension to sixty lighthouses was planned in 1881, but never realized after trials in 1883 and 1884 of gas, oil, and electric lights. The trials found that the arc light had poorer penetrating power in haze and fogs than oil or gas lights.[65] By 1901, of 1,100 lights around the British coast, only five were lit by electricity.[66]

JOSEPH HUME INTERVENES AGAIN

Parliament received several petitions in 1845 for reducing lighthouse dues and supporting them directly from central funds as a public service. There were no complaints about the quality of the lights, but Hume was led to convene another select committee, its membership including Viscount Palmerston and George Clerk, vice president of the Boards of Trade. Several engineers gave evidence, including James Walker for Trinity House, and Alan Stevenson. Three other engineers described ways of designing lighthouse structures, but as in 1834 the committee did not attempt any systematic technical inquiry. Faraday was not called as a witness, although his contribution to improving the ventilation of burners and installing lightning conductors was praised.

The committee again recommended a single board, and added that all lighthouse expenses should be defrayed out of the public purse.[67] In 1849 Henry Labouchere, president of the Board of Trade, tried to pass a bill for a government board of management, but failed given the strong adverse Scottish and Irish interests.

In a pragmatic solution, Trinity House, as a private charitable corporation, was given a public role that would otherwise have been performed by government. In any case, given the close relationship between Trinity House and the Board of Trade, there was increasing supervision by government. That increased further when in 1850 the funds were transferred to the Board of Trade.[68] In 1854 financial control of the three boards was placed in the hands of the Board of Trade by the Merchant Shipping Act. This act also stipulated that Trinity House had control of approval of new projects over the other two boards.

THE LIGHTHOUSE AS A PUBLIC GOOD

Coastal lighthouses are effectively a monopoly. As with railways, government stepped in to ensure the public interest, which was defined both in terms of reduced cost to trade and of public safety.[69] Private owners had not invested in the latest technologies, and only after greater state intervention did the system improve.[70] Centralization, it was argued, would lead to economies. In this case, as with the Poor Laws, the aim was public control but at one remove from government.[71] Nationalization, or public ownership, would avoid profit being made by those who did not contribute to the trade of the country at the expense of those who did.[72] In this case, nationalization redistributed the dues within the private sector from the lighthouse owners to shipowners and therefore, it was argued, to the public through reduced trading costs.

British manufacturers were not able to produce lenses to the quality of the French until Chance Brothers employed a Frenchman, George Bontemps, in 1848. Faraday reported to Trinity House in 1854 that the company had created a product that could match that of the French. The role of the lighthouse as a public good created the need for state-sponsored programs of research and design when it came to the optical sciences and technologies used in their illumination.[73] The Board of Trade supported many trials of technologies over the following years, taking on the role of a state funder of applied science to improve lighthouses. Nevertheless, funding constraints limited speedy adoption of new technologies.[74]

In 1856, following the Merchant Shipping Act 1854, which gave the Board of Trade responsibility for colonial lighthouses, Faraday was appointed to the additional role of scientific adviser to the Board of Trade. The act had also placed the lighthouse function of Trinity House and the Scottish and Irish commissioners under the Board of Trade, although in practice they retained some freedom. Despite the regulations implemented in recent shipping acts, shipwrecks and loss of life continued. Between 1852 and 1860, 10,336 ships, about one in every two hundred, were destroyed in collisions or wrecked on the coasts. Attention turned again to coastal signals. The time was ripe for a royal commission, perhaps also instigated with an eye to the Board of Trade securing further control.[75]

ROYAL COMMISSION ON LIGHTS, BUOYS, AND BEACONS, 1858–1861

In 1858, under the Earl of Derby's Conservative government, the royal commission on harbors of refuge was already underway when a second was established to look at the complementary question of lights for navigation, compared to those in other countries. Tragedies in 1859, as the commission started its work, brought the challenge home. It was a bad year for shipping. Material losses amounted to £1.5 million and many lives were lost. The *Royal Charter* was smashed on the Anglesey rocks and the *Pomona* near the coast of Ireland, drowning 870 in these two wrecks alone.[76]

The commission did not report until 1861, but it went into far more technical detail than any previous inquiry. Chaired by Rear Admiral William Baillie Hamilton, the other commissioners were the naval captain Alfred Ryder, the chemist John Gladstone, and two members of local marine boards. This time there was a wealth of scientific evidence. Gladstone offered a theoretical paper and suggested experiments, and Airy sent in a long letter after observing lens systems. Reports from Faraday were received, including on the electric light, and observations were communicated by Chance, and by David and Thomas Stevenson of the Scottish lighthouse dynasty.[77]

Most of the evidence was acquired through visits, questionnaires, and reports, but David Stevenson and the electric light experts William Cutler and Holmes among others later gave oral evidence. Some fifty scientific and technical experts wrote in response to questions from the commissioners, among them Brewster, John Herschel, William Rankine, Fitzroy, William Hallows Miller, Thomas Robinson, and the Earl of Rosse.[78]

The commissioners found the science of lighthouse illumination "in a transition state, and capable of further development."[79] The last coal light, at St. Bees on the Cumbrian coast, had only been extinguished in 1822. Now there were different oils available, the option of gas, and the first electric light was undergoing trials at South Foreland. New wick and lens systems were being developed. Nevertheless, the commissioners recognised that French lights were of better quality. This was a report designed to inform practice, and the findings were given in technical detail. They included suggesting the further use of fog signals, a matter which would much exercise Faraday's successor as adviser, John Tyndall.

The commissioners pointed out that, unlike the case in Britain, "The Lighthouse Boards of foreign countries generally include engineers, hydrographers, and professionally scientific men."[80] The omission of such expertise they regarded as leading to inefficiency and missed opportunities, although they praised the example of the Scots for the way in which they employed an engineer. That British anomaly would remain, through weight of tradition. So would the complexity of institutional arrangements. The commission's recommendation to create a new board called the Trinity Commissioners for Lights, directly responsible to Parliament and that would have expert members, including the astronomer royal, was not taken up. Nor were light dues abolished, a course the commissioners urged, as had every previous select committee, and they remain even in the twenty-first century.

THE ARRIVAL OF JOHN TYNDALL

Faraday's work on lighthouses in the early 1860s, punctuated by the royal commission, continued as before. He was to be found examining proposals for lime lights, overseeing the installation and testing of electric lights, and working with Chance to improve the manufacture of high quality glass and optical systems.[81]

To these were added a new attention to acoustic fog signals. The initiative came from the Irish astronomer Thomas Robinson, who had provided evidence on them to the royal commission. At the British Association meeting in Manchester in 1861, a committee chaired by Robinson was established with members including Charles Wheatstone and John Gladstone, who had been one of the royal commissioners. Following the sinking in fog in 1863 of the *Anglo Saxon* off Cape Race in Newfoundland, where permission for the installation of a fog signal had been refused by the Board of Trade, the British Association lobbied Thomas Milner-Gibson, president of the Board of Trade, for an investigation into their potential. Faraday was persuaded, reluctantly given his declining health, to trial at Dungeness the fog signal invented by the American Celadon Daboll.[82] But by 1865 it was clear that Faraday was becoming too ill to continue.

From 1836 to 1883 there were only two scientific advisers to Trinity House and the Board of Trade: Faraday and Tyndall. Tyndall, Faraday's protégé at the Royal Institution, took over Faraday's responsibilities in 1865 and was formally appointed scientific adviser to the Board of Trade and Trinity House on Faraday's death in 1867. Furthermore, as the Board of Trade sought more control over the lighthouse organizations, he was appointed adviser to the Scottish and Irish boards. This did not please the Scots, especially when they later discovered that they were paying part of his salary, and throughout his time as scientific adviser Tyndall had a robust relationship with them. His relationship with the Irish was generally better, not least given his lengthy defence of the gas lights developed in Ireland by John Wigham, but also because he was Irish.

OIL, GAS, AND ELECTRICAL LIGHTS, AND FOG SIGNALS

When Tyndall became scientific adviser to the Board of Trade and Trinity House, the mechanisms of government were different from those at the start of Faraday's tenure. An efficient administrative machine under Farrer was in place, the lighthouse system was nationalized, and what mattered was technological improvement. Different types of oil, whether the vegetable colza oil or mineral oils from petroleum, vied with gas and electricity as sources of light. Different burners, lenses, and lantern designs were in competition.

The engineers of the three lighthouse boards had their own preferences. Tyndall, who was no diplomat, had the task of adjudicating between the competing options in his reports for the Board of Trade and Trinity House, and for overseeing experimental trials.[83] He could call on others, such as the chemist William Valentin at the Royal College of Chemistry.

Tyndall saw himself as an objective, evidence-based adviser. The broad administrative arrangements were settled, and no further nineteenth-century legislation troubled the institutional relationships. The issues were technical, scientific, and economic. One problem for Tyndall was how to judge the merits of different lights. How important was intensity, reliability, ease of use, or cost? In the end, although based on science, these were not scientific judgements alone, even though Tyndall imagined that he could make them. Whereas Faraday had been concerned with the reliability of lights, Tyndall emphasised their power and penetration.

Cost was a particular constraint, both for installations and for experimental trials. The president of the Board of Trade was not always a cabinet position, reducing its influence. And Farrer, permanent secretary to the Board of Trade from 1867 to 1886, was known for his parsimony. Sums available for experimental research were never large, nor were they available consistently, arguably revealing a misunderstanding by government of the nature of scientific research.[84]

Tyndall, like Faraday, also had to deal with the professional pride of the established engineers, such as David Stevenson in Scotland and James Douglass at Trinity House. The engineers had decades of practical experience on their side and were not easily influenced by the theoretical approaches of the scientific newcomers. The Scots had resisted Brewster's dioptric lens apparatus for two decades after its introduction in France.

Tyndall's work on fog signals from 1874 to 1884 was at least uncontroversial, and led to the booming horns now encountered around foggy coasts. It illustrates one aspect of the connection between fundamental and applied research. Most of Tyndall's scientific research was aimed at answering questions generated in his search for understanding the laws of nature. He took the common Baconian view among the practitioners of science that fundamental understanding would inevitably lead in time to applications for society, but that no one should constrain the investigator in his choice of research area.

The work on foghorns was unusual, in that Tyndall's extensive observations while floating in the Trinity House yachts off the Kent coast listening to bells, whistles, trumpets, horns, and guns led him to an understanding of how the composition of the atmosphere affected the transmission of sound.[85] It is an example of research to solve a practical problem generating new theoretical understanding rather than the reverse. Both Faraday and Tyndall promoted systematic processes of experimentation, but their part-time positions and primary interests elsewhere reduced their impact. They did not have the institutional and administrative bases of Airy at the Royal Observatory at Greenwich and the Admiralty, or of Abel at the Royal Military Academy at Woolwich and the War Office.

The dispute over oil and gas burners encapsulates the problems both of coming to common scientific judgements and of the significance of nonscientific factors.[86] Wigham's gas light had been installed at Howth Bailey off Dublin in 1865. Tyndall reported in 1869 that although gas might be more expensive than mineral oil, its ease of handling, its distinctiveness in fog, and the ability to vary its light intensity easily made it a preferable choice. He recommended its increasing use in Ireland, perhaps expecting that the other organizations would see its value. But the Scots disagreed, as did Trinity House when they carried out their own tests.

Tyndall was authorized to conduct a series of trials, and although they confirmed him in his view of the benefits of gas over oil, because of the control over power and ease of use, Trinity House was not convinced. The disputes continued into the 1880s. Joseph Chamberlain became president of the Board of Trade in 1880, and sought to resolve the quarrels by asking Trinity House to organize definitive experiments to compare oil and gas lights in 1882. They took place at South Foreland, with the additional comparison of an electric light. Tyndall objected to the constitution of the Illuminants Committee established to oversee the trials, on the grounds that it contained three engineers known to be opponents of gas, two of whom were brothers and one the patentee of the oil lamp. When he failed to have the membership of the committee changed to his liking, he resigned. The committee attempted to carry on, but when Robert Ball, scientific adviser to the Irish commissioners, also resigned the whole effort collapsed. Tyndall released his emotive account to the press, and received much support. But the politicians managed to convey the impression that he was being unreasonably partial to an Irish inventor, although Wigham was in reality a Scot working in Ireland, and smoothed over the troubles.

What exercised Tyndall, and led to his resignation, was his perception of the system within which his advice was received. It did not place enough weight on sustained scientific research to inform decisions. Writing later, when he published his account in three long polemics in the periodical press, he recalled a statement made by the royal commissioners in 1861, which had described the Board of Trade as "a department of the Government, whose president changes with the Government, whose members are not selected for their knowledge of Lighthouse Illumination, and who have not necessarily any officers specially instructed in that subject."[87]

Tyndall genuinely felt, with some justification, that Wigham had been treated unfairly by biased officials and engineers with a financial interest. In the event, the Board of Trade and Trinity House may have been more swayed by their estimates of cost. Despite the greater power of the gas light, oil was good enough and cheaper.

Wigham continued to promote his system. He asked Michael Hicks-Beach, president of the Board of Trade from 1888 to 1892, to establish another committee. He wanted the committee to review his invention to include Tyndall.[88] Hicks-Beach rejected Tyndall as being too partisan, and asked George Stokes, as president of the Royal Society and now an MP, to chair a committee of fellows and report.[89] Stokes asked Lord Rayleigh and William Thomson to join him. Their report in 1890 gave some preference to electricity, acknowledged some benefits of gas over oil, but decided that oil was to be preferred over gas for simplicity and economy. They suggested that all three should remain in use, depending on circumstances. It was a diplomatic solution of which Tyndall would have been incapable, although Wigham disputed it at length.[90] Bruised by their experience with Tyndall, Trinity House did not appoint a successor until 1896, when they appointed Rayleigh. The Board of Trade did not replace him at all.

TOWARD THE TWENTIETH CENTURY

Until this point Britain's lighthouses had been isolated structures, standing without easy communication with each other or with central agencies. Electricity and the telegraph changed the possibilities.

In 1887 the Board of Trade established a committee to examine the desirability of electrical communication between light vessels and lighthouses and the shore, with the aim of saving lives at sea. The engineer and shipowner Alfred Holt was one of the nine members of the committee, chaired by the third Earl of Crawford and Balcarres, politician and astronomer. As chairman of the Marine Committee for the Mersey Docks and Harbour Board, Holt was considering telegraphic connection to the *Bar* lightship in the River Mersey. The committee initially recommended an extended trial with the *Sunk* lightship near Harwich.[91] Eighteen months later, the connection to the *Bar* lightship had still not been made, and the committee concluded on the basis of the trail with the *Sunk* lightship that electrical communication was too expensive to be justifiable at the time.[92]

The proposition was reexamined in 1892, when a royal commission was established to inquire into which lighthouses and light vessels might be connected with the telegraphic system. The nine commissioners included William Preece, engineer-in-chief and electrician to the General Post Office. Their aim was to find ways, both practical and economical, to aid communication with vessels in distress and to issue storm warnings.

In the circumstances, appointing a commission was a practical strategy given the number of government departments with an interest, including the Board of Trade, the General Post Office, and the Admiralty, in addition to the three lighthouse bodies. The commissioners reported five times over the next five years, during which time they identified the installations that should be connected either by aerial wires for those on shore or by cable for those on islands or piles. They also explored noncontinuous connections, required for remote places such as the Fastnet Lighthouse. Light vessels were particularly problematic as the cables taking telephone wires would sometimes snap, but the noncontinuous system also failed, with "electric energy being almost entirely lost in the sea."[93]

Just over the horizon was a new technology. As the commissioners completed their work in 1897, Guglielmo Marconi was registering his patent for a radio communication device. The commissioners noted a successful trial across the Bristol Channel, and imagined that the method would support communication with remote light vessels and lighthouses. In 1901 the first wireless signals were transmitted across the Atlantic, ushering in a technology for the twentieth century and a powerful solution to the problem of communicating with remote vessels and installations.

These examples from the end of the nineteenth century illustrate the integration of expert advice into the improvement of systems for supporting shipping and safe navigation. Some of these improvements were purely technical, but legislation and regulation also drew heavily on scientific expertise. Here, familiar factors constrained the extent and speed of regulatory change, in particular the laissez-faire approach to private industry and the related emphasis on individual and institutional responsibility.

PART V

INDUSTRY

2

FACTORIES, NUISANCES, AND THE HOME OFFICE

THE HOME OFFICE WAS CREATED IN 1782, WHEN ONE OF THE TWO SECretaries of state was made responsible for domestic and colonial affairs. With the creation of the position of secretary of state for War in 1794 and its extension to secretary of state for War and the Colonies in 1801, the home secretary was left with responsibility for domestic affairs. The department thereby gathered an increasing number of functions with the growth of government during the nineteenth century. Many of these involved intervention and inspection, in areas as diverse as factories, mines, explosives, food quality and adulteration, salmon fisheries, dissection of bodies, cruelty to animals, vivisection, and nuisances caused by things such as smoke, industrial emissions, and graveyards.[1] Scientific evidence informed policy and practice in all these areas. The department was also responsible for policing and prisons, which are not addressed in this book since technical advice was relatively unimportant for their development.

One factor that shocks today, given the huge size of bureaucracy in a modern economy and society, is the small number of staff in nineteenth-century civil government departments. Even in 1848 there were only twenty-two permanent staff in the Home Office, and the home secretary could and would scrutinize all the daily mail. That rose to thirty-three by 1870 and thirty-six in 1876.[2] Like so much of government, the Home Office was staffed through personal networks. Recruitment by open competition was not introduced until the 1870s. Victorian administration was lean and personal. Access to senior politicians by advisers and inspectors could be straightforward, for those who managed to gain entry in the first place, although individual ministers showed different levels of interest and initiative.

FACTORIES

The Factories Act of 1833 is generally taken to herald the introduction of the major inspectorates, which would subsequently bring so many people with scientific, engineering, and medical expertise into influence on Parliament and government.[3] Although some rudimentary attempts at inspection had been envisaged in earlier factory legislation, and the practice had been introduced into Irish prisons, the political and public visibility of this new departure makes it stand out. What this episode also reveals is the impact of the extensive evidence produced from the perspective of social science and medicine, even if that evidence was partial, unsystematic, and designed to support the beliefs of its originator, Edwin Chadwick.[4]

By the 1830s, legislation limited some hours of work for children, but there was still widespread agitation in the emerging industrial areas for a ten-hour workday for workers of all ages. Against this were arguments about the freedom of workers to contract their labor and the rights of parents to take responsibility for their children. Prevailing models of political economy held that interference of the state with business activity would be damaging.[5]

When the Tory MP Michael Sadler brought forth a ten-hour bill in 1832, many medical men gave evidence to the committee established to scrutinize it, as they had to a previous committee in 1816.[6] Whether they were local practitioners or from the medical elite, they were almost all on the side of increased regulation. In 1816 the medical witnesses included Ashley Cooper, Anthony Carlisle, and Gilbert Blane. Blane was an eminent naval physician, who provided valuable advice both to the Royal Navy, based on careful statistics, and later to the Home Office on conditions in prisons and convict ships. It was Blane who in 1795 authorized lemon juice to be distributed throughout the navy to combat scurvy.

Although the committee of 1816 put selected statistics to the witnesses, designed to show that very few children were dying at work, the medical men argued from limited observations and anecdotal evidence that working conditions and hardships led to stunted growth and poor health, which in turn could lead to a variety of diseases, even if the mechanisms were unknown. Blane was rightly sceptical of the dubious figures put to him by the committee.[7]

By 1832 the medical men were even more in evidence. They included ten who were fellows of the Royal Society, among them Anthony Carlisle, Benjamin Brodie, Mark Roget, Joseph Green, and Charles Bell. All were outspoken in their belief that the extensive hours of work by children in poor factory environments was deleterious to their health, and some drew direct comparison with the conditions of colonial slavery.[8]

The select committee unearthed a host of abuses. When Sadler failed to return to Parliament at the 1832 election, the evangelical Tory Lord Ashley, driven by humanitarian considerations, took over the bill. Opposition from manufacturers led to the establishment of a royal commission, established by the government to report speedily with the aim of heading off the bill. In the middle of his work on the Poor Laws, Chadwick was called on as one of three central commissioners to serve on this royal commission on the employment of children in factories,[9] which sat amid labor unrest and strikes around the country.

The three central commissioners—Chadwick, Thomas Tooke, and Thomas Southwood Smith—were assisted by twelve others. They were charged with collecting information "as to the employment of children in factories, as to the effect of such employment on their morals and bodily health and as to the propriety and means of curtailing the hours of their labour." Tooke was an economist and free trader, and a founder with David Ricardo, Robert Malthus, James Mill, and others of the Political Economy Club. Southwood Smith, a Unitarian, had studied medicine in Edinburgh and became physician to the London Fever Hospital. In 1832 he lectured over Jeremy Bentham's body, and prepared his skeleton, which now sits in University College London. A close colleague of Chadwick, he worked on many subsequent public health matters, and fell from influence only when Chadwick lost authority in 1854.[10]

Southwood Smith drafted the medical questions for gathering information and Chadwick the rest. Beside Southwood Smith, three other commissioners were medical men. Although they produced much evidence of the impact of social and economic conditions on health, Chadwick downplayed it. He claimed that it had no statistical rigor, and that many different causes of disease were advocated, which might not be related to factory work at all.

The commissioners' recommendations led to the Factories Act 1833. It required a reduction to nine working hours for children under thirteen and no working for children under nine. It introduced requirements for children's education, and an inspection system to assure compliance, to encourage greater uniformity locally, and to protect those seen as vulnerable. Although the medical men were in demand to give evidence in this period, they were not well represented as inspectors as regulation increased. That indicates a priority by government, reflecting the needs of industry, on employment conditions rather than wider health and environmental considerations. No medical man became a factory inspector, and only James Kay became an assistant Poor Law commissioner.[11]

As with later railway legislation, central inspection did not seek to remove corporate management responsibility from the factory owners.[12] Four factory inspectors were appointed, with some support from manufacturers who perhaps saw them as a means of preventing unfair competition or of freezing out smaller businesses.[13] The inspectors could enter factories employing children, direct safety and sanitary matters, and investigate the provision of education for children. They had legal powers of enforcement, although they were rarely used and were removed in the Factories Act 1844.

The four first factory inspectors were Whig appointees of whom only one—the Scottish geologist, merchant, and educational reformer Leonard Horner—had prior experience of mills and textiles. He had been one of the twelve assistant commissioners on children's employment in factories, and was Charles Lyell's father-in-law. In that sense these inspectors were atypical of those who followed in areas such as railways, mines, and shipping. They found the act difficult to enforce, but under pressure from manufacturing interests the Whigs passed no amending legislation.

Eventually the Tory Home Secretary James Graham succeeded in passing the Factories Act 1844. Influenced by reports from the inspectors, this tightened limits on working hours for women and children, and strengthened educational provisions. It also brought in some health and safety regulations, enabling the state to determine aspects of the working environment in these private businesses. Although Prime Minister Robert Peel's government still resisted reducing the workday to ten hours, that limit was finally introduced for women and children in the Factories Act 1847. Despite concerns that it would reduce working-class wages and lead to lower economic performance, the moral argument for limiting work hours on the grounds of improving the health and strength of workers, and of tending to domestic and social harmony, prevailed. Adult men were considered to be responsible for their own interests.

Whether coincident or not with the passing of the Factories Act 1844, it was the collapse of a cotton mill in Oldham, and of part of a prison building at Northleach, that spurred Graham to commission Henry De La Beche and the engineer Thomas Cubitt to report to him.[14] The collapse of the mill had killed twenty people and injured many more. William Fairbairn had presented a report to the inquest, and suspicion fell on the cast-iron beams, in particular their removal red-hot from the sand when cast, which while saving time would lead to brittleness. De La Beche and Cubitt had the iron analyzed by Robert Angus Smith at the Royal Manchester Institution, and sought the views of Eaton Hodgkinson and Fairbairn on the best form of cast iron. Angus Smith's experiments indicated that little was known about the effect of chemical composition of cast iron on its appearance and strength. The commissioners recommended a proper examination, with experiments at Woolwich under the

Board of Ordnance, and proposed that wrought iron should be used in preference wherever possible. That conclusion was reinforced by the commission established to investigate the Dee Bridge collapse in 1847 (see chapter 7).

Concerns about working conditions in other industries led to similar inquiries. For example, Seymour Tremenheere, the first inspector of mines, was commissioned to examine the employment of women and children in bleaching works in 1854. The testimony he took from factory managers and workers, including poignant contributions from young children themselves, included medical perspectives. Two medical men argued for the harm to health from the long hours of work, and Tremenheere noted two petitions to Viscount Palmerston, the home secretary, signed by dozens of medical men and clergy that made the same case.[15]

Throughout the rest of the nineteenth century, the provisions of factory acts were extended to other areas of manufacturing and trade, particularly following a royal commission from 1862 to 1867. This one included the results of an extended investigation by the physician and nutritionist Edward Smith on the food of the poorer laboring classes.[16] Not until 1874 was the ten-hour workday clearly established, and at the same time the minimum age for full-time employment was set at fourteen.[17]

Since most of this legislation relates to employment conditions and rights, its dependence on scientific advice is limited, although some medical advice was brought to bear and a medical inspector was appointed in 1898.[18] But its location in the Home Office ties it to wider Home Office responsibilities that have scientific, engineering, and medical dimensions, and that were already apparent in relation to manufactories. These relate to the concept of nuisance.

SMOKE AS A NUISANCE

The legal construct of nuisance played a major role in discussions of social conditions and public health. Dating back to the reign of King Henry III, the law of public nuisance enabled action to be taken against inconveniences to the public in general, whereas the law of private nuisance protected individuals against interference with their comfort and the enjoyment of their land. That comfort and enjoyment was threatened by pollutants from industrialization and from animal and human waste. Steam engines and factories belched out smoke and noxious fumes, which traveled on the wind. Rivers carried toxic chemicals and untreated sewage. This pollution affected both the landowners on their rural estates, where it was carried from tall chimneys or by the river system, and the urban working class crammed together in cities. Those with the power and money to take action were the landed gentry, well represented in Parliament, who saw the quality of their rural way of life threatened. It was their ethos of stewardship of the land that played a significant role in tempering some of the excesses of free market industrialization.[19] They used Parliament and the courts to protect their estates, and they called on scientific and engineering expertise to help.

Resorting to the law was no quick panacea, and infrequently used. It was expensive, and difficult to prove cause and effect. Local courts and local boards of health might include the very people creating the nuisance, or their friends. Several cases in mid-century demonstrate that the economic importance of industrial activity influenced outcomes in favor of manufacturers, although this was not always the case.[20] The development of legislation throughout the nineteenth century is a story of the gradual introduction of regulations and inspection, informed and influenced by scientific evidence, in the context of strongly held principles of noninterference in private businesses, individual responsibility, and local government.[21]

The problems were already apparent early in the century. The reforming Conservative Michael Angelo Taylor, the MP for Durham, proposed and chaired a select committee on steam engines and furnaces in 1819 and 1820.[22] The committee called on medical men to describe the damage to health and life, and on inventors and manufacturers to demonstrate that smoke abatement was technically feasible. Several witnesses described furnaces that could consume their own smoke, greatly reducing pollution. They included the young engineer Josiah Parkes, who was in partnership with his father and brother in a worsted mill in Warwick. He explained how his method of burning the smoke, an improvement on James Watt's "smoke-consuming" furnace, was both more economical and avoided contamination of the bleached cloth stored nearby. He described how his method was being applied to other steam engines. Its success was attested by several other witnesses, including members of the committee who had viewed installations. Parkes stated that William Wollaston, Humphry Davy, Thomas Brande, and other "scientific gentlemen of the greatest eminence" had visited his works and given his invention "the most unreserved approbation."[23]

Taylor brought forth a bill to enable local prosecutions of steam engine owners, with powers to impose costs and require remediation. Although it passed, members of Parliament from mining, smelting, and machining industries managed to have those exempted.[24] In subsequent years, several towns brought in local acts through Parliament to attempt to deal with the continuing problems. Bodies such as improvement commissions were enabled by local acts to require removal of a nuisance, but it was a patchwork with no national legislation, and in practice opposition from particular interests often reduced the effectiveness of laws. The only other option was recourse to the common law with all its associated problems. By the 1840s, with local acts applying to few towns, the Liberal MP William Alexander Mackinnon, shocked by the choking and dirty state of the atmosphere in towns compared to the country, initiated a select committee to consider the need for national legislation.[25]

Mackinnon's select committee on smoke prevention met in 1843. It was clear to the committee that smoke, caused by incomplete combustion, could be much diminished if not entirely prevented by inventions such as those which increased airflow to burn it.[26] As in 1820, it was argued that this would be more economical too, at least in most cases, and might even be extended eventually to coal-burning fires in private homes.[27] In reality, given the multitude of different furnace designs and modes of operation, good abatement was difficult to achieve.

Whereas Taylor's committee had examined mostly engineers, inventors, and manufacturers, Mackinnon's also called many scientific experts to give evidence, including Brande, Michael Faraday, Andrew Ure, and Neil Arnott. They set out the chemical principles and knowledge in detail, while Parkes and other engineers and inventors described the various possible technical solutions and existing patented systems. The responses of Brande, Faraday, Ure, and others went beyond the scientific evidence, when they expressed the view that the government should legislate to compel steam-engine owners to remove smoke. Ure was a firm advocate of free trade and the factory system but also an industrial inventor, so perhaps his desire to support inventions won out.[28] The focus at this point was on smoke as a nuisance, as there was no clear evidence about its impact on health. Indeed, there was a widespread belief that coal smoke could be beneficial by fumigating miasmas.

Mackinnon sought to bring in a bill to make it illegal for owners of many furnaces to create opaque smoke for more than a specified time each day, but Prime Minister Peel, judging the extent of parliamentary opposition, would not give the government's support.[29] Mackinnon struggled on, and a further select committee chaired by him in 1845 took additional testimony and examined the implications of the bill.[30] The scientific and technical evidence was regarded as proven, and Fairbairn added to it by summarising a report he had recently completed for the British Association, in which he was enthusiastic about the technical and economic possibilities of smoke abatement.[31] The committee urged compulsion for smoke abatement of stationary steam engines. Even then, lobbying from owners of iron, copper, coal works, and distilleries resulted in them being exempted.[32]

Mackinnon withdrew his bill, on the grounds that a matter of such importance should be left to the government. The government still did not act to legislate, but noting increased public interest, reflected in the pages of the *Times*, it appointed two commissioners to investigate. Home Secretary James Graham approached the Earl of Lincoln, commissioner of Woods and Forests, who instructed De La Beche, director of the Museum of Economic Geology, and Lyon Playfair, the museum's chemist, to submit a report. This report again regarded the scientific and technical case as proven. It concentrated on situations in which local acts were failing to reduce smoke, and on the grounds for exemption of businesses such as iron manufacture and distilleries.

De La Beche and Playfair did not confine their recommendations to scientific matters. They argued that although furnaces "should be brought within the provisions of legislative enactments, the progress of important branches of our national industry should not, on the other, be impeded."[33] In their view the obstacles to smoke abatement were primarily legal and political, not scientific and technical. In reality, the lack of easily controlled technologies and the difficulties of producing robust scientific evidence that would stand up in court offered ready loopholes for manufacturers.

Although the government claimed that it could not do anything, as the evidence was inconclusive, Mackinnon reintroduced his bill. Finally prompted to action, the new Whig government included a smoke-control article in the Town Improvement Clauses Act 1847. It remained unsatisfactory given its reliance on diverse local administrations.

Wider legislation on nuisances was brought in through the Nuisances Removal and Diseases Prevention Act 1848. By this time, nuisance laws were also being invoked to deal with insanitary conditions and disease, under the threat of cholera, as well as unwholesome meat. The local post of nuisance inspector, or sanitary inspector, featured in the Town Improvement Clauses Act 1847 and the more comprehensive Public Health Act 1848.[34] London was included in the Smoke Nuisance Abatement (Metropolis) Act 1853 at the instigation of the home secretary Viscount Palmerston, after John Simon, as medical officer of health for the City of London, had managed to get a smoke clause written into the City of London Sewers Act 1851. The new abatement act included emissions from steamboats on the Thames. Viscount Palmerston then ordered an enquiry by the General Board of Health into the effectiveness of abatement measures.[35]

In 1855 the Nuisances Removal and Diseases Prevention Act obliged every local authority to have a sanitary inspector with powers to enter premises and complain to justices who could order abatement. The Local Government Act 1858 transferred responsibility for towns to local boards of health, still excluding metals, mining, and potteries. Further pressure resulted in a clause in the Sanitary Act 1866 that stipulated that a local board must act as a nuisance authority. Little of significance happened thereafter. The only major changes before the revising and consolidating Public Health Act 1936 were those in the Public Health Act 1875. This allowed authorities to bring actions against polluters outside their district if the smoke created a nuisance within it, but again gave exemptions to the mining and metals interests. Enforcement remained sporadic and limited.

Private coal fires received attention in the 1880s, when Lord Stratheden and Campbell regularly introduced bills to curb smoke in London.[36] They were as regularly rejected by the government, which claimed that such intervention would be impossible, given people's enjoyment and expectation of open fires and the costs of conversion to closed stoves burning anthracite or coke.[37] Toward the end of the century, gas was increasingly touted as a "smokeless" fuel, but although its burning created less smoke than coal, its production simply transferred problems of pollution to the immediate environments of the gasification plants.[38] Science had little more to add to any of this during the nineteenth century. It took the Great Smog of 1952, after decades of further agitation, to usher in the Clean Air Act 1956, which at last initiated effective smoke controls.[39]

THE ALKALI ACT 1863

The Alkali Act 1863 may be seen as the first significant act regulating industrial pollution, discounting the ineffective act on steam engines in 1821.[40] But attempts and means of abating nuisances, including smoke or other pollution from manufacturing businesses, had a long prior history. The gas industry is an early case in point (see chapter 10). Later on, the Nuisances Removal and Diseases Prevention Act 1855, which particularly dealt with disease-related public health matters, included some elements of industrial pollution as statutory health hazards. It was not strong or specific enough to deal with major polluting industries.

The chemical industry had been established in Merseyside by 1823, when James Muspratt built a factory for the production of soda, used in making glass, soaps, and textiles.[41] Its clouds of pollution wafted over the surrounding countryside, poisoning crops and animals with hydrochloric acid. Building higher chimneys merely spread the problem over a larger area. The landowners downwind of factories approached the Earl of Derby in 1862, and he called for a select committee of the House of Lords.

In asking for the committee, Lord Derby drew parallels with the current investigation in the House of Commons on utilizing sewage for agriculture, which he said was otherwise "poisoning our rivers and streams."[42] He pointed out that there was limited remedy in law for individuals or the public at large. These were either through obtaining damages, which the wealthy Muspratt had paid on several occasions, or through a discretionary power under the Public Health Act 1858 to seek to prevent the creation of a nuisance. Neither had been effective, although Lord Derby stressed that he had no intention of offering "the slightest impediment to the great manufactures of the country." Instead, he looked to technological solutions that might be used in the best manufactories, and which laggards could be obliged to adopt. Many manufacturers recognized their responsibility to reduce harmful emissions if possible.

Lord Derby construed this as an issue beyond party politics, although the vocal complainants were landowners defending their property interests. His committee was balanced between Conservatives and Liberals, but landed rather than manufacturing interests were well represented. Two of the fourteen members had scientific backgrounds, but the technical detail which the whole committee explored is notable in its depth. The scientific members were the Liberals Lord Belper, who had been chief commissioner of railways between 1846 and 1848, and Lord Talbot de Malahide. Giving evidence to the committee, one witness recollected John Henry Vivian having claimed to have solved the problem, although at considerable expense, at his copper smelting works in Swansea in the 1820s. With Davy, Faraday, and Davies Gilbert, he had introduced a water-washed flue system to remove the acidic gas.[43] William Gossage described how he had employed a similar process to remove hydrochloric acid from his chemical works. He believed that an inspector could sample such a system to ensure that all the acid was being removed, without interfering with the manufacturing process, and argued that independent inspection was essential.

Gossage had claimed that the only process for which there was a certain remedy was in alkali works, not in smelting, and argued that legislation for others should await the discovery of suitable methods. The questions became one of the definition of what a noxious vapor was, and what remedies were available. Many medical men, including Philip Holland, medical inspector of the Burial Department for the Home Office, stated that vapors from works in their areas were injurious to health and that legal means of reducing them had proved ineffectual. The committee took scientific evidence from figures such as Playfair, Wilhelm Hofmann, Edward Frankland, and John Percy. Hofmann and Frankland had recently visited alkali works to examine their systems for acid removal. Playfair defined as noxious the vapors emitted from soda, sulfuric acid, ammonia, and smelting trades, and asserted that the first three could be prevented by chemical means. He thought soda manufacture created "the monster nuisance of all."[44] The question was whether it would be economic to prevent emission, so that compulsion by law would be feasible without damaging the businesses. Playfair recommended letting manufacturers use the best method they could where it was believed possible, and imposing penalties for the escape of noxious gases. The manufacturers broadly agreed. They could see that some action was going to be necessary.

The committee thought that the law on nuisance should be made more uniform throughout the country, perhaps by making provisions of the Smoke Prevention Act universal, but that alkali and chemical works required special legislation. Following the advice of manufacturers and scientific experts, they proposed that no particular method for the removal of noxious gases should be specified, but that an inspection system should be introduced, with imposition of a substantial penalty if gases escaped.[45]

The Alkali Act 1863 was brought in by Lord Stanley of Alderley. It set a standard for emissions and created the post of an inspector who could bring actions for damages in the county courts. Unlike factory inspection, which concentrated on conditions of employment and reported to the Home Office, this technical inspectorate reported to the less prestigious Board of Trade. In 1873 responsibility moved to the new Local Government Board.

THE ALKALI INSPECTOR

The first inspector was the Manchester-based chemist Robert Angus Smith. He would serve until he died in post in 1884. The young Angus Smith had attended Thomas Graham's lectures in Glasgow in the late 1820s, studied in Justus Liebig's laboratory in Giessen from 1839 to 1841, and worked as Playfair's assistant in Manchester from 1843 to 1845. Playfair then asked Angus Smith to help with his work on the royal commission on the state of large towns and populous districts, and he came to know people like Hofmann and Chadwick. Playfair may have recommended him for the post.[46]

The approach that Angus Smith took toward industry was supportive and consultative rather than punitive. The enabling nature of the legislation, developed with the involvement of the industry, soon reduced emissions of hydrochloric acid to legally acceptable levels. It led to innovations in works and chimney design, and the exploration of new profitable ways to use the trapped gas, such as making bleach for textiles. Despite concerns in other settings that regulation would tend to inhibit innovation, this approach of setting limits but not specifying the technology had the reverse effect. The technical expertise of Angus Smith and the other inspectors, and the implications of ever-developing knowledge and technologies, made this flexible approach an effective way of proceeding. Prosecution was resorted to only in clear cases of neglect. A flexible approach was also a consequence of the wider social and economic pressures. Angus Smith and his colleagues, with limited budgets, had to steer a course between protecting public and private interests in environmental terms, and taking account of the profitability of manufacturing businesses.

Angus Smith used his annual reports to note successes and to raise concerns. Among them was the fact that the act only covered hydrochloric acid, and at a level that was not encouraging further reduction. Angus Smith started to look at atmospheric pollution on a wider scale. The Public Health Act 1872 placed him under the new Local Government Board, with John Lambert as its permanent secretary and John Simon in place as medical officer of health. While the Alkali Act 1863 had aimed at the protection of property, this move led also to a focus on public health, although Angus Smith, still based in Manchester, was not in close contact with Simon in London. This public health work included involvement with the Cattle Plague Commission, the Metropolitan Sanitary Commission, and the Mines Commission. Nevertheless, his chemical approach to understanding disease, based on Liebig's concept of ferments, was never entirely superseded by his later recognition of germ theories and the significance of Louis Pasteur's work.[47] While his quantitative results on the analysis of air and water may have been questionable in many cases, he was respected by many at the time, including William Farr and Florence Nightingale. Even allowing for some inaccuracies, his identification of "acid rain" can be seen as a founding contribution to "chemical climatology."[48]

In 1872 Angus Smith argued for sulfur and nitric acid works to be placed under similar inspection to alkali works. That call was heeded to a degree when the president of the Local Government Board, George Sclater-Booth, brought in the Alkali Act (1863) Amendment Bill of 1874, which extended inspection to wet copper works and required alkali manufacturers to use the "best practicable means" to prevent escape of all other noxious gases. Angus Smith's reports and advocacy persuaded Lambert that more action was needed. Even Queen Victoria complained of the noxious fumes at Osborne House, which came from a cement works and therefore escaped the act. A petition to the House of Lords stimulated a request in 1876 for a royal commission, which Home Secretary Richard Cross established.

Chaired by Lord Aberdare, and with the South Shields industrialist and Liberal MP James Stevenson representing the manufacturing interest, the nine members of the Royal Commission on Noxious Vapours included the chemists Frederick Abel, Alexander Williamson, and Henry Roscoe. Other chemists—including William Odling and Chandler Roberts, chemist to the Royal Mint—gave evidence, as did Angus Smith and his subinspectors. The commissioners were "to inquire into the working and management of works and manufactories from which sulphurous acid, sulphuretted hydrogen, and ammoniacal or other vapours and gases are given off, to ascertain the effect produced thereby on animal and vegetable life, and to report on the means to be adopted for the prevention of injury thereto arising from the exhalations of such acids, vapours, and gases."[49] It is revealing that the terms refer to "animal and vegetable life," the property of the landed class, and not directly to citizens.

The commissioners visited major manufacturing centers and questioned nearly two hundred witnesses. Many doctors and medical officers of health stated their views about the damage to health from local works, and farmers complained of damage to crops, trees, and cattle. The recommendations proposed extending regulations to require inspection of many other types of chemical works, to cover the emission of acids containing sulfur and nitrogen, and for the Local Government Board to have the power to set standards or to require the adoption of the best practicable means for preventing the escape of noxious gases.[50] While Lambert and Simon supported local inspection, in line with the recent extension of local responsibilities, Roscoe and Stevenson argued in the final report that only central authorities should have the power to initiate prosecution, since otherwise ratepayers' money might be used to protect private property.[51] Most of the industrial witnesses supported further central inspection and limits on more emissions achieved through "best practicable means."

After a few false starts the new act of 1881, largely composed by Angus Smith at Lambert's instigation, put many of the recommendations of the royal commission into law. It fixed standards for sulfur and nitric acids, placed many new works under the test of "best practicable means," and gave the Local Government Board the power to extend regulations to other works as soon as suitable regulatory measures could be developed. The inspectorate was increased in size, as was the number of works to be inspected, quadrupling to around one thousand. In addition, local authorities could appoint resident chemical inspectors. A consolidating measure, the Alkali Works Regulation Act 1906, set the basis of vapor regulation for the future. This act remained in place until the Control of Pollution Act 1974.

Angus Smith's undoubted success in catalyzing gradual improvement, and his recognition that an emphasis on technological invention was required to reduce emissions rather than on punitive inspection and regulation, should not blind us to the limits of what he achieved. One only has to read *The White Slaves of England*, published by Robert Sherard in 1897, to recognise that dire conditions still prevailed in the alkali industry, among many others.[52] Those conditions may not have been Angus Smith's responsibility, which was concerned only with the effects of emissions outside the works, but still at the end of the nineteenth century, so many people were slipping between the gaps in the factory acts, alkali acts, nuisance laws, and public health acts.

This period at the turn of the century finally saw the appointment of factory inspectors with specialist medical expertise, and in 1893 the appointment of women, who were also able to become sanitary inspectors.[53] The development of regulation of these enterprises over the century illustrates the importance given to commercial issues, and to the assumed personal responsibility of the employee to take the appropriate protective measures, rather than to considerations of public health.

EXPLOSIVES

The effective regulation of explosives, which became a Home Office responsibility, is primarily due to one person, the Royal Artillery officer Vivian Dering Majendie. Major Majendie, as he then was, was the architect of the Explosives Act 1875. Appointed inspector of gunpowder works by the Home Office in 1871, he became chief inspector of explosives from 1875 until his death in 1898.[54] Majendie had served in Crimea and during the Indian Mutiny, and spent ten years at the Royal Laboratory in Woolwich where ammunition was made and tested.

The legislation Majendie inherited was based on the Gunpowder Act 1860, which had been passed after an explosion at a factory in Birmingham killed twenty-five people. Places making and storing gunpowder and other explosives needed a licence, issued by a magistrate, and the home secretary could authorize inspection. The act was flawed, since licences could not be time-limited nor suspended. After a fatal and devastating explosion in Erith in 1864, Home Secretary George Grey commissioned Colonel Edward Boxer, superintendent of the Royal Laboratory in Woolwich, to report. He concluded that more restrictions were required, including the sort of "inspection and control as that authorised by act of parliament in the case of the floating magazines in Liverpool," but nothing was done.[55]

At the same time, new types of chemical explosives were being developed, such as guncotton and nitroglycerin, the explosive in dynamite. It was the fatal explosion of a cartload of dynamite in 1869 that led to the Nitro-Glycerine Act 1869, which required licences for the substance. The Liberal home secretary Henry Bruce referred the question of licensing to Simon, medical officer of the Privy Council, who in turn sought the help of the chemist William Allen Miller. Both Miller's report in 1870, and an earlier report by Abel, the War Office chemist, laid out the danger of transporting dynamite unless it was handled with care.[56] More explosions led Bruce to instruct Majendie to inspect explosives factories across the country. Majendie concluded that the Gunpowder Act was inadequate to protect the public and that a qualified government inspector was needed.[57] He gave detailed proposals for legislation in 1872, reinforced them with a call for a single act of Parliament in 1873, and urged an approach in 1874 that was flexible enough to respond to new developments.[58]

Bruce did not manage to introduce a bill, but the incoming Conservative home secretary Richard Cross established a select committee to inform legislation. Majendie's evidence and proposals formed the focus of the hearings, and his advocacy for flexible legislation was backed by Colonel Charles Youn-ghusband, superintendent of the government gunpowder factories. Technical evidence was also provided by Abel, by Alfred Nobel, the manufacturer and inventor of dynamite, and by the chemist August Dupré.

Nobel, despite his commercial interest, recognized the need for an inspector to have some individual powers—for example, for determining the minimum separation between buildings and for safe packing methods. Dupré, educated in Germany and now naturalized, had been asked in 1873 to advise the Home Office. He had worked with Majendie and the coroner to investigate an explosion of guncotton in Stowmarket in 1871.[59] The engineer George Bidder, who had been a member of the War Office committee on guncotton and lithofracteur, an explosive mixture containing nitroglycerin, added his views on safe storage and transport. The discussion with all these witnesses was technical, as they explored the characteristics of individual explosives and the manner in which the trade could be continued with adequate safety.

Despite concerns from manufacturers about flexible and uncertain powers that might be given to inspectors or made through Orders in Council, the select committee agreed with Majendie and recommended comprehensive legislation for making, storing, transporting, and using explosives for civilian purposes based on a classification by Abel.[60] A massive explosion of gunpowder in a boat on the Regent's Park canal in October 1874 gave additional impetus. It killed three people and had been caused by sheer carelessness, as Majendie's report demonstrated.[61] Only by luck was far greater damage averted. Cross introduced a bill in February 1875 that soon became law on a nonpartisan basis. It split responsibility between local authorities for the licensing and inspection of small fireworks factories, and the Home Office for all other factories and storage depots. Majendie and his deputy were both from the Royal Artillery, and subsequent inspectors were recruited in the same way. That parallels the recruitment of Royal Engineer officers to be railway inspectors (see chapter 7).

For the next twenty years, Majendie produced annual reports suggesting improvements and investigated accidents. In specific cases a committee would be established to report. For example, the Home Office appointed a committee on the manufacture of compressed gas cylinders in 1896, consisting of five fellows of the Royal Society, including Dupré, which proposed a range of regulations.[62]

The explosives inspectors, increased to three in 1881 and four by the end of the century, were supported by Dupré, who tested explosives at Woolwich. Majendie badgered the local authorities into effective action and improved the situation in factories and magazines. He was sensitive also to the needs of industry, even if the powerful Nobel Explosives Company complained about his actions. His view, not uncommon among the industrial inspectors, was the same as that of Grey, who had argued that restrictions on trade could only to be justified by the necessity of providing for public safety. In other words, as Majendie also declared, "All State interference in industrial life is absolutely contrary to the first principles of economical science, and should only be resorted to as exceptional and as work of necessity. It is a bad thing in itself, and should only prevail to cure a worse thing curable in no other way."[63] Majendie's impact can be assessed by looking at overall accident figures. They remained roughly constant from 1876 to 1896, while the number of factories and magazines more than doubled.[64]

Colonel Sir Vivian Majendie, as he became, died in harness of a heart attack in 1898 after an exhausting cross-examination by a select committee on petroleum. He represents the ideal type of nineteenth-century inspector, operating within an effective model of regulatory legislation. He believed in minimal government intervention in private business except on the grounds of safety, and was able to use tact, discretion, and the flexibility of the act, of which he had been a prime mover, to improve social outcomes.

VIVISECTION

The question of vivisection, using animals for medical research, offers a further glimpse into Home Office responsibilities toward the end of the nineteenth century, and a specific example of regulation. It illustrates clashes between social values on the one hand and scientific and medical opinion on the other. Like the disputes over the Contagious Diseases Acts, which took place over a similar period,[65] arguments about vivisection were linked to strong agitation by women.

Britain had long had a movement against cruelty to animals. An act to prevent the cruel and improper treatment of cattle was passed in 1822, brought in by the Irish MP Richard Martin, who then established the Society for the Prevention of Cruelty to Animals in 1824.[66] Children had to wait until the 1880s for an equivalent organization. Martin's act was updated in 1835, when dogfighting was outlawed, and again in 1849, setting a maximum penalty of £5 and compensation of up to £10 for the offences of beating, ill-treating, overdriving, abusing, or torturing animals. While the legislation was driven by straightforward notions of sparing cruelty to animals, there was also a medical dimension. Giving evidence to a select committee on cruelty to animals in 1832, William Youatt, a veterinary surgeon and lecturer on veterinary medicine at the University of London, stated that dogfighting was not only cruel but was also likely to lead to an increase in rabies.[67]

Several events brought the question of vivisection to public attention in the 1870s. Among them were the publication in 1873 by John Burdon

Sanderson of his edited *Handbook for the Physiological Laboratory*, a demonstration of the injection of alcohol into two dogs at the meeting of the British Medical Association in 1874, and the work of David Ferrier on stimulating the cerebral cortex of animals to produce specific bodily movement. The first revealed actual practices in medical schools and laboratories, and the second resulted in an unsuccessful prosecution of three doctors involved.[68] Ferrier's work challenged those who rejected a materialist account of the mind, with its implications for religious beliefs.[69] Like the other examples, it also generated feelings of disgust and outrage among many people.

Two bills were introduced into Parliament in May 1875. The first, in the House of Lords, was informed by leading anti-vivisectionist Frances Power Cobbe.[70] It required all vivisections to be undertaken at registered premises open to inspection, and a personal licence for anyone carrying out an experiment without anaesthetics. The second, by Playfair in the House of Commons, developed with the help of Thomas Huxley, Charles Darwin, and Sanderson, allowed legalization of all painless experiments and required licensing for painful experiments without the use of anaesthesia, leaving the definition of what constituted painful in effect to researchers. Playfair's bill also drew on a code of practice that had been agreed by the British Association in 1871, which was a recognition that at least some self-regulation was necessary. Sanderson and the physiologist Michael Foster in Cambridge, had been instrumental in drawing it up. Neither bill succeeded, although their positions were not far apart. The forthcoming royal commission would polarize positions and make accommodation more difficult.

The royal commission was established, at the instigation of Queen Victoria, by Prime Minister Benjamin Disraeli's Conservative government in June, "To inquire into the practice of subjecting live animals to experiments for scientific purposes, and to consider and report what measures, if any, it may be desirable to take in respect of any such practice."[71] The seven commissioners were chaired by Viscount Cardwell who, among many government posts, had been president of the Board of Trade and most recently secretary of state for War. Others included the Liberal MP and recent architect of education and ballot reform William Forster; John Karslake, who had recently been attorney general; the surgeon John Erichsen; the journalist and anti-vivisectionist Richard Hutton; and Huxley. Erichsen was appointed a fellow of the Royal Society soon after the commission reported, making four of the seven commissioners fellows. The scientific and medical establishment had considerable influence on the process. Home Secretary Richard Cross, informed by the fourth Earl of Carnarvon, had aimed to represent all opinions but with a balance in favor of vivisection. Cardwell and Forster were both vice presidents of the Royal Society for the Prevention of Cruelty to Animals, which, although opposed to some aspects of vivisection, took the pragmatic strategy of advocating stricter regulation rather than total abolition.[72]

Fifty-three witnesses were called, from both sides of the argument, but the eminent physicians, surgeons, and medical researchers who gave evidence were overwhelmingly on the side of minimal regulation. They included Joseph Lister, James Paget, Sanderson, and Simon.

Huxley did not attend for the first seven days of evidence, detained by lectures in Edinburgh, but as soon as the anti-vivisectionists started appearing he was almost invariably present. He did miss the controversial evidence given by a close colleague of Sanderson, the bacteriologist Emmanuel Klein, who when asked if he had any regard for the sufferings of animals replied, "No regard at all."[73] That created an outcry and a surge of opposition, and may have led to a tighter regulatory structure than otherwise would have been the case.

Huxley for one thought Klein's testimony would be damaging. Darwin, at his own instigation and encouraged by Huxley, made a cameo appearance to bolster the case after Klein's evidence, in support of the British Association's proposed code of practice.[74] The commission endorsed a regulatory structure that allowed experiments for human and veterinary purposes, and for educational demonstrations under anaesthetic.

The Cruelty to Animals Act 1876, brought in by Cross after extensive lobbying and negotiation, defined an administrative process to regulate vivisection. It allowed vivisection only for original, useful purposes, with a licence from the home secretary.[75] Laboratories used for experiments needed prior approval, and animals could only be tested without anaesthetic or used for demonstration with special dispensation from the Home Office. This left the home secretary with considerable discretionary power, which could only be effectively exercised with scientific and medical advice.

The first inspector appointed under the act, part-time, was George Busk, a close friend of Huxley. Busk, naval surgeon and zoologist, was a safe pair of hands for the medical researchers. He held the post for ten years, reporting to the Home Office annually on the names of the licensees, the authorities recommending and granting licences, the places of experimentation, the total number of experiments, and the number that he believed involved inflicting pain or suffering.[76] He was succeeded by Erichsen, one of the original commissioners supportive of vivisection, who served until 1900.

The act may be seen as a political compromise, as it pleased neither side. Researchers thought it was too restrictive. Some licences were refused by the home secretary and there is some evidence that it did constrain research.[77] The Physiological Society was founded in 1876 in part to monitor the impact of the act on research. In this period of the 1870s, the science of physiology, led by Sanderson and Foster, was just becoming established. Without government funding for research, and without the opportunities for earning money through consultancies, like the chemists, it was a challenge to develop the discipline and to make the case for its value and for the need for vivisection, against traditional medical practices. The Association for the Advancement of Medicine by Research was founded in 1882 with similar objectives. It gradually worked itself into a position in which it was involved in administering the act to the benefit of the research community.[78]

Anti-vivisectionists thought the act was too lenient. They believed it unjustifiable to inflict pain on any animal, and that such procedures tended to dehumanize those undertaking the research. The contrary argument was that the benefits to human and animal welfare outweighed the pain, although some medical men disputed the real gains in knowledge and practice that had been achieved. In consequence, far from putting the matter to rest by the time-honored use of a royal commission, the debates became more intense.[79] Simon, having just left the civil service, was vocal in his opposition to the act, which he thought imposed restrictions on vivisection crippling to science.[80] Others, including James Stansfeld and the seventh Earl of Shaftesbury, president of Cobbe's abolitionist Victoria Street Society, had wanted stronger measures.

Matters came to a head when anti-vivisectionists prosecuted Ferrier in 1881. That attempt was unsuccessful, but Ferrier's work was cited again when a bill for the Abolition of the Practice of Vivisection was brought to Parliament in 1883. In a long speech opposing the bill, Playfair defended the act, arguing on utilitarian grounds that "It is not the mere, or even continuous, infliction of pain which is an offence against moral law, but the unnecessary infliction of pain without an adequate motive to benefit mankind by the act." He did not want a bill that "would drive English physiologists to foreign countries, or to make them work secretly to evade an unjust law, and thus brand as criminals men whose whole object is to ameliorate the condition of suffering humanity."[81] His view prevailed.

The Anatomy Act 1832 had given the Home Office responsibility for inspecting the controversial use of human bodies for medical training.[82] Now, more than forty years later, the public image of medical practice, teaching, and research came again under vivid parliamentary scrutiny. The elite of the medical community achieved most of what it wanted, and limited any severe damage to its freedom to research, but the uneasy relationship with the public continued. Opposition to vivisection remained. A second royal commission revisited the question from 1906 to 1908, reporting in 1912. But that proved to pose no challenge to the growing political influence of experimental medicine, and the act was not revised until 1986.

HOME OFFICE RESPONSIBILITIES
AT THE TURN OF THE CENTURY

The Home Office, like the Board of Trade, gathered a disparate range of responsibilities with the increase in centralizing legislation during the nineteenth century. Those addressed in this chapter have been factories, smoke and industrial pollutants as nuisances, explosives, and vivisection. Mining, which was a Home Office responsibility until 1920, is considered in chapter 10. Fishing, which moved to the Board of Trade in 1886, is considered in chapter 6. Local boards of health, a Home Office responsibility from 1858, moved to the Local Government Board in 1871. Responsibility for graveyards also moved to the Local Government Board in 1900.

Pressure for centralizing legislation invariably came up against both resistance to interference in private businesses and to demands for the delegation of local powers, two of the potent political forces that shaped the ways in which experts had to operate. The influence of scientific, engineering, and medical expertise was constrained by these powerful forces. Yet by the end of the nineteenth century, the experts were embedded on the inside and gaining sway from the outside, in an increasingly bureaucratic administration that required specific expertise both for regulatory purposes and for policy advice.

10

COAL, GAS, AND ELECTRICITY

IT WAS COAL THAT POWERED THE STEAM ENGINE AND BRITAIN'S INDUS-trial rise. The seas were ruled by the Royal Navy, which protected fishing into the bargain, and this combination of powers lay behind Britain's ability to secure and maintain its empire.

Coal mattered, to the extent that in the mid-nineteenth century a major royal commission examined several geologists over alarming possibilities of the exhaustion of coal reserves.[1] Coal mining was a private business, unlike in some continental countries, so legislation came up against the familiar notions of the primacy of the market, private property rights, personal responsibility, labor contracts, and free trade. But mining was also a technical activity, informed by engineering practice and scientific knowledge. Legislation throughout the century reveals, as in many other areas, both the limits of that practice and knowledge, and the limits of the influence of advisers on politicians with particular values and defending powerful interests.

Up to 1835, legislation was concerned primarily with the protection of mining property against sabotage by miners angry at their conditions of work.[2] As mining expanded, stimulated by the development of canals, steam engines, and then the railways, the risks to miners became increasingly visible in tragic accidents. Humphry Davy's invention of a safety lamp in 1815, and inventions of lamps by others, gave some protection. But they also led to the development of deeper mines, so that a select committee on the coal trade in

1829 noted evidence that loss of life since the introduction of the lamp was no better than before.[3]

It was the Whig MP Joseph Pease who instigated and chaired a select committee on accidents in mines in 1835. Pease, the first Quaker to sit in Parliament, had in 1830 become the largest owner of collieries in South Durham.

The witnesses called included the inventor and engineer Goldsworthy Gurney, George Birkbeck, George Stephenson, and William Clanny.[4] Clanny and Stephenson both described the safety lamps they had invented. Several witnesses, including Gurney, Birkbeck, and Stephenson, discussed mine ventilation.[5] Their views differed, in that while Stephenson recommended the common method of a furnace at the bottom of the shaft to create an upward current of air, Gurney and Birkbeck worried about the danger of explosions. Birkbeck suggested mechanical extraction of air while Gurney proposed using high-pressure steam to create an air current. David Reid, who gave evidence to the select committee on the ventilation of the Houses of Parliament a month later, was not called by this committee. His work on ventilating the new Houses of Parliament using a furnace was eventually taken up by Gurney when Reid was fired in 1852.[6] Birkbeck was in demand on this subject, as he outlined his ideas to the select committee on ventilation of the House of Commons fifteen days later.[7]

Although the committee recognized the damage to property and trade caused by frequent explosions and by other causes such as noxious gases and flooding, they claimed that it was the interests of humanity that exercised them. They put the explosions down to accumulations of firedamp, and so examined experiments on many lamps, demonstrated by Jonathan Pereira, a lecturer in chemistry at the London Hospital Medical School.

The findings pointed to the importance of ventilation, safety lamps, and good maps and plans. The committee recommended that all fatal accidents should be reported to the home secretary, and that coroners and juries should have "some fit and proper person or persons" to advise them. The long arm of Davy hung over the committee's deliberations. One of the committee members reported Davy telling him that "he considered the system of ventilation as perfect as it could be," and agreed with the suggestion that "all that was required at the hands of science was the means of producing sufficient light for the workmen without endangering explosion."[8] The committee left the evidence to speak for itself and urged mine owners to be more responsible. No legislation followed.

It was Lord Ashley who broke the logjam when he rose in Parliament in August 1840 to call for an inquiry into the employment of children in mines.[9] This proposal was not based on safety or science, but on what Ashley saw as the damage to children's educational and moral development. Four commissioners were appointed—the factory inspectors Leonard Horner and Robert Saunders, Thomas Tooke, and Thomas Southwood Smith. Horner has been recognized as the most influential of the first factory inspectors.[10] He was also a notable geologist. Southwood Smith, a physician, had already worked with fellow medical colleagues Neil Arnott and James Kay on reports for the Poor Law commissioners. Such was the public interest that ten thousand copies of the vast report of this commission were printed in 1842, more than any other in the century.[11] It brought conditions to light that could not be ignored. The scenes described were appalling, shocking Victorian society. Women worked almost naked. Children as young as four worked alone and in the dark for hours on end. Fatal accidents were widespread, although the commissioners did not think that mines could be rendered perfectly safe by any means yet known.[12]

As soon as the first report was published, Ashley, in an impassioned speech, brought in a bill to address some of the problems, supported by Home Secretary James Graham.[13] The Mines and Collieries Act 1842 prohibited employment of women and of children younger than ten in the mines, and the operation of steam engines by children under fifteen. Seymour Tremenheere, an inspector of schools, was appointed the first inspector of mines, with a role similar to the factory inspectors, to examine conditions of employment.[14] He was a Whig, favorable to employers and hostile to trades unions.[15]

THE MEN OF SCIENCE ARE CALLED

The political imperative behind the Mines and Collieries Act 1842, as Ashley and others saw it, was to counter the danger of the moral degradation of the populace who were subjected to conditions in the mines that led to depravity. Safety was a secondary consideration, and in any case the current view was that science could not do much more for it. Tremenheere was not expected to inspect mines underground, and did not do so. Many miners were resistant to underground inspection, fearing interference in their livelihoods.

One group in the mining community had taken matters into its own hands. Following an explosion in the Wallsend Mine in 1839, which killed fifty-two people, the local MP Robert Ingham established the South Shields Committee to inquire into the accident. The committee's report was published in 1843,[16] and its secretary made clear to a House of Lords select committee in 1849 that they had undertaken this work not with the purpose of interfering in the mines, but "merely to investigate the matter scientifically, and put ourselves in communication with scientific and practical men throughout the Empire."[17] The report amounts to a systematic literature review, enhanced by correspondence with Stephenson, Gurney, Birkbeck, Thomas Graham, and others. Covering safety lamps, ventilation, scientific instruments, maps and plans, medical treatment, and scientific education, the committee ended with a call for an advisory system of inspection with the option of appeal to a judicial body, should inspectors' recommendations not be put into practice. Ingham was out of Parliament when the report was published, which may account for it not being actively pursued.

It was a dreadful explosion in Haswell Colliery on September 28, 1844, when ninety-five people were killed, which formally brought in the scientific experts.[18] The miners had only recently returned to work after a lengthy strike that had generated much public sympathy. The explosion reignited that interest. The magistrate conducting the inquest, the Conservative Thomas Maynard, was eventually persuaded by the lawyer for the bereaved, the Chartist and trade unionist William Roberts, to agree to an inspection of the mine. Roberts petitioned Prime Minister Robert Peel to send two government representatives to the inquest. Peel suggested Michael Faraday, who had recently undertaken an inquiry for the Ordnance Office into an explosion at the Waltham Abbey Royal Gunpowder Factory, and Babbage. Charles Babbage declined and the geologist Charles Lyell agreed to attend, while insisting that he and Faraday be accompanied by a mining engineer. The choice of a scientific team may have been made as a consequence of the significance of scientific findings in the select committee report of 1835.[19]

Faraday and Lyell visited the mine and took part in the inquest. Although they agreed with the expected verdict of accidental death, they had seen enough in the mine to know that safety considerations were not paramount. While the presence of Faraday and Lyell had been helpful to the government in adding credibility to the verdict at the inquest, and in absolving the mine owners of charges of liability, their subsequent report to James Graham turned out to be more difficult to deal with.[20] Although admitting their lack of practical knowledge, Faraday and Lyell suggested how firedamp had created such a huge explosion, and proposed solutions. They also highlighted the dangers of coal dust. To complement any practical solution they urged better scientific education of colliery workers, so that miners understood the need for particular safety precautions and did not, for example, light their pipes underground from safety lamps.

The Committee of the Coal Trade, an organization of the mine owners, took exception to aspects of the suggestions for drawing firedamp away from the mine, and Faraday responded with further observations. The dispute soon reached ministers. Not wishing to reject the report and raise public anger, nor to commit to implementing any recommendations and offend the industry, the government tabled the report during a debate on the controversial Maynooth Endowment Bill, the passing of which led to William Gladstone's resignation. Although Peel was one of the more supportive politicians toward science, he was not averse to burying advice when politically expedient. The clear recognition of coal dust as a potential danger was for the moment buried with it.

This investigation illustrates two significant features of scientific advice, in the manner in which Faraday, who was largely responsible for the report, shaped it. The first is that Faraday and Lyell placed much of the responsibility for safety on the miners rather than the mine owners, although they recognised that the owners were "guided by considerations of present profit."[21] As far as they could see, the ventilation at Haswell was one of the best. The implication was that the owners had done what they could, and that the lack of education and care taken by the miners was a significant factor in causing the accident. The second was Faraday's suggested system for additional ventilation of the parts of the mine. This was based on an understanding, although neither Faraday not anyone else provided clear evidence for it, that the gas that exploded had accumulated in the "goaves" formed when rocks above the worked seam collapsed.

Faraday's proposed system of ventilation of the goaves, and his suggestions for means of working that would minimize collapse, were criticised by the mine owners as ineffective and expensive. Faraday's opinion carried credibility with the government and the public, helping to exonerate the mine owners, but his scientific analysis and proposed solution came up against the practical expertise of those engaged in mining and was found wanting by them.[22]

THE MUSEUM OF ECONOMIC GEOLOGY

The geological museum had been established at Charing Cross in 1835 at the suggestion of Henry De La Beche, and opened in 1837. De La Beche was the first director of the Ordnance Geological Survey, created in 1835 under the Board of Ordnance.

In 1838 De La Beche persuaded the mining geologist Thomas Sopwith to urge the government to establish a collection of mining records, given the lack of landowners' knowledge of geology despite its economic significance.[23] A committee including Sopwith, De La Beche, Lyell, and William Buckland, with the MP Charles Lemon and the Marquess of Northampton, resulted in the formation of the Mining Record Office, opened in 1839.[24] A keeper of mining records was appointed in 1840 and the voluntary deposit of mining plans was encouraged.[25] The following year the office became the Museum of Economic Geology. It moved to Jermyn Street in 1851, becoming the Museum of Practical Geology when the Government School of Mines and Science Applied to the Arts was opened with De La Beche as its director. The Royal College of Chemistry merged with it in 1853, and it became the Royal School of Mines in 1863.

Responsibility for the Geological Survey had been transferred to the Department of Woods and Forests in 1845 under the Geological Survey Act. So it was the Earl of Lincoln, first commissioner, who wrote to De La Beche in 1845 at the request of the home secretary, to ask him to carry out a further inquiry with Lyon Playfair into accidents in mines. Playfair had just become chemist to the Geological Survey, where Edward Frankland was appointed his assistant.

It was clear that the government thought legislation might be necessary to try to reduce the escape of gases into mines and to improve ventilation. De La Beche and Playfair reported a year later to the new first commissioner, Viscount Canning.[26]

This inquiry stimulated De La Beche and Playfair to try to resolve a disputed question about the constitution of dangerous gases in mines. Through chemical analysis, and supported by Thomas Graham, who had come to the same conclusion, they found that it consisted almost entirely of firedamp. While also commenting on the explosive nature of coal dust, they concentrated on the difficulty of removing firedamp safely. Furthermore, to reduce deaths following explosions, they suggested that people entering a mine after an accident should breathe through a bag containing Glauber salts or lime to absorb carbon dioxide that might otherwise asphyxiate them. Playfair, De La Beche, and Warington Smyth, the mining geologist to the Geological Survey, had visited several mines after disasters caused by explosions. They commented on poor practice in operating safety lamps and on the dangerous use of candles, but decided that electric lights were not sufficiently developed to be used safely. Then they turned to ventilation. They thought that rules could not be laid down in legislation for removing firedamp since the practice would depend on the nature of each pit. So, they recommended an extension of the discretionary powers of "properly qualified persons" in different districts to form a "system of judicious inspection."

A few months after this report, George Grey asked De La Beche and Playfair directly for further advice following fatal explosions in Warwickshire and Lancashire.[27] They obtained permission for Warington Smyth to investigate, and his report again highlighted the importance of different local conditions and practices in the mines.

Disasters continued to happen. Two months later, in March 1847, Grey wrote to Viscount Morpeth, who had succeeded Viscount Canning, to request that De La Beche inquire into an accident in Barnsley that had killed seventy-three people.[28] He took Warington Smyth with him. Notable in these reports is the statement that mine owners were cooperative, providing information "without any attempt at concealment." The prevention of accidents was in their business interests, and for many a matter of common humanity. The report yet again recommended inspection "rightly conducted."[29]

COAL MINES INSPECTION ACT 1850

Still, the government took no action, despite a petition to Parliament by sixty thousand colliers in June 1847, and two attempts to pass bills to empower the inspector to direct colliery owners to alter their systems of ventilation and to compel the use of safety lamps.

But in 1849, after yet more accidents, George Cornewall Lewis, under-secretary at the Home Office, wrote at Grey's request to John Phillips and John Kenyon Blackwell to ask them to inquire into the state of collieries and ironstone mines, and specifically into their systems of ventilation.[30] Phillips was a geologist and assistant secretary of the British Association. Blackwell was familiar with mines. This request followed a report earlier in the year by Tremenheere on the explosion in Darley Main Colliery that had killed seventy-five people. Smyth had also visited at the request of the government, and he and Tremenheere had identified poor ventilation practice as the problem. Tremenheere told the home secretary that the inquest jury and many of the witnesses had expressed the view that underground inspection on behalf of the government should be delayed no longer.[31] Phillips and Blackwell gave extensive practical suggestions for improving ventilation, and highlighted the need for proper education of miners and supervisors.[32]

The issue had also been brought to the attention of the House of Lords after a series of petitions, including one from Gurney, who had a personal interest in promoting his steam-jet scheme of ventilation. A select committee examined many witnesses, including Gurney, Tremenheere, De La Beche, Playfair, Clanny, and the physician and inventor Neil Arnott.[33] This committee heard proposals for several inventions for better ventilation of mines but pronounced itself unable to judge between their technical merits. Nor did it take a position on the advantages of different safety lamps. It pointed out that accidents could be caused by means other than explosions and recommended that inspectors should have the right to enter mines, but not to take responsibility from mine owners by ordering specific work or enforcing penalties.

This time the government, rather than a private member, introduced a bill to allow underground inspection of mines. The Earl of Carlisle, as Lord Morpeth had now become, brought it forward.[34] He met some opposition from the usual suspects, such as the mine-owning Earl of Lonsdale in the House of Lords and Colonel Charles Sibthorp in the House of Commons, and also from Benjamin Disraeli as being "hasty and ill-conceived,"[35] but the bill passed in quick time. The Coal Mines Inspection Act 1850 established the collection of fatal accident statistics and the underground inspection of mines, required accurate plans to be kept. It also expanded the inspectorate to five, including Tremenheere, to cover the entire country. Two more were appointed in 1852. Whereas the first factory inspectors were either intelligent laymen or partners in textile firms, the new mine inspectors all had some practical experience of mining and were preferred by the mine owners to the theoretical men of science for that reason.[36]

A FLURRY OF SELECT COMMITTEES

With six inspectors attempting to cover around two thousand collieries and many more pits, it was evident that the effects of inspection would be patchy at best. But the inspectors' reports did bring the truth about accidents to notice and influenced parliamentary responses.[37]

Three select committees—in 1852, 1853 and 1854—highlighted the problem of limited inspection, questioning people including Gurney, Robert Stephenson, and the inspectors Joseph Dickinson and Herbert Mackworth. The 1852 committee argued for more inspectors and for a board of "scientific and practical men" to guide and instruct them.[38] James Mather, secretary of the South Shields Committee, proposed this combination of practical and scientific men. His view, harking back to the Haswell inquiry, was that "Even Professors Faraday and Lyell, than whom there are no abler men in Europe as scientific men, have committed errors, not in principle, but in its application, which practical men would not."[39] The 1853 and 1854 committees felt able to make technical recommendations about ventilation, although they contradicted each other.

A group of inspectors, including Dickinson and Mackworth, drew up rules for collieries, which they submitted, with the permission of Viscount Palmerston, to a deputation of coal owners and workmen.[40] It was this recommendation on rules, including adequate ventilation, that became law in the Coal Mines Inspection Act 1855, tightened in 1860. A further accident at the Hartley Colliery led to the Coal Mines Act 1862, which prohibited single shafts. The absence of a second shaft had prevented rescue, trapping 204 miners who all died. It remains one of the worst mining disasters in England.

ROYAL COMMISSION ON MINES, 1862–1864

In 1862 a royal commission was appointed to look into all mines to which the 1860 act did not apply, such as metalliferous and slate mines.

The commissioners, balanced politically, were chaired by Lord Kinnaird, a Whig who was close to Viscount Palmerston. Philip Egerton, the palaeontologist and Conservative MP, was joined by three Cornish MPs. With the emphasis on health, Edward Greenhow, physician and epidemiologist, and Philip Holland, burials inspector for the Home Office, were also appointed. Robert

Angus Smith reported evidence on the physiological effects of carbon dioxide, by observing people in an airtight chamber.[41] He concluded that carbon dioxide was more physiologically harmful than he had thought, stating also that it "almost always comes in bad company."[42] Other reports on ventilation and atmospheric analysis were added by the toxicologist Alfred Swaine Taylor and Albert Bernays, a chemist at St Thomas's Hospital. William Farr provided mortality statistics.[43] The commissioners recommended practical changes based on their understanding of ventilation in coal mines where mortality was lower, excepting accidents. Other conclusions were shaped by medical evidence, including causes of poor health among miners, and by statistics on accidents.

While the report may have had some influence in practice, no legislation followed despite three more select committee investigations, until eight years later in 1872 when the Coal Mines Regulation Act 1872 was passed.

Legislation was initiated in 1869 by Home Secretary Henry Bruce in Gladstone's first government. But it took several years, under pressure of other business and the need for detailed consideration, before a bill could be passed, strongly supported by Disraeli.[44] The 1872 act allowed for more inspectors, made requirements on ventilation and safety lamps more stringent, and required mine managers to hold certificates of competency. The equivalent consolidating act for factories came in 1878.

ROYAL COMMISSION ON ACCIDENTS IN MINES, 1879–1886

Despite the new regulations, dreadful accidents mounted up. Around a thousand miners were killed each year in the 1870s. Although the number of deaths had remained relatively constant while employment and tons of coal extracted doubled, which gives an indication of the impact of inspection, the effect on public opinion could not be ignored. Public pressure after several major disasters led to another royal commission on accidents in mines in 1879.

The commission's task was to inquire whether "the resources of science furnish any practicable expedients not now in use, and are calculated to prevent the occurrence of accidents, or limit their disastrous consequences."[45] This commission brought in the full weight of the scientific experts, who had been largely absent from committees since the 1850s. Of the nine members appointed, five were fellows of the Royal Society. They were Warington Smyth, who chaired the commission; John Tyndall; Frederick Abel; Robert Clifton; and the third Earl of Crawford and Balcarres. Three represented the employers. Thomas Burt MP, trade unionist and one of the first working-class members of Parliament, represented the employed. Their inquiries were so extensive that although they made an interim report in 1881, it was not until 1886 that they submitted their final report.[46]

One issue the commissioners picked up on was the potential danger of coal dust explosions. Faraday and Lyell had warned of this in 1845, in the report that was buried by Peel. The danger had been more recently identified by the French in 1867 and 1875, and in England by William Galloway. He had shown that small amounts of firedamp could ignite coal dust explosively, and others had since shown that a mixture of coal dust and air alone could explode. Abel had followed up these results and summarized extensive investigations after an explosion at Seaham Colliery in 1880 had killed 164 people.[47] These studies showed not only that coal dust was explosive but that small amounts of dust in mines could also trigger explosions of gas when suddenly heated.

In the first phase of work the commissioners visited twenty mines and interviewed sixty-nine witnesses, including mine inspectors, colliery managers, and mining engineers. Their examination was exhaustive, from collapses of walls and roofs, the most common cause of deaths, to the emission of firedamp and other gases, explosions, the influence of coal dust, methods of illumination, ventilation, blasting, signaling, and lifesaving after accidents.

Their interim report in 1881 pointed to the need for further experiments, on questions such as the pressures of gas in strata, the use of different safety lamps in diverse gas and colliery conditions, the development of a more sensitive gas detector than the Davy lamp, on coal dust, on electric lighting, and on methods of bringing down coal without using sparks or flame. They established that the two main causes of death in mines were falls of the roof and walls and explosions of firedamp. Both were intimately connected with the illumination of workings, hence the commission's emphasis on safety lamps and the possibility of electric lighting.

In the second phase, the priority was experimental work, much of which related to the lamps. More than half their final report was devoted to experiments on 250 different lamps. Other lines of experiment included the extent of flame projection in blown-out powder shots in the presence and absence of coal dust, under Abel at the Royal Engineers in Chatham. They also examined means of bringing down coal without explosions, and the use of Edward Liveing's gas indicator to detect gases to which safety lamps were not sensitive. Galloway was asked to experiment with using water in conjunction with explosives. The aim was to reduce sparking and keep the temperature below that at which gases could combust and explode. The findings left a rich resource of information on the use of explosives, safety lamps, lighting, and gas in mines.

Following the report, Lord Salisbury's Conservative government passed the Coal Mines Regulation Act in 1887. It tightened up employment and safety practices across the board. The scientific evidence led to extensive modification of the thirty-one general rules in force and the addition of eight new ones, particularly concerning safety lamps. The work of the commission proved to be a positive step in the decades-long process of changing societal norms of employment and safety practices.

COAL DUST AND THE COAL MINES REGULATION ACT 1896

Ever since Faraday and Lyell's work in 1845, the dangers of coal dust had been noted. The royal commission's report in 1886 had emphasised the challenge. But it still remained. So in 1891 a royal commission was appointed "to inquire whether there are any practicable means of preventing or mitigating any dangers that may arise from the presence of coal dust in mines." The commission, chaired by Joseph Chamberlain and including Lord Rayleigh, reported in 1894, refining the findings of the earlier commission. The consequential Coal Mines Regulation Act 1896 aimed at preventing explosions, particularly from coal dust. It gave the home secretary powers to modify rules on the storage and use of explosives in mines and the use of safety lamps.

Throughout the century the impact of inspectors was constrained since there were still only forty at the end, their salaries and budgets were small, and they had limited powers. Although they could instigate prosecution if mine owners did not comply, this was difficult in practice, since local magistrates could be influenced by mine owners. Nevertheless, their collection of statistics on the causes of accidents and of occupational diseases amounted to a solid body of evidence that informed miners themselves, mine owners, and later inquiries.

Inspection had to contend with the prevalent belief against intervention in private enterprise unless such a response became politically unavoidable.[48] The inspectors operated by advice and suggestion. Tremenheere approved of this advisory approach, which many saw as guarding against the danger of shifting responsibility from the mine owner to the state. Dickinson, who became chief inspector in 1884, held the same view. Steadily but slowly legislation was introduced that aimed to improve employment conditions and to reduce accidents. Expert advice offered only gradual improvement, against the political and economic position and power of the mining companies.

GAS AND ELECTRICITY

Gas lights first flickered in Pall Mall in 1807. A form of gas lighting had been invented by William Murdoch for the firm Boulton & Watt in the 1790s, which won him the Royal Society's Rumford Medal in 1808. The firm developed a process for generating the gas from coal, in works that belched out smoke and produced noxious waste products, including contaminated lime, corrosive acids, tar, and ammoniacal liquids.

But it was the Gas Light and Coke Company, using a method invented by

Frederick Winsor, which managed to establish itself in 1812, following a select committee that received objections from Murdoch over his rights of invention.[49] The consultant chemist Frederick Accum, who worked for the Gas Light and Coke Company, gave evidence claiming that Winsor's method was better than Murdoch's.[50] Davy appeared unavailingly for the defense, stating that Murdoch had been given the Rumford Medal as the Royal Society had decided that he had been the first to establish the process.[51] The Gas Light and Coke Company installed gas generation plants in many towns, often close to rivers for removal of waste products. The plant in Westminster was soon causing trouble. Local residents successfully took court action for damages under nuisance laws for the effects of effluent running in a local sewer. The company then built an underground sewer, but that just removed the pollution directly into the River Thames.

As more companies were formed, and opposition became more vocal, even the industry's consultant Frederick Accum called in 1819 for parliamentary action.[52] Public pressure had its effect, as Parliament started to include clauses in bills incorporating gas companies that forbade them from dumping effluent into sewers or watercourses. The remedies were rarely effectual as there was no inspectorate or effective means of enforcement, and the companies were prepared to pay occasional damages. Equally, there was no incentive to try to provide scientific evidence of the linkage between effluent and fish death. Gas companies could deny that they polluted, could argue that pollution was not as bad as some maintained, and could claim that there was no evidence that their waste was killing fish. The royal commission on the metropolitan water supply did report in 1828 on the damage done by gasworks in addition to steamboats, but no effective legislation resulted.[53]

A ROYAL SOCIETY INVESTIGATION

While pollution from effluent and smoke could to some extent be ignored, explosions were another matter. The explosion at the Westminster works of the Gas Light and Coke Company in 1813, although it killed no one and caused little damage, nevertheless caused alarm.

Home Secretary Lord Sidmouth wrote to Joseph Banks, asking him to establish a committee to examine and report. Banks's committee included Colonel William Congreve, an expert on explosives, William Wollaston, Thomas Young, and John Rennie.[54] Young was asked to research the impact of the explosion by comparing it with an equivalent force of gunpowder, which he calculated to be between five and ten barrels.

The committee reported from its experiments that fire could not be conveyed along closed tubes, a gratifying discovery in terms of safety, and confirmed the company's finding that the explosion had been caused by leakage due to inattentiveness. It suggested that gasworks should be distant from other buildings and that storage gasometers should be small and numerous.[55] With later experience, this advice appeared overcautious. The Gas Light and Coke Company was duly allowed to proceed to illuminate London.

By 1823 there were still concerns about such operations and a select committee looked into them as a bill was brought in. The committee took into account the Royal Society's views and two recent reports by Congreve on the state of gasworks in London. Congreve's reports, made in 1822 and 1823, had been ordered by Sidmouth and Peel, respectively, as home secretaries. In the first, Congreve laid out a wide range of possible causes of explosions in gasworks and from leakage from pipes and buildings.[56] The second was a comprehensive financial and technical audit of the gas companies.[57] The financial calculations were made with a view to establishing fair prices and to giving the government information about policy alternatives, such as the use of oil gas rather than coal gas for lighting, which he suggested might be beneficial for expanding fisheries, "viewed as a principal nursery for seamen."[58] Congreve, the member for Plymouth, would have been aware of the importance of seamen as a source of men for the navy. He proposed regulations to limit the size of gasometers and their distance from each other and from houses, to surround pipes with clay to help prevent leakage, and to limit companies to separate areas to remove the difficulty of identifying responsibility for leaks and nuisances.

Davy and Wollaston gave evidence to this committee, in addition to the engineer John Millington, professor of mechanical philosophy at the Royal Institution. Davy, from what he had seen at the Westminster works, thought the danger of explosions there was low. But he was careful to say that more experimental evidence would be needed before he could pronounce with any certainty, or on any other location, and he was not prepared to give any opinion on proposed regulations in the bill before Parliament. Wollaston backtracked on some of the recommendations of the original Royal Society committee. Like Davy, he suggested that more experiments were needed, and repeatedly stated that he did not have the knowledge to comment on specific questions asked, such as setting limits on the size of gasometers or their distance apart. He did then agree with the view put to him that one could rely on the interests of the company to take all possible precautions, rather than on regulation or parliamentary legislation, as did Millington.

The committee concluded that "the danger likely to arise from Gasometers and Gas Works is not so great as has been supposed, and that therefore, the necessity of interference by the legislative enactments . . . does not press at the present period of the session."[59] Despite that, they noted recommendations about the design and siting of gasometers, and thought that the secretary of state should have further powers to carry into effect improvements for the safety of the public. Given that half-hearted and somewhat self-contradictory approach, the less than helpful evidence from Davy and Wollaston, and the fact that individual gas company acts already included powers of inspection under the Home Office, it is unsurprising that the bill did not pass.

This episode reveals Davy and Wollaston being very circumspect about their knowledge, and unwilling to comment on the proposed regulations in the bill. Even the original Royal Society report, to which Congreve and Wollaston but not Davy had been party, came in for implicit criticism by the committee, since the basis of its comparisons between gasometer and gunpowder explosions was unclear to them. It is interesting to compare Davy's reticence here with his forthright approach at the same time to protecting the navy's ships without extensive experimental trials, which resulted in the problems described in chapter 3.

Two years later Peel asked the Royal Society to advise on whether supervision and control by the Home Office remained necessary.[60] Although the report criticized aspects of management, such as ventilation at some works, gasholder construction, pipelaying, and the use of safety lamps, the Royal Society thought that serious accidents were improbable and lamely suggested "a propriety of occasional superintendence."[61]

MONOPOLIES AND METERS

Much gas legislation was concerned with regulating monopoly or near-monopoly undertakings in the public interest.[62] That could cover both financial aspects and technical dimensions—for example, to deter fraud and to assure quantity and quality of supply. There were also clauses in relevant acts to penalize companies polluting rivers, but as the royal commission on salmon fisheries noted in 1861, "The want of a competent authority to enforce such regulations in the interest of the public renders them nugatory."[63]

The metropolis was often a testing ground for developments. A select committee on the supply of gas to the metropolis, which sat in 1858 and 1859, surfaced a whole set of issues relating not only to financial returns and monopolies but also to the illuminating power of different supplies, the reliability of different meters, and the lack of standard measures of volume.[64]

An act for regulating the measures used in sales of gas was passed in 1859, but immediately required the services of George Airy, the leading adviser to government on questions of weights and measures, to pronounce on its effectiveness in preventing fraud. According to Airy, the act had been passed as a private member's bill without the government noticing that it required the establishment of standards.[65] This was a matter for the Exchequer Office of Weights and Measures, and Airy was approached for advice by Thomas Spring

Rice, now Lord Monteagle, comptroller general of the exchequer. Monteagle was clearly under pressure to deliver the standards, as on Charles Wheatstone's recommendation he also wrote to Frankland, then lecturer on chemistry at St Bartholomew's Hospital, and to George Lowe, a gas company manufacturer. Frankland had carried out analysis of coal gas as a consultant in 1851.[66] Wheatstone, who had suggested that Monteagle would need a whole new department to manage the work, had stressed that both were fellows of the Royal Society, and both replied with complementary information.[67]

Airy nevertheless held the main responsibility. He brought in the chemist William Hallowes Miller and submitted a full report. He then had a weight, a cubic foot measure, a gas transfer mechanism and a test gas holder designed, made, and lodged with the Exchequer standards. Certified copies were to be stamped and sent to the chief magistrates in London, Edinburgh, and Dublin.[68]

THE GAS REFEREES

Under the Metropolis Gas Act 1860, companies were given monopoly powers within their boundaries, but their dividends were limited and they were obliged to supply residents. They had to provide illumination and gas pressure to given standards, although there was no effective provision for monitoring their adherence.[69] Several clauses relating to purity were included following evidence to the select committee on the bill, which was given by Henry Letheby, professor of chemistry at the London Hospital and chemical referee to the City of London Corporation for the quality of gas.[70]

The operation of this act was reviewed by a select committee in 1866. Local gas examiners provided evidence of the variable quality of gas, and the committee concluded that gas companies in London provided lower illumination at a higher price than elsewhere and generated excess funds. They recommended increasing the required illumination and reducing the maximum price.[71] The committee also found that purification of gas was imperfect and that sulfur compounds were injurious to pictures, metals, and leather. Robert Baxter, solicitor for the gas companies, declared that the gas companies supported the creation of a chemical board of three members to analyze the purity and illuminating power of gas, funded by the industry but reporting to the home secretary.[72] Letheby proposed that each gas company should be required to establish a testing station and that each local authority should appoint testing officers, with a chief analyst reporting weekly.[73]

These recommendations led to the City of London Gas Act 1868. This act and its successors provided for three gas referees, with responsibility for overseeing the quality of gas produced by the burgeoning and lucrative gas industry. There was a view that the local authorities in London should be able to establish their own supply, if the companies could not agree on a quality and price satisfactory to consumers. The outcome was an act that established regulations for price and quality, but did not permit supply by the public City Corporation. The companies won the battle for business, but at the cost of increased regulation.

While gas analysts or referees had been appointed on a local basis for many years, the creation of a new group of three gas referees gave greater central control. Like the alkali inspectorate, they reported to the Board of Trade. They were required to be "competent and impartial persons, one at least having practical knowledge and experience in the manufacture and supply of gas." They could specify both the methods of testing and permissible limits of impurities, a remarkable power, with gas at all times to be free of sulfuretted hydrogen.[74] Letheby was appointed chief gas examiner. The first appointees as gas referees were Robert Patterson, John Peirce, and Frederick Evans.[75] Patterson, educated as a civil engineer, was a journalist with an interest in chemistry. Peirce was an engineer, as was Evans, who worked for the Chartered Gas Company and was the only referee with practical experience of gas establishments.

THE GAS REFEREES AT WORK

The gas referees soon reported on their activities, which were aimed at improving and standardizing the quality of gas, for which they set out protocols to be followed at testing stations.[76] Comparing illuminating power, defined by the "candle," was one problem, which required new photometers to be developed. The referees also compared different gas burners, so that they could advise consumers on which were the most efficient.[77] Impurities were a major challenge. According to the referees, there was no satisfactory process for completely removing sulfur, the levels of which could not be accurately determined anyway. Since the referees had the power of being able to set the limits for sulfur, that was a stumbling block.

The passing of the City of London Gas Act 1868 gave a direct impetus to the development of photometric techniques, approaches to chemical analysis, and methods of purifying gas. The "sulfur question" led to a new understanding of the nature of sulfur compounds in gas and how to minimize or remove them.[78] Ammonia removal was a more straightforward problem, not least because it gave a profitable by-product, and the referees were able to reduce allowed limits substantially.[79]

When Evans resigned in 1870 he was replaced by William Pole, an engineer who had managed a gasworks in his youth.[80] Pole is one of the many lesser-known scientific experts so active behind the scenes. He had worked with Robert Stephenson on calculations for the Menai Bridge, with James Rendel advising the Italian government on harbors, and with James Simpson

to establish the Lambeth Waterworks Company's works. From the late 1850s he was drawn extensively into government circles, advising the War Office on iron armor and on the Martini-Henry rifle, and acting as secretary to royal commissions on railways, on metropolitan water supply, and on pollution in the River Thames.

CHANGING THE GAS REFEREES

By 1872 the Board of Trade had decided that the work of the original gas referees was nearly complete.[81] The major remaining problem was the presence of sulfur in gas, as the smell from gasworks was a cause of frequent complaint. Having considered abolishing the office of gas referees, the board realized it could not do this under the 1868 act. It decided instead to reconstitute the referees as a scientific body, including a chemist and someone with practical knowledge and experience in the manufacture and supply of gas.

John Pierce and Patterson left, and the two were replaced by the chemist Augustus Vernon Harcourt and the physicist John Tyndall, making all three referees fellows of the Royal Society. Harcourt, nephew of one of the founders of the British Association, had previously worked with the referees on removing sulfur from gas.[82] Tyndall, who had recently revealed new findings on photochemistry and the scattering of light, was already well known to the Board of Trade in his role as scientific adviser on lighthouses, where he had developed knowledge of different coals and the production and burning of gas. Their role was part-time, their pay £200, about a third of the previous fees.

The gas referees, as before, were to inspect gas company works and processes of manufacture, both for routine matters and in response to complaints about nuisances, and to prescribe and certify the mode, location, and apparatus for testing gas purity. They would also provide evidence to select committees on gas bills.

Much of the focus of the referees was on the measurement and removal of unwanted impurities such as sulfur and ammonia, and on the standardization of the illuminating power of burners. Photometry was critical to this work, and a feature of meetings over many years. The Board of Trade also sought independent advice. For example, a committee established in 1879—consisting of Andrew Williamson, William Odling, and George Livesey—recommended that the use of candles in setting photometric standards be discontinued.[83]

Tyndall, suffering from ill health, was replaced in 1892 by Arthur Rücker, organizer of his retirement dinner at the Royal Institution in 1887. Pole continued in post until 1899, and Harcourt served for forty-five years.

Scientific advice continued to be required for setting standards. A committee chaired by Odling and including the three gas referees Pole, Harcourt, and Rücker, as well as William Dibdin and Frankland, reported on photometric standards for testing coal gas in 1895.[84] In addition to these technical matters, the Home Office picked up on the potential danger of carbon monoxide poisoning from water gas. A committee was convened with Lord Belper in the chair, aided by Henry Cunynghame of the Home Office; Henry Parsons, assistant medical officer of the Local Government Board; the physiologist John Scott Haldane; and the chemist William Ramsay. Parsons, Haldane, and Ramsay all submitted reports, backed up by evidence from witnesses, including several engineers and medical officers, chemists such as Frankland and Henry Roscoe, and John Tatham, statistical superintendent of the General Register Office. The committee recommended that any poisonous gas should be made with a pungent smell, that regulatory powers should limit the proportion of carbon monoxide allowed in the gas, and that there should be better reporting of causes of deaths by coroners.[85]

The powers of the gas referees were extended in 1920 to the whole United Kingdom. The Office of the Gas Referees was then abolished in 1938 and its responsibilities taken over by the Board of Trade. The gas referees had served their function of giving public confidence at a time when governments were not seen to make consumer interests a priority. The scientific experts had proved able and willing to act as trusted intermediaries, embedded in state bureaucracy.

ELECTRICITY

The harnessing of electricity fascinated the nineteenth century public.[86]

The electric telegraph, developed by William Cooke and Charles Wheatstone in 1837, was soon employed on the railways. Its insulation was perfected over the years, so that by the 1860s instantaneous communication was possible between London and any part of Great Britain.[87] Later still, in 1876, Alexander Graham Bell's invention of the telephone offered enhanced means of communication. By 1885 there were about twelve thousand subscribers to the telephone service, which was a public undertaking of the General Post Office following nationalization of the telegraph companies in 1870.[88] Public electricity supply was only established in the 1880s, at first for lighting and then for power.

Between these major developments of the telegraph and public electricity supply, it was the extension of communication by electrical means that occupied the interests of the state. The most visible, or rather invisible, example of this is the submarine cable.[89] The significance of these cables to the government is demonstrated by the knighthood awarded to William Thomson in 1867 for his contribution to their success, and his elevation to the peerage as Baron Kelvin in 1892.

PUBLIC AND PRIVATE ELECTRICITY SUPPLY

Public and domestic electric lighting was a late nineteenth-century arrival. After the initial use of arc lighting from mid-century, Thomas Edison's announcement of the invention of the incandescent light in 1879 and the first International Electrical Congress in Paris in 1881 set the direction. In early 1882 a central generator was set up at Holborn Viaduct to supply the area, although it was abandoned in 1886 after heavy financial losses. Street lighting was offered at the price of gas, despite its greater expense to produce, and supplies were taken by hotels, restaurants, and other buildings. This facility, and Edison's patents, were then bought by the Edison Electric Light Company, Ltd., which was established in 1883.

Electric lighting was developing at a time when there were growing complaints about the quality of gas lighting, and of the dangers and adverse environmental effects of gas generation. The House of Commons took an interest in the subject when individual towns started to explore the use of electricity. A select committee chaired by Playfair in 1879, stimulated by proposed bills for electric lighting in Liverpool and elsewhere, examined whether municipal corporations should be allowed to adopt schemes for lighting by electricity and the conditions under which private companies should operate.[90]

At this early stage in development of the technology, the committee gave much attention to scientific and engineering aspects. Eighteen of the twenty-four witnesses had scientific or engineering expertise. Tyndall gave an introductory lecture and demonstration to the committee on the science, to which William Siemens, William Thomson, and the electric specialist John Hopkinson added their perspectives. Thomas Keates, consulting chemist to the Metropolitan Board of Works, described his work with Joseph Bazalgette on lighting the Thames embankment. Engineers included James Douglass, engineer-in-chief to Trinity House; Thomas Stevenson, engineer to the Board of Northern Lights; William Preece, electrician to the General Post Office; and George Deacon, who had been aboard the *Great Eastern* to lay the Atlantic cable in 1865.

The scientific witnesses pointed out that electric lights, which were already used in some lighthouses and public places, avoided the production of the "vitiated air" generated by gas lights, and that electrical systems could be used to distribute power as well as lighting. The sanitary benefits of electricity over gas had become a banner around which many rallied. Tyndall took the opportunity to remind the committee that the scientific researcher should not be constrained by the immediate need to produce practical results. Playfair asked him, "Did Professor Faraday, as a man of science, think it his duty to apply his discoveries to the practical purposes of producing electric heat, or electric

light, upon an economical scale?" Tyndall answered, "No; but with the pre-
science of a true man of science, he predicted that his results, which were
exceedingly feeble in the first instance, would receive their full development
hereafter."[91] The committee did not want to hold back any such development
and thought that local authorities should have the power to authorize the lay-
ing of wires for public lighting. But it did recognize the danger of monopoly
powers of private companies, and suggested that they might be given permission to operate for a limited period, after which the municipal authority
would have an option to take control.

Companies vied for this market, raising memories of bubbles in railway
and submarine telegraph shares earlier in the century. Many private bills were
brought forward for new private companies, existing gas companies, and
municipal authorities to supply electric light and power. Petitions opposing
these bills, concerns about speculation, and the consideration of powers for
companies to dig up the roads to lay cables led to a select committee to examine general legislation.[92]

The primary issues now were political and commercial rather than scientific, and many of the scientific and engineering witnesses had financial
interests.[93] Frederick Bramwell, a shareholder in the Edison Company, gave
evidence on behalf of the Electric Lighting Company bills, as did William
Spottiswoode as president of the Royal Society,[94] Siemens as a representative
of his company, John Lubbock as director and shareholder of a company, and
Hopkinson. The experts disagreed over questions of costs, with Hopkinson,
Spottiswoode, and Siemens giving figures that differed by a factor of four.

Chamberlain, as president of the Board of Trade from 1880 to 1885, was
concerned like many before him to limit the powers of monopolies in the
public interest. As he had done with the gasworks in Birmingham, he was also
open to bringing enterprises into public ownership by local government. The
outcome was legislation that granted companies twenty-one years of regulated
activity, after which the businesses could be bought by the local authority.
Bramwell, Spottiswoode, Siemens, Hopkinson, and Lubbock had all argued
for this figure, doubtless in a coordinated move with the companies in which
most of them had an interest.[95] Their interests did not go unnoticed, as Lubbock in cross-examination was asked, "Possibly that may tinge your mind a
little with regard to the conditions upon which they should be bought up,
and the length of time and so on." He replied, "I must leave others to judge of
that."[96]

The Electric Lighting Act 1882 allowed companies to break up the streets
for cables and for local authorities to spend and borrow money. The Board
of Trade had regulatory powers that it could exert through licensing and the
use of orders. The following year a select committee examined the working of

the act in practice following various complaints. Hopkinson, electrician to the Edison Company; Bramwell; and Lubbock, at that point chairman of the company, just before the merger of the Edison and Swan Companies, were examined about costs and competition. Details of provisional orders bills were amended by the committee, but contrary to expectations, the envisaged growth in electricity supply did not materialize. The years following passage of the 1882 act coincided with a general economic downturn. However, Bramwell, Lubbock, and others retrospectively blamed the government's restrictive conditions when they appeared before a House of Lords select committee in 1886 to review the operation of the act.[97]

The amended Electric Lighting Act 1886 extended the time that private companies could operate before possible public purchase to forty-two years. Coincident now with an upturn in the economy, the electrical industry grew again. Even so, despite Britain's technological capability, it lagged behind others such as the United States. It is arguable that the prudent British desire to balance the interests of the public and business communities at a time of rapid technological change, of arc lights to incandescent lighting and of direct to alternating current, constrained innovation to a degree.[98] Continuing concerns about the slow development in Britain led to a further amending act in 1909, and eventually to the establishment of the Board of Electricity Commissioners in 1920, to try to accelerate the development of electricity supply.[99]

As in other areas of private business, scientific and technical advice was important in informing regulation, but political and economic factors set the major constraints. In mining, the primary factors influencing the extent and speed of regulatory legislation were the instinctive desire not to interfere with private businesses or with the freedom of individuals to enter into contracts of employment, allied with the evident limitations of scientific knowledge and engineering expertise in helping ensure safe conditions for working. Gas production and distribution offered different challenges, as gasification plants were in urban areas, monopolies emerged, and distribution networks directly served public streets, businesses, and homes. Here, both national and local systems of regulation for quality, price and public safety were the response. That brought technical experts center stage both in regulatory roles and as influencers of further legislation. The later development of electricity generation and supply is more analogous to gas production and distribution rather than mining and coal production. For both gas and electricity, the contested ground was economic, between private and public ownership, as local authorities often sought public provision. Coal mining remained a private business throughout the nineteenth century, giving a different context to the provision of expert advice.

PART VI

SOCIAL CONDITIONS AND PUBLIC HEALTH

11

WATER, SANITATION, AND RIVER POLLUTION

PUBLIC HEALTH IS A MATTER OF POLITICS, NOT A STRAIGHTFORWARD question of applying medical, scientific, or engineering knowledge. Not least when such knowledge is contested between specialists in the same field. The public health choices that governments and local authorities made in the nineteenth century were informed by expert advice, often persuasively and sometimes dangerously. But they were not determined by it.[1] While the scientific experts, the engineers, and the medical men were all involved, the provision of water supplies and sanitation systems became a major nationwide exercise in engineering.[2]

WATER IN THE METROPOLIS

A good water supply and system for removing waste are prerequisites for sound public health. Prior to the nineteenth century, villages might be supplied by a parish pump and larger towns by private companies. Most cesspits were private, with nonhuman waste flowing in open street drains. In 1815 London households were permitted by their sewer commissions to connect their sewage drains to the main sewers, adding much human waste to the animal and industrial waste in the sewers. The sewage was not treated before flowing straight into rivers or the sea.

There was limited or no competition for water supply. That changed in London in the early nineteenth century. Parliament incorporated several water companies between 1806 and 1811, in competition with the existing ones. It was assumed that competition would benefit the consumer. But once the companies had invested in separate systems, and fought each other to a commercial standstill, they agreed to withdraw to specific areas and became local monopolies. Prices rose and customers lost out. Similar problems arose in towns elsewhere. Few local authorities possessed their own waterworks, and in London there was no single metropolitan authority that might have developed one.

The challenges were primarily political and economic, in the context of technological change, heightened by the fact that dozens of members of Parliament had interests in private water companies.[3] Nevertheless, the engineers and the medical practitioners were rapidly drawn into the debate, and were committed to helping. The 1828 charter of the Institution of Civil Engineers even included the objective of "removing . . . noxious accumulations as by the drainage of towns and districts to prevent the formation of malaria and secure the public health."[4] Complicating the issue was the oppositional nature of the debates. Experts were called on both sides of arguments, as agitation for public ownership or replacement of polluted supplies led to legal hearings as well as parliamentary committees and commissions.

In 1821 Parliament took a hand by establishing a select committee on the supply of water to the metropolis, when a substantial increase in water rates which was in prospect led to a public outcry.[5] About forty people were called to give evidence, including the engineers to the water companies. While wringing its hands as to the predicament inadvertently created by unfettered private competition for this public utility, the committee recognized that some restrictive regulations were necessary to protect the public; however, it avoided recommending increased charges for the apparent improvements in amount and quality of supply. The grounds were that it would be necessary "to require further aid from men of science, and practical knowledge in the construction of works; and to consume more time in the investigation" to do this.[6] The judgement could not be made "without the assistance of very skilful and experienced engineers, unconnected with the parties concerned."[7] But the committee had called no independent engineers to give evidence. It failed to tackle the main issue at all, and conveniently blamed a lack of independent scientific and engineering advice.

Following complaints about water quality, stimulated by a pamphlet titled *The Dolphin*, which pointed out that the Grand Junction Waterworks Company had its intake next to the notorious Ranelagh sewer, Parliament returned to the subject in 1827.[8] The Radical MP Francis Burdett, who had helped publicize the pamphlet, pressed for a parliamentary inquiry. The government established a royal commission to look into the River Thames as a source of water.[9]

The three commissioners chosen were the country's leading civil engineer,

Thomas Telford, with the physician Mark Roget and the chemist Thomas Brande. Their inquiry was circumscribed, as the home secretary, the third Marquess of Lansdowne, explained, to a description of the quantity and quality of the water supplied by the water companies, in other words to the wealthier rather than poorer classes.

The commissioners, appointed under George Canning's government, reported in April 1828 to Home Secretary Robert Peel in the Duke of Wellington's government. They found no significant problems with the quantity of water available and supplied. Quality was more problematic, with suspended filth often not removed by settling. Contamination from industries such a gasworks, sewage, and churning by steamboats and tides made Thames water dirty and stinking. Fish were unable to survive in it.

As well as examining engineers, the commissioners asked John Bostock, chemist and professor of physiology at Guy's Hospital, to undertake water analysis. His evidence illustrates the shortcomings of chemical analysis at this time.[10] The description of types of solid contaminants is qualitative, and the analysis of substances in solution, which the commissioners recognized could be important, is limited. Bostock recommended filtration through sand and charcoal. A test with a filter of sand had produced water that "was perfectly free from visible impurities and had lost all unpleasant flavour or odour."[11] Chemistry in this sense was irrelevant. A common sense definition of what was potable was sufficient. Only later in the century did chemical analysis develop sufficiently to challenge this concept of purity.[12]

The dangers were also attested by many medical men, including Henry Halford and Benjamin Brodie, who sent in their views on samples of water from the Grand Junction intake.[13] They could do no more than make general statements, on the basis that such water was likely to upset the balance of health of individuals drinking it. While the chemists were not yet able to detect and specify the chemical constituents unambiguously, let alone associate them with specific health outcomes, the medical men were equally constrained by the prevailing philosophy of constitutional medicine as one of balance of health, or restoring "balance" or normality to the affected body. The association of specific diseases with specific causes came later, with the development of statistics, epidemiology, and microbiology.

The commissioners, recognizing the limitations of the evidence, concluded by recommending filtering to remove solid matter and suggested that other water sources should be sought. They also wanted the water companies, as monopolies, subjected to "some effective superintendence and control."[14] In making such a political statement they were arguably exceeding their brief to examine "the salubrity and description of the water itself, and the quantity in which it is furnished," while making generous use of the subsequent permission to make "such observations upon the whole of the subject as cannot fail to occur to gentlemen so highly qualified."[15]

This royal commission highlights the contested nature of water analysis, which continued for many decades.[16] Analytical chemistry was developing fast, as industries such as mining, textiles, metals, gas, and alkalis grew. Spas promoting mineral waters for medical purposes contracted chemists to reveal their compositions. Despite the discrepancies that emerged between different analyses, the chemists managed to establish a measure of authority in the reliability of their conclusions, which spilled over into consultations on the public water supply.

The reason the commissioners had appointed Bostock was because the analyses submitted to them by various chemists were inconsistent. They declared him "eminently qualified,"[17] and indeed he had published *On the Purification of Thames Water* in 1826. The further problem with the public supply was its contamination by putrefying matter. Bostock was not the only one to point this out. William Lambe also submitted an analysis of Thames water, arguing that organic contaminants were dangerous, and that their detection and nature had escaped the techniques of chemical analysts.[18]

In further responses to petitions, Parliament established select committees in 1828, 1834, and 1840, but none concluded that an alternative supply of purer water would justify the expense. The 1828 committee recommended that Telford be commissioned to make a plan for supplying the metropolis with wholesome water.[19] Telford was employed by the Treasury and produced his report in 1834, seven months before his death. He proposed to draw from the River Verulam in the north and the River Wandle in the south, with additional extraction from the River Lea.[20] A select committee examined the report but had no time to come to any conclusions, given quarrels between the ageing Telford and his assistants.[21]

The House of Lords further examined the subject in 1840. It likewise laid out evidence obtained from engineers, two medical men, and Richard Phillips, chemist and curator of the Museum of Economic Geology, but also advanced no conclusions. Telford himself had realized that the project would probably be too expensive.[22]

Throughout the nineteenth century, the water companies were innovating and improving their services. The challenge, especially in London, was the huge increase in population, from less than a million in 1801 to 2.3 million in 1851 and 4.5 million by 1900, allied to the multitude of different responsible local authorities. Steam-driven pumps were developed. Wooden pipes were replaced by cast iron, and continuous, high-pressure supplies gradually introduced. Following the royal commission report of 1828, James Simpson introduced sand filtration at Chelsea Waterworks in 1829, which was extended to other areas in the 1840s.[23] Intakes were moved over time upstream of sewage outlets. These were welcome changes, but water companies were subject to incessant complaints about monopolistic practices.

By the 1840s waterworks acts generally had a scale of maximum charges related to rates, and a maximum level of dividends. The Waterworks Clauses Act 1847 imposed a duty on companies to supply water on demand at constant pressure, with local boards of health able to initiate action to remedy deficiencies. Some improvements preceded government regulation, which in hindsight was sometimes contrary to improving public health. For example, the belief of key figures at this point, such as Edwin Chadwick, William Farr, and John Simon, that fever was caused by a miasmatic atmosphere rather than by germs transmissible in particular ways, led to the consequences of untreated sewage being flushed into the Thames, the source of so much drinking water. Had John Snow's theory of cholera transmission been accepted earlier, things might have been different.

SANITATION

There were undoubted advances throughout the century in reducing nuisances, but they were painfully slow. The emphasis of public health policy was on the provision of clean water and the removal and treatment of waste, exemplified by the massive and expensive construction of the Victorian water and sewage systems. It was the ultimate technological fix for a host of social problems.[24] How did it happen?[25]

The driving force in mid-century was Edwin Chadwick. Like many of the reformers, he was influenced by Jeremy Bentham's ideas for a managed system of government, and his utilitarian argument that public policy should aim at maximizing the happiness of the most people.[26]

Chadwick's time as secretary of the Poor Law Commission was not a happy one, as his single-minded drive to eliminate relief to the able-bodied poor outside the workhouse led to massive public resistance. That was exacerbated from 1837 by the severe financial depression that spread across the northern manufacturing districts. It was an era of riots, strikes, and the rise of Chartism.

The Whig politician Robert Slaney pressed for action in 1840, and established a select committee on the health of towns, "with a view to improved sanitary regulations for their benefit."[27] This was an issue that affected all classes, since although pauperism itself was confined to the poor, there was concern that disease among the poor could easily spread to more prosperous areas.

Nor was it just a matter of disease. Physical degradation, it was argued, led to discontent, drunkenness, crime, and moral degeneration. Social stability require that it be combated, in which sanitation could play a role. Following his previous committee in 1838 on the education of the working class, Slaney aimed to put selected expert opinion on the record.[28] The witnesses started with Thomas Southwood Smith and Neil Arnott, who made reference to their recent report on the sanitary state of the laboring classes, work that had been commissioned by Chadwick. They were followed by many local medical officers and sewage representatives, surveyors, an architect, and the builder Thomas Cubitt. Much statistical information was provided on the living conditions of the poor.

The medical men such as Arnott claimed that fever and ill health arose from foul conditions and rotting matter. Southwood Smith declared that it was "not possible for any language to convey to the mind an adequate conception of the filthy and poisonous condition in which large portions of all these districts constantly remain."[29] Chadwick put this in its most extreme form in 1846, when he stated to a parliamentary committee that "all smell is, if it be intense, immediate acute disease, and eventually we may say that, by depressing the system and rendering it susceptible to the action of other causes, all smell is disease."[30]

The committee concluded that "evils of a most extensive and afflicting nature are found to prevail, affecting the health and comfort of vast bodies of their fellow-subjects, and which might be removed or much lessened by due sanatory [sic] regulations."[31] The members pressed the need for legislative intervention. They recommended general acts to facilitate proper building, to provide effective sewerage, and for other improvements. Local boards of health, with inspectors to enforce sanitary regulations, would report to a central board of health or to the home secretary. Already, political concerns around social unrest were being seen as a medical problem, soluble through interventionist legislation, infrastructure, building, and engineering.

Chadwick was not happy with aspects of this report. Rather than Slaney's emphasis on a general building act to regulate the quality of housing, his focus was already on water and sewerage.[32] In his view, concentrating on house sanitation alone ran the risk of simply increasing filth nearby if it could not be swept away by an effective water and sewage system. By this time Chadwick was sidelined in his official role as secretary to the Poor Law Commission. But after a motion in the House of Lords, the Poor Law Commission was tasked with a full inquiry into the sanitary conditions of the working classes, and Chadwick set to work. Following a typhus epidemic in London in 1838, in 1839 he had asked his medical experts—Southwood Smith, Arnott, and James Kay—to undertake local surveys on causes of fever, as its incidence was increasing Poor Law claims. Their reports, and survey data from questionnaires to

Poor Law medical officers in London, led to his monumental *Report on the Sanitary Condition of the Labouring Population of Great Britain* (1842).

These three men all became significant figures. Southwood Smith was physician to the London Fever Hospital, a gentle man to whom Bentham had left his body for dissection.[33] Arnott, also a friend of Bentham and of John Stuart Mill, had written a major book on physics and was recognized as an expert on ventilation. Kay, later James Kay Shuttleworth, a Manchester physician, would make a name for himself as secretary to the Committee of Council on Education. The collection of studies by Southwood Smith, Arnott, and Kay identified "filth" as the major cause for concern. They recommended better ventilation of houses, the removal of refuse that caused noxious atmospheres, regular water supply, and better drainage and sewerage.[34] What none of them highlighted were factors seen as predisposing causes by much medical opinion, such as lack of clothing, heat, and food, indeed poverty itself.[35] It was the so-called exciting causes, those that seemed to be causal and that came to be associated with "filth," that took center stage.

Chadwick and his medical advisers held a theory of disease that put it down to a foul atmosphere created by decomposing waste, a miasmatist theory. With regard to fever, Arnott saw the diseases of various types as a poisoning, a malaria from decaying filth. Southwood Smith was of the view that the atmosphere caused plague. They distinguished between contagious and epidemic diseases, with the former involving a "specific animal poison" that would affect anyone exposed.[36] For those, predisposing causes such as foul living conditions were not important. Epidemic diseases, by contrast, resulted from "a condition of the air," from emanations of decay, and for which practices like the removal of wastes and better ventilation could inhibit transmission. People exposed must succumb, but how quickly would depend on their condition of life.

Indeed, the supposed mechanism of disease was not a major issue in deciding sanitary approaches. Contagionist and miasmatist explanations did not necessarily lead in different directions, and everyone could agree that removing filth would tend to improve health. But it was argued that there was a need to improve the physical environment for all, or disease would spread to all, including the higher classes. Poverty was incidental to the policy. What mattered were the physical conditions that exposed people to the tainted air. The arguments were backed up by dubious statistics.[37]

Their recommendations therefore focused on the physical causes of fever and on "proper sanatory measures."[38] In effect this marginalized discussion of the broader reasons for infant mortality and ignored endemic diseases such as tuberculosis, which had less immediate political salience. This approach disregarded the occupational and economic causes of disease, and of destitution.

That view was challenged. The Scottish expert in social medicine, William Alison, raised the wider issues at the British Association meeting in Glasgow in 1840, and had debated the previous year with the Reverend Thomas Chalmers on whether food or moral discipline was the road to Scotland's salvation. His criticism of the Poor Law policy was that it drove people to starvation to avoid the taint of the workhouse, leading to disease and death. Chadwick ignored Alison's evidence and selectively quoted medical and statistical evidence in his report.[39] Whereas for Alison poverty led to disease, for Chadwick disease led to poverty. Cesspools, piles of filth, and stagnant drains became the problem, not destitution. This view prevailed despite the fact that Farr's first annual report had attributed some disease and death to starvation.[40]

Chadwick had managed to survive the fall of the Whig government in 1841, and his sanitary report arrived under Peel's new Tory government. James Graham was at the Home Office, worried about revolution, as 1842 saw the Plug Plot Riots and strikes spreading across the country. The sanitarian approach was now ascendant. Chadwick relied on what he saw as the science, allied to efficient administration and enforcement. So, he recommended a reliable water supply, drainage, removal of wastes, recycling of sewage, and local authorities with boundaries coincident with drainage basins.

Informing these practices were the work of people like John Roe, a sewer engineer who had designed an integrated system with smaller egg-shaped tile pipes for unimpeded flow, and James Smith of Deanston, who advocated flushing liquid sewage directly to fields as a fertilizer (see chapter 5).[41] It was imagined that sale of this fertilizer would pay for the sanitary works. This was an approach that helped uphold the political status quo. Public health became an environmental policy, based on science and technology, because the political alternatives were too difficult.

The *Sanitary Report* can be seen as an ideological manifesto rather than an empirical survey of conditions affecting health.[42] It offered a way of averting revolution and combating Chartism by making minimal changes that were acceptable to established interests, which included those of the scientific experts. It was a means of combating moral degeneration through improving the environment. It was thought that improved conditions would lead to better awareness of the need not to overreproduce, so that people would produce offspring more responsibly and raise their wages in the process, vindicating the factory system.

Chadwick was unimpressed by the current state of medical knowledge and the inability of medical men to cure so much disease. He argued that prevention was the preferable approach. Whatever the precise causes of disease, removing the filth and decaying matter from houses and streets would surely reduce the problem, thereby decreasing the burden on taxation and increasing the availability of people to work. Chadwick's experience was that epidemics resulted in an increased charge on the poor rates, a particular problem when the rate could not be legally applied to removing the nuisances. He was turning the challenge of improving public health into one of engineering and efficient administration, not one of medicine, and certainly not one of the nature of employment, the level of wages or the ability to buy food. Such "scientific" advice found ready political purchase.

ROYAL COMMISSION ON THE STATE OF
LARGE TOWNS AND POPULOUS DISTRICTS

Chadwick's *Sanitary Report* was a major stimulus to the establishment, in 1843, of the Royal Commission on the State of Large Towns and Populous Districts. Its remit was to inquire into the "Causes of Disease" and the "best means of Promoting and Securing the Public Health," especially with respect to the "Poorer Classes."[43] Its terms of reference and recommendations, shaped by Chadwick under the eye of Graham, became the basis for legislation in public health throughout the 1840s and 1850s, and a subsequent commission in 1869 would reaffirm its findings.

Chadwick now sought to line up people with engineering and scientific expertise to sit on the commission and give evidence. The commission was packed with them. The political commissioners included the fifth Duke of Buccleuch, the chairman and lord president of the Privy Council, and the Earl of Lincoln, both Tories. Lord Lincoln was first commissioner of Woods and Forests between 1841 and 1846, and was active in seeking scientific advice from Henry De La Beche and colleagues at the Museum of Economic Geology, which was accountable to his department. To them were added George Graham, the registrar general, and the Whig Robert Slaney. Four were medical men, at least by education. They were David Reid, who was working on the ventilation of the Houses of Parliament;[44] the surgeon James Martin; Richard Owen; and Lyon Playfair, still only twenty-five years old and personally invited by Peel.[45] Owen, an anatomist and palaeontologist, is best known for coining the term *dinosaur*, but he was thoroughly committed to sanitation. He wrote to Chadwick, "I would rather achieve the effectual trapping of the sewer-vents of London than resuscitate graphically in Natural History records the strangest of the old monsters which it has pleased God to blot out of his Creation."[46] The engineers included Robert Stephenson and Smith of Deanston, and De La Beche was also appointed. Chadwick had nominated Owen,

Stephenson, Smith of Deanston, and De La Beche.[47] His further suggestions of Michael Faraday, George Airy, and the physician James Clark were not taken up by Graham.

The first report in 1844 was based on evidence from architects, engineers, sewers commission officials, and reports from several medical men, statisticians, and the water engineer Thomas Hawksley, who had built a much admired high-pressure water supply system in Nottingham.[48] The second report in 1845 laid out the results of visits by commissioners to many parts of the country to report on conditions firsthand.[49] Playfair, in his report on Lancashire, in which he was assisted by Robert Angus Smith, developed a mathematical calculation of the savings supposedly to be achieved by a sanitary approach due to reduction in death and disease.[50]

Few of the commissioners had any direct professional experience of the matters under examination. Stephenson took no part in later deliberations, but he and the others largely supported or at least acquiesced to Chadwick's plans. The evidence of Southwood Smith, also a commissioner on the employment of children in mines and manufactories, was key. For him, a filth poison emitted from the decaying waste caused fever and, indirectly, most other diseases and social problems. Removing the filth was the best way to solve the challenges. Arnott and Reid argued for better ventilation, and Reid for better physical conditions of food and warmth, but the focus on filth won through. The physician and medical statistician William Guy gave forensic evidence on the incidence of tuberculosis and of suggested remedies for this pervasive disease, emphasising better ventilation, but it seems to have barely registered with the commissioners.[51]

The young and enthusiastic Playfair, protégé of Peel, was keen to make his mark. He had based himself in London from 1845, with a post at the Museum of Economic Geology in Craig's Court, before moving on to the new School of Mines in Jermyn Street with De La Beche. Here he was also drawn into sanitation matters, inspecting Buckingham Palace and Eton College. His examination of Buckingham Palace revealed such horrors that the government never disclosed the results to Parliament.[52] Playfair's extensive report on Lancashire betrayed his view of the working class as both victims of their conditions and wilfully bad in their behavior. The campaign for sanitation was, in part, one to improve the morals of the lower orders through technology. It was intended both to improve lives and to reduce class conflict.

The commission's first report laid out the groundwork of evidence, analysis, and recommendations that would be made concrete in the second.[53] The proposal was a regular water supply at high pressure, linked to an integrated sewerage system that could remove filth speedily and continuously, recycling it profitably for agricultural use, allied to a means of financing major improvements over the long term. It was clear that the only remedies to health it would consider were structural. The emphasis was on engineering, law and administration, not medicine.

The second report set out a seventy-five-page plan for practical and legislative action, which could be traced back to the *Sanitary Report*.[54] Standing firmly on the science, as he saw it, Chadwick had told Graham that there could be no objections to the principles of the *Sanitary Report* as they derived from the "law of gravity."[55] This second report did not take a view on whether destitution was a cause or consequence of fever, but argued that one could endorse sanitary reform without needing to discuss the influence of poverty and distress on mortality and fever. It contained more extensive and robust statistical evidence than previous reports.[56] The suggested framework for legislation that emerged in the report was based on coordinated municipal engineering, locally managed to national standards, with a local medical officer of health to inspect and report on sanitary conditions, disease, and mortality.

Chadwick had marshalled the engineers, medical men, and scientific experts to achieve the outcome he wanted. He, De La Beche, and Playfair reiterated many of these conclusions when they were called before a select committee on private bills, chaired by Joseph Hume in 1846. This committee pointed to the inconsistencies, expense, and poor quality of legislation under the current system. It recommended greater use of public general acts, the use of inspectors who could provide independent evidence for committees on private bills, and consolidating legislation.[57]

The select committee's report reveals the complexity not only of the diversity of local authorities responsible for different matters—such as lighting, paving, water, sewage, and cemeteries—but also the challenge in such circumstances of putting any scientific and technical advice into effective and consistent practice. Throughout the century, the pace with which expert advice could be assimilated and employed was inextricably intertwined with the complexity of local government arrangements, and of the degree of influence or control that could be exerted on them by central government when it tried to do so.

THE GENERAL BOARD OF HEALTH

While the report of the commissioners was being finalized, in December 1844, the voluntary Health of Towns Association was formed as a cross-party initiative to lobby for change, with the first Marquess of Normanby and Southwood Smith as leading individuals. Lord Normanby, as home secretary in the soon to be outgoing Whig ministry, had unsuccessfully introduced a public health bill focusing on sanitary improvement in 1841. The secretary was Henry Austin, a young engineer. He was the brother-in-law of Charles Dickens, and helped

win the influential Dickens over to supporting Chadwick, whose role in Poor
Law administration Dickens had roundly criticized.[58] Many others gathered
under this banner, including Viscount Morpeth, Lord Ashley, and Simon. So,
public health became focused on sanitary improvement, which meant technologies to remove decomposing matter and hence disease. It was left at this
stage to independent people such as John Snow and William Budd, along with
Farr, to investigate the aetiology and epidemiology of particular diseases (see
chapter 12).

Lord Lincoln, Peel's commissioner of Woods and Forests, introduced a far-
reaching bill, but it failed when Peel's government fell in 1846. The task now
devolved on the Whigs and therefore on Viscount Morpeth, the incoming
commissioner of Woods and Forests.[59] It was Viscount Morpeth, a member of
the Health of Towns Association and an advocate of sanitary measures, who
introduced public health bills in 1847 and 1848, with cholera advancing in the
direction of Britain. In 1846 the private bill procedure had been changed to
require expert assessment of proposed projects, and the government passed
the first Nuisances Removal and Diseases Act 1846, under the threat of cholera, which enabled local authorities to take action on the basis of statements
from two qualified medical practitioners. For the first time, there was a statutory approach to public health hazards, rather than relying on the courts and
existing nuisance laws, although many towns had developed bylaws on the
basis of public health, preceding this national legislation.[60]

Experts were becoming integrated into the system, and public health issues
into nuisance laws. That process continued with the Public Health Act 1848.
This act gave towns powers to make sanitary improvements and established
the General Board of Health to support and supervise them. It was a compromise between local democracy and central control, in a similar manner to
the Poor Law Amendment Act 1834, although the metropolis of London was
omitted from the act due to political opposition. Unlike the new Poor Law
Board, which was run by a minister responsible to Parliament, the General
Board of Health was not directly responsible to Parliament. That emphasized
its need to work by persuasion.[61]

Having survived the fall of the Whigs in 1841, Chadwick also survived
the fall of the Tories in 1846. Indeed, he found himself in a strong position.
The Poor Law Commission, from which he was anyway semidetached, was
abolished in 1847 in a move catalyzed by the Andover workhouse scandal. It
became the Poor Law Board, with new members, and was made more accountable to Parliament than the previous commission. In the process, Home Secretary George Grey, at the instigation of Prime Minister John Russell, promised
Chadwick a place on a commission arising from a health of towns act.[62]

This became the General Board of Health, following the passing of the

Public Health Act 1848.[63] The board was chaired by Viscount Morpeth, with Chadwick as the prime mover and Lord Ashley as the third member. It was Viscount Morpeth, the emollient politician, who smoothed the way for the abrasive Chadwick where he could. Viscount Morpeth became the seventh Earl of Carlisle in 1848 and would remain in post until 1850. Southwood Smith soon joined them, initially in response to the emerging cholera outbreak.

The General Board of Health could set up local elected boards of health, with consolidated powers over water supply, drainage, sewerage and sewage treatment, roads, and new construction. These had to appoint a surveyor and could employ a medical officer. The new boards could take on long-term debt to finance infrastructure, based on an inspector's report. The inspectors were all engineers. Smith of Deanston became one of the first inspectors under the act, joined, among others, by Dickens's brother Alfred and Charles Babbage's son Benjamin. Their role was to cajole the local boards into action, by setting out their observations of sanitary horrors and their statistics and discussing them with the local authorities. They could also undertake private commissions for towns to act on their advice. While this was happening, Chadwick's focus had turned to the sewers.

SEWERS

Before industrialization, sewers were constructed to carry off surface water. Cesspits accommodated human waste, which would then be carted off as "night soil." But from the early 1800s sewers had to deal increasingly with industrial wastes and sewage. London, the rapidly growing metropolis, became a place of contention as the several localized Commissions of Sewers tried to tackle the emerging problems.

By 1834, nearly twenty years after households had been permitted to connect their waste drains to the main sewers, public complaints were sufficient to instigate a select committee on metropolis sewers.[64] The committee was largely satisfied with the service provided by the commissions. The focus in their report was on administrative matters, including the ability of commissions to raise money, and powers to improve sewers and to require the connection of new houses. They urged more coordination between the commissions, but without recommending an overarching board for the metropolis.

Given the state of scientific and medical evidence about the connection between sewage and disease, the committee was unable to make any suggestions. Several medical men attested to the injurious effects of sewage on public health, but without being able to be specific. One witness, Peter Fuller, backed up by the Reverend William Walker, a mathematician and chaplain of St George's Hospital, suggested a solution to the problem of foul air, which was thought to be dangerous to health. As the committee reported, "It has been proposed by medical men of considerable eminence to purify the air in the Main Sewers by building Furnaces, at intervals, along their course, and by closing some of the Gullyholes, and providing others with Traps."[65] Fuller had consulted Faraday and Brande, and the committee then asked Faraday to appear before them. He gave skeptical but qualified approval to the plan, insisting that it would require experiments. The committee did not recommend any action and the idea was given up.

Although there were inevitably problems with London's sewer system at such a period of change, it was Chadwick who turned up the heat on them. He saw the commissions as a barrier to the introduction of his integrated London-wide system of water supply and sewage management. His trenchant criticisms, often exaggerated and untrue, came to a head in the late 1840s.[66] Chadwick had lost his place on the abolition of the Poor Law Commission, but Prime Minister Russell put him in charge of the Metropolitan Sanitary Commission, to recommend actions.[67] That resulted in the creation of a single Metropolitan Commission of Sewers, which had responsibility for the whole of London apart from the City. It was riven by dissent, both by conflicting personalities and by competing engineering solutions for the drainage of London, resulting in Chadwick's removal as a commissioner. The Metropolitan Commission of Sewers, which had become independent of the Board of Health after Chadwick's removal, continued with limited success and its functions were absorbed into the new Metropolitan Board of Works in 1855.[68]

DEMISE OF THE BOARD OF HEALTH

Chadwick's political support faded in the early 1850s, after the departure of the Earl of Carlisle in 1850, followed by Lord Ashley moving to the House of Lords in 1851 on becoming the Earl of Shaftesbury. Chadwick's vision of integrated works of constant high-pressure water supply using small bore tubular earthenware pipes, with flushed sewage removal and recycling of sewage as fertilizer, was based on a scientific and theoretical approach. The experienced engineers responded to their city and town clients with more pragmatic and nondoctrinaire solutions, based on their engineering experience, and often favoring big brick-built sewers. Chadwick's waning influence, and his growing antagonism toward the most influential engineers such as Thomas Hawksley and Joseph Bazalgette, who disagreed with his approaches to engineering sewers, led to his own removal from the General Board of Health in 1854.

In the event, scientific theory and the pursuit of administrative power in the hands of Chadwick met engineering expertise, allied to the power of the water companies and the vestries, and lost. Chadwick tried to have Playfair appointed to the board, but the job of president fell to Benjamin Hall, a thorough opponent of Chadwick, whose defeat was complete.[69] He had nevertheless left a substantial legacy. Between 1848 and 1854 Chadwick and his engineering inspectors had brought more than two million people under the Public Health Act.

The General Board of Health was formally abolished in 1858, along with the local boards of health, which became local boards for government and infrastructure administered by the Local Government Act Office. The Privy Council Medical Department was charged with overseeing health and disease. In reality, the Local Government Act 1858, brought in by Conservative home secretary Charles Adderley, proved far less decentralizing than his rhetoric suggested.[70] Over its thirteen years of existence it confirmed more than 1,600 separate local schemes, mostly for water supply and drainage. The engineers were now prohibited from private practice, and therefore appeared more impartial and authoritative than in Chadwick's time.[71] They had considerable delegated authority since ministers, while taking formal responsibility under the mid-Victorian doctrine of "ministerial responsibility," could not be involved in all responses given the huge increase in demands from local government.[72]

The central power was increased after the passing of the Sanitary Act 1866 (see chapter 12). In practice this was welcomed by the smaller local authorities, which lacked their own expertise. Technical authority in government was substantial, but was called on here from outside the center. That contrasts with the approach of Simon in the Medical Department of the Privy Council, who took initiatives himself. In 1871 the Local Government Board was created to oversee the separate functions of poor relief, health, and municipal engineering.

The development of the governance and administration of the water and sewage systems in the metropolis and across the country illustrates the ultimate importance of political over scientific principles. Granted that some intervention was necessary to secure the public good, how was this to be achieved? John Stuart Mill's views on the London water supply are instructive. His position was that powers of central government should be checked by local institutions, even at the cost of some inefficiency, as a defense against excessive interference in personal liberty. So, while the monopolistic practices of water companies should be curbed, the administration of any regulation should be in the hands of a suitable local authority, supported by central information and advice.[73] That had not been Chadwick's view, and it is this tension that runs through central government intervention and inspection throughout the century. In this case, because there was no London-wide democratic body, Mill supported giving powers to a central board. He would later introduce the first bill to create an elected London County Council, which was not realized until 1889, coincident with the abolition of the Metropolitan Board of Works.

RIVER POLLUTION AND SEWAGE

The disposal of sewage, and broader questions of water pollution, now took center stage.[74] The political demise of Chadwick had left Hall in charge at the General Board of Health, and in July 1855 he moved to become first commissioner of Works in Viscount Palmerston's government. In this role he saw through the Better Local Management of the Metropolis Management Act 1855, known as Hall's act, which established the Metropolitan Board of Works. It was the first London-wide body, though it would still have to deal with the smaller units of local government.[75] The board was concerned to ensure that waste was separated from water flowing into the Thames. It proposed to build intercepting sewers running for five miles below London to deposit the waste further down the river. The Great Stink of 1858, when Parliament itself was almost uninhabitable, spurred Benjamin Disraeli to ask the board to act immediately. This lengthy development, under the board's chief engineer, Bazalgette, opened officially in 1865 but took until 1875 to complete. Not until the late 1880s was significant sewage pollution of the Thames ended.[76]

Questions of supply were inevitably connected to those of pollution, and discussions about what substances might be injurious to health or even cause particular diseases. By the mid-1850s, dangerous disease-causing agents were increasingly believed to be specific organic ferments that could enter another body where they would cause further putrefaction and decay. Rather than a specific germ theory, this was a fermentation, or zymotic, theory derived from Justus Liebig, which had considerable influence on British medical thought.[77] It led to the idea that it was the process of decomposition that caused disease, like an internal rot that was caused by an external rot. This took the emphasis away from specific chemical pollutants, even if some of those might also be injurious to health.[78]

Wilhelm Hofmann, a pupil of Liebig and now professor of chemistry at the Metropolitan School of Science and chemist to the Museum of Practical Geology, was a key advocate of this theory, and he had the ear of government. He made a report on the metropolis water supply in 1856 with his student Lyndsay Blyth, lecturer on natural philosophy at St Mary's Hospital.[79] Hofmann and Blyth suggested that the organic nitrogen content would be a good indicator of the potential danger from disease-causing ferments.[80] They were not able to develop this approach, which had to await Edward Frankland in 1867, but it implied that water could not be shown to be definitively safe or harmful, merely more or less potentially dangerous.

In their complementary report, the engineers William Ranger, Austin, and Alfred Dickens stated that "it is a most difficult thing to discover, by any chemical process hitherto adopted, the precise amount of these poisonous ingredients which may produce such injury to the system; while, on the other hand, it is impossible to say how small a quantity of such matter might not, under certain circumstances, be injurious."[81] The burden of almost impossible proof was therefore starting to fall on the water companies to show that their water was safe.

The extent to which pollution of the Thames exercised Parliament is illustrated by the select committee which sat in the summer of 1858, just as the Great Stink arrived. It was brought about in response to Goldsworthy Gurney's plan for purifying the Thames.[82] Gurney was well-known to parliamentarians, as he had been engaged to develop the heating and lighting of the new Houses of Parliament.[83] His scheme involved discharging all the sewage destined for the Thames directly into the river below the low water mark, so that the bad smells, which he thought were created by exposure to air and heat, would be prevented. That would then require that gases building up in the sewers across London be burnt in situ. Witnesses such as Bazalgette were not impressed, and nor was the committee. The members were skeptical that sewage mixed with water would no longer smell. They also thought that the burning of sewer gases by numerous furnaces was impractical, although the concept was supported by chemists William Maughan and Lewis Thompson. John Lawes offered suggestions for deodorizing, recycling, and purifying the solid sewage. The committee wanted the sewage either taken well down the river beyond the metropolis before it was discharged, or deodorized and only the "purified liquid part of it" discharged into the river. The report noted the suggestions of several engineers, including Hawksley, George Bidder, and James Walker, that embanking the Thames would increase the scour of the stream and prevent accumulation of mud.[84] Embankment would follow, but for other reasons.[85]

FIRST ROYAL COMMISSION ON THE BEST MEANS OF PREVENTING THE POLLUTION OF RIVERS, 1865–1868

By the mid-1860s, pollution of the Thames and other rivers was still a major political issue.[86] Lord Robert Montagu, who had chaired the select committee in 1864 on utilizing sewage from the metropolis (see chapter 5) brought forth a River Waters Protection bill in 1865. He withdrew it on the second reading, when Home Secretary George Grey and others objected that it was too far-reaching and would damage industry, and on the understanding that the Sewage Utilization Bill would pass.[87] The government then appointed three commissioners to inquire into river pollution, the first time the word *pollution* is used in the title of an official committee or commission report. It was not used in the title of a local act until 1867 or of a general act until the Rivers Pollution Protection Act 1876. Brought forth under Disraeli's government, this was the first consolidating measure to counter river pollution, though of limited effectiveness. This royal commission had a rocky and lengthy life in two phases.

In the first phase, up to 1868, the three commissioners were the engineer Robert Rawlinson,[88] inspector in the Local Government Act Office; the agricultural chemist John Way, once a student of Thomas Graham; and the agriculturalist John Harrison. They were given a large brief: to inquire how far the use of rivers in England and Wales for carrying off the sewage of towns, and the refuse arising from industrial processes, could be reduced without risk to public health or serious injury to industry; how far such sewage and refuse could be utilized or rendered harmless before reaching rivers; and to examine the effect on the drainage of lands and towns of obstructions to the natural flow of rivers caused by mills, weirs, locks, and other navigation works. They got as far as investigating the river basins of the Thames, Lea, Aire, and Calder before disagreements between them led to the commission being revoked and reformed. Their conclusions, insofar as they reached them, included creating a Conservancy Board for each watershed area, with penalties for allowing sewage or other injurious industrial refuse from entering rivers.[89] The report on the Thames resulted in an act in 1868 that strengthened the powers of the Conservancy Board of the Thames to prevent sewage and other injurious matter from entering the river and allowed it to fine offenders.[90]

By now, the focus in terms of public health was on organic matter in the rivers, particularly associated with sewage, not least given the arrival of cholera in London again in 1866. With the Thames passing through Oxford, the commissioners took the opportunity to interview several of Oxford's most eminent men of science. Henry Acland, Regius professor of medicine, said that no one knew when water once poisoned became safe again, and argued that "therefore sewage should be kept out of rivers altogether."[91] Benjamin Brodie, professor of chemistry, maintained that "the poisonous quality does not depend upon the absolute quantity of the organic matter. We have no measure of the poisonous influences of water except upon human health and life."[92] In other words, the chemists could not detect it. Charles Daubeny, professor of botany at Oxford University, agreed that chemists could not determine the quantity of organic matter that might be harmful, although he thought filtration might help make the water safer.

The commissioners agreed with arguments like these in recommending the prevention of any sewage reaching the river system, impractical though that might be. Rawlinson and Way reiterated this view in their evidence on the Lea Conservancy Bill in 1868, when they disagreed with the chemical analyst Alfred Swaine Taylor. He continued to argue on the basis of inorganic analysis that the water was safe, and denied "the existence of what cannot be discovered."[93] The policy question at issue was whether or not one should take a precautionary approach. Despite the apparent difficulties of concluding anything useful to policy, water analysis remained important.

SECOND ROYAL COMMISSION ON THE BEST MEANS OF PREVENTING THE POLLUTION OF RIVERS, 1868–1874

The new commission was again composed of three people: William Denison, previously a royal engineer; John Morton, an agriculturalist and journalist; and Frankland. Frankland had reached a position of notable influence and received a salary of £800 for this commission, exemplified by his concurrent involvement with the Royal Commission on Water Supply, which examined and quoted his evidence in particular detail. That influence was increased when Denison died in 1870 and was not replaced. It is possible that Frankland owed his position on the commission to Treasury savings to be achieved by using him as analyst for both.[94] Frankland was the prime mover as commissioner, and central to questions of the quality of water supply for the next two decades.

Like Simon and Farr, and Chadwick earlier, Frankland used his position of influence with government to advocate for specific policies. Developing a belief that tiny particles, whether "germs" or putrefying matter, were the agents of disease but undetectable and probably not removed by filtration, he devised an analytical approach that sought to determine the extent of prior sewage contamination in water supplies. Frankland had recognized during the cholera outbreak of 1866 that he could not detect the "choleraic poison" by chemical means, nor show whether purification techniques were effective.[95] Recognizing that chemical analysis could not determine unequivocally whether or not water was safe, he sought ways of taking into account its history of likely contamination.

Frankland extended his work in the six reports of the new commission. These covered the Rivers Mersey and Ribble, a patent process for sewage treatment, pollution by woollen manufacturers, Scottish rivers, pollution by mining and metal manufacturing, and domestic water supply.[96] Over several years, Frankland gathered a comprehensive and systematic bank of evidence through surveys, interviews, and the work of his analytical assistants. He used this to inform his advocacy for standards for levels of substances in effluents and of processes to recycle industrial wastes. The reports were influential in establishing effluent standards, even if those were not defined in law.[97] Frankland was not afraid to present his results in a form that promoted public discussion and anxiety about the level of contamination of their water. While Charles Tidy, Henry Letheby, and William Odling sought to reassure the public, Frankland was convinced that the public needed protection and that water supplies could be engineered to remove the dangers. These experts, and others, battled out their positions in front of the Royal Commission on Water Supply from 1867 to 1869.[98]

In reaching this position of influence, Frankland managed to see off challenges from various quarters about the reliability and meaning of his analyses. Convincing in public debate, holding the prestigious position of professor of chemistry at the Royal School of Mines, and a member of the nine-strong and well-connected X Club, the powerful inner group of metropolitan men of science, he was well placed to repel opposition.[99] That opposition came particularly from James Wanklyn; William Crookes, editor of the *Chemical News*; Tidy; and Letheby. Tidy and Letheby were both consultants to London water companies. The case of the chemists was weakened in 1882 when the Medical Department of the Local Government Board declared that chemical analysis could not identify water that had caused typhoid.[100] Judgements of potability had to be made in future on the basis of a wider range of factors.

The detection of organic matter in water became the battleground for disputes both between the analysts and in front of commissions and committees. In 1856 Hofmann and Blyth had stressed the need to determine the proportion of nitrogen in water that was organic. This would indicate the presence of material that might cause disease, as William Budd thought, or indicate disease itself, as Arthur Hassall thought. They devised no analytical procedure. But by working with Henry Armstrong, Frankland devised a method of determining organic carbon and nitrogen content in 1868, by oxidation of the residue from an evaporated sample of water. The ratios were used to indicate prior contamination.[101] At the same time, his former student Wanklyn developed a method of measuring nitrogen by oxidation of protein in the water. Although Frankland's technique was deemed better, many preferred Wanklyn's as it was simpler to carry out. So while Odling and Andrew Williamson among others preferred Frankland's method, others such as Letheby, Way, Angus Smith, William Allen Miller, and Augustus Voelcker preferred Wanklyn's.[102] Wanklyn attacked Frankland acrimoniously, but in the process managed to alienate many of his colleagues.

At this stage there was no agreed scientific understanding of infection and putrefaction, or of the nature of any underlying living material. The idea of specific reproducing "germs," promoted by Louis Pasteur, vied with zymotic theories of fermenting matter and with the possibility of spontaneous development of life. The belief that disease germs were living particles, advocated by Frankland, tended to put the existing supply in question. Against that, Letheby's view that poisons from nonliving putrefying matter were significant worked the other way, and he defended the companies' water.[103]

The core of Frankland's case was that the danger in water supplies was prior sewage contamination. Whereas those who believed in the zymotic theory of putrefaction reassured committees and the public that natural processes and sufficient dilution by water would render them harmless, Frankland was developing an understanding based on germ theories. On that basis, he argued, such living matter might be longer-lasting, and he believed that he could detect the possibility of prior contamination with this organic matter chemically. The only safe water was water that had not been polluted by sewage or manure.[104]

The consequence of taking this approach was a strong precautionary view. Proving a negative, the absence of potentially dangerous material, was all but impossible, especially since the actual nature of the dangerous material was unknown and would later turn out to be bacteriological. But Frankland's authority was substantial, and government listened. Tidy succeeded Letheby in 1876 and added to the opposition, with Crookes and Odling, to Frankland's conclusions.[105]

Frankland, who was a consultant to the General Register Office, a post he retained when the Local Government Board was created in 1871,[106] also had to deal with the new post of London water examiner, which had been established under the Metropolis Water Act 1871. The appointee was Colonel Francis Bolton, an engineer, but Frankland managed to outmaneuver him and continued publishing his reports for the registrar general, even when Bolton's position was transferred to the Local Government Board in 1875. Frankland's consultancy extended to the royal residences, when he analyzed the water supplies to Windsor, Sandringham, and Balmoral in 1874 and 1875.

RIVERS POLLUTION PROTECTION ACT 1876

One result of the royal commission was the Rivers Pollution Protection Act 1876, the first to attempt a general remedy of the problem and to extend it beyond the nuisance laws, which had proved unable to resolve the environmental problems.[107] Like so many other initial attempts at regulation it was weak and relatively ineffective, although George Sclater-Booth, president of the Local Government Board, stated that no loan for sewerage works was now being sanctioned unless provision was made for the purification of the sewage.[108]

Although the act prohibited the discharge of sewage and of manufacturing and mining pollution, there were loopholes. Existing sewage channels could be used if the "best practicable and available means" were deployed to render the sewage harmless. Likewise, the "best practicable and reasonably available means" were to be applied to mining and manufacturing emissions.[109] Proceedings against a polluter required the sanction of the Local Government Board, which had to have regard to the industrial interests that might be affected.

Nevertheless, local authorities such as the Cardiff Union Rural Sanitary Authority could use the framework of the act, short of formal prosecution, to force improvement in their own areas.[110] Angus Smith was appointed as the first inspector, adding to his role as alkali inspector. The act was strengthened in 1893,[111] before the question of sewage disposal was visited by another royal commission in 1912.

Once the problem of pollution had been admitted, the solution on a large scale depended on engineering. As the royal commission on river pollution was finishing its work, Home Secretary Richard Cross appointed John Hawkshaw to report on purifying the River Clyde. His report set out schemes for a large-scale sewage system, based on gravitational flow and irrigation on land. He rejected existing filtration systems because the solid would be difficult to dispose of, and dismissed the proposal by Crookes for chemical treatment, the ABC process of his Native Guano Company, for being financially unviable.[112] Hawkshaw was optimistic about the feasibility of processes to clean up industrial effluent on site to make it harmless.[113]

In parallel, the Local Government Board commissioned a review of methods for treating town sewage. There were three methods available, in the shape of sewage farms, land filtration, and precipitation by chemical processes. Rawlinson, the chief engineer, and Clare Read, MP, one of the board's secretaries, undertook the investigation. They visited sites around the country, listed almost five hundred patents concerned with sewage precipitation and chemical treatment, and noted reports such as Voelcker's on the value of sewage as a fertilizer. Their conclusions show the technical and scientific challenges remaining.[114] They found that no precipitation method did more than separate the solid, although they regarded that as an improvement, but also that no use of precipitated solid as a manure could cover its costs. Land irrigation they found impracticable in many cases, since it had to be a continuous process regardless of agricultural need. More effective technical solutions would await biological treatment and, not until 1915, chlorination of water supplies.

BIOLOGICAL SEWAGE TREATMENT

William Dibdin, who had given evidence to the Royal Commission on Metropolitan Sewage Discharge, and served from 1882 to 1897 as chief chemist to the governing bodies of greater London, was instrumental in developing biological treatment. He had started in the 1880s with chemical precipitation, using a combination of lime, alum, and iron sulphate, and deodorization by permanganate. This approach was reviewed by a committee consisting of Odling, Frederick Abel, August Dupré, and Alexander Williamson, who were not enthusiastic.[115] Although the scheme was put into effect, criticism led

Dibdin to explore the biological option, while the board asked Henry Roscoe to take over the chemical process.[116]

In 1889, when the London County Council replaced the Metropolitan Board of Works, Dibdin reclaimed responsibility for sewage treatment. Following a study of effluent filtration by the council's engineer Alexander Binnie and Benjamin Baker, Dibdin and Binnie experimented with filtration. A continuous flow of air, using Dupré's idea, was used to support the life processes of bacteria that were thought to purify the sewage. This started out as more of a political than a technical solution to the problem, by demonstrating that the authorities were taking action, even if the scientific evidence for its effectiveness was lacking.[117] There was no direct proof that microorganisms would purify the sewage sufficiently to make it harmless. Only later did the understanding emerge of why artificial filtration was more efficient than land filtration or other methods.[118] Dibdin resigned in 1897 to establish a successful sewage treatment consultancy with George Thudichum, with whom he had worked on the process.

A royal commission on sewage disposal reviewed these artificial methods from 1898, under pressure from local authorities who claimed they did not have enough land for sewage farming and wanted the government to issue loans for treatment by artificial means alone. The commissioners noted that at the time of earlier investigations, "the science of bacteriology was in its infancy." They commissioned both chemical and bacteriological investigations of sewage treatment, agreed that artificial methods had improved in effectiveness, and recommended that the Local Government Board should support them.[119] This it did following the report.

The outcome was a move to sewage treatment rather than recycling, and a failure to manage in an integrated manner urban wastes and rural land use. The early Victorians held to the ideal of recycling matter to its rightful and useful place in the environment. By the Edwardian period, when recycling sewage as fertilizer had not proved economic, that principle was no longer seen as important.[120]

INFECTION AND DISEASE

UNDERSTANDING OF DISEASE CHANGED ENORMOUSLY DURING THE NINE-teenth century. For much of the century there were overlapping theories, broadly contagionist or environmental, the latter including a belief in miasma, or foul atmospheres as causative of fevers, although there was much intersecting of categories.[1] Contagion implied a mode of transmission, and miasma some emanation dispersed in the atmosphere. Belief in contagion favored quarantine as a protection, which could generate political resistance on the grounds of affecting trade or personal freedoms. For example, Parliament established a select committee in 1819 that declared the plague contagious and reaffirmed quarantine arrangements.[2] The miasmatic approach regarded the "miasma" as a toxic substance present in the air of unhealthy places.[3]

A further important dimension was the concept of predisposing causes, such as social conditions, which might be sufficient in themselves to trigger disease. The philosophy of medicine emphasized maintaining health or balance, rather than treating diseases as caused by specific external agents. The causes of diseases were not clear, until the gradual acceptance of germ theories later in the century, nor were questions of what predisposed one person to succumb rather than another. So there was much debate over methods of countering disease and infection.[4]

THE ARRIVAL OF CHOLERA

The waves of cholera that arrived in Britain in the nineteenth century had a major impact on approaches to public health. People in Britain were used to living with endemic and epidemic diseases. Whether it was smallpox, tuberculosis, scarlet fever, typhoid, or "jail" fever (typhus), disease was a fact of life. What set cholera apart was the speed at which it could strike, and the violence and horror of its mode of death. It was as feared as the bubonic plague or yellow fever.

With cholera arriving in St. Petersburg, Russia, in June 1831, the Privy Council wrote to Henry Halford, president of the Royal College of Physicians, asking him to recommend physicians to serve on an advisory board, thus giving the government authoritative cover for any policies. The composition of the board was similar to one established in 1805 when yellow fever had struck Gibraltar,[5] with six fellows of the Royal College of Physicians and representatives from Customs, the navy, and army. The Privy Council requested advice on the modes of transmission of cholera, by humans or by materials, on means of combating it, and on rules for quarantine.

The board, taking a contagionist approach, advised that the disease was infectious and that quarantine for people or materials should be fourteen days, though there was no certainty about its mode of transmission. It also recommended that local boards of health should be established across the country to manage the identification and isolation of cases and to clean and ventilate places of infection. This was the first legal recognition of a health hazard, and required powers of coercion that were politically difficult at such a time of reform.[6] The use of chlorine gas as a disinfectant was explored, with the board taking evidence from medical men and from Michael Faraday, who later undertook experiments on the effect of treating smallpox vaccine with chlorine, finding that it appeared to inactivate it.

The dreaded cholera reached Britain in October 1831, starting in Sunderland, just as the new rules and regulations agreed by the Board of Health were being published. Many local boards of health were quickly established, soon amounting to more than a thousand.[7] But the central advisory board was abolished on the grounds that the government needed a dedicated full-time board to meet daily. This became the new Central Board of Health, of which William Pym, superintendent of quarantine at the Privy Council Office, was a key member.[8] The physicians were asked to form a voluntary advisory committee. The government thereby kept tight control, reduced its costs, and was perhaps disappointed when the advisory committee recommended that quarantine be maintained at fourteen days rather than the ten that the government was seeking on commercial grounds.

The new Central Board of Health appointed a central medical staff to deal with the local boards and to implement national sanitary regulations and nuisance abatement powers through local authorities. For the first time, medical practitioners were required to testify that the nuisance was injurious to health, although local powers to remove the nuisances were limited. That reflects the continuing reluctance of central government to interfere in matters concerned with private property.[9]

The legal powers were tightened up in the Cholera Act 1832, as the disease made a relentless advance, killing five thousand by mid-1832. The coercive approach to quarantine was tempered to one of persuasion, with more attention given to sanitary approaches such as fumigation and the cleansing of homes. But the local boards were never given explicit powers to remove nuisances, such as slum conditions, which might lead to cholera transmission. Nor was it clear who would pay. For example, there was no power to use the poor rate. Then, as the epidemic abated, the pressure reduced. The Central Board of Health was abolished at the end of 1832. Public health was not yet seen to require a permanent national organization. It was to be left to local administrations, until the threat of a further assault of cholera in the following decade.[10]

During this time, the balance of influential opinion in Britain turned away from contagion and quarantine and toward the idea of the origins of disease in morbid atmospheres or miasma, produced by filth and decomposing organic matter.[11] The public health solution was therefore the introduction of sanitary approaches, as described in the previous chapter, despite strong opposition from many in the medical community who held contagionist views.

THE MEDICAL COMMUNITY IN
THE EARLY NINETEENTH CENTURY

Law, the church, and the military were respected occupations by the start of the nineteenth century. Compared to these, except for the elite physicians and surgeons, medicine had lesser prestige. Most medical men were correspondingly less visible. They treated individuals, and occasionally advised public inquiries, but they did not lead or decide in public affairs in the way that was expected of lawyers and churchmen. Nor was the medical community a united one. It lacked political power. That was only attained in substance after the reforms to the profession in 1858, and it coincided with a change in emphasis in attributing disease to scientific rather than to social causes. At the same time, many in the profession lost sight of awareness of the social factors underlying disease, for which poverty stood for so many.

The views that medical men held changed from disease as imbalance between humans and the environment to disease as invasion, often by specific germs, although there was continuous tension between these perspectives.[12] Many related factors shaped the public and political debate about what to do. The medical evidence and knowledge about disease was disputed within the medical community, as was the significance of poverty and disadvantage. There was disagreement about which diseases were contagious, and in what ways, leading to conflicts over policies such as the imposition of quarantine. And belief in freedom of choice for the individual came up against demands for measures of compulsion for the greater societal good. The politicians, as ever, had to try to balance these tensions.

It should be borne in mind, while considering the status of the medical fraternity, that the position of the scientific practitioners was lower in some respects. Following a notorious court case, after the Severn and King Sugar Factory had burned down in 1819, the judge declared that the many chemists, who had given conflicting evidence on both sides, could not be described as professional in the manner of lawyers and doctors. They were therefore ineligible for their expenses to be paid by order of the court.[13] Early in his career, Faraday was one of the witnesses alongside many fellows of the Royal Society. Just as with evidence from medical men, it would be clear to members of parliamentary committees and to commissions that certainty could not necessarily be provided by scientific evidence, and that they would have to make judgements about who was authoritative and whom to trust.

At the beginning of the nineteenth century, there were three primary medical communities.[14] Social supremacy went to the physicians, who would attend rich private patients in their homes, brought together in the premier Royal College of Physicians of London and the Royal College of Physicians of Edinburgh. Surgeons, represented by the Royal College of Surgeons of London and the Royal College of Surgeons of Edinburgh, came next. The elite surgeons had comparable status to the physicians, and would also administer to the rich, but many surgeons were seen as craft practitioners, with their historic association with barbers. At the bottom came the apothecaries, increasingly becoming general practitioners to the poorer classes, collected together in the Society of Apothecaries. They in turn had a developing turf war with the chemists and druggists, in the trade of preparing drugs and medicines. These bodies had their own charters, and could license members to practice, although that power was limited. During the nineteenth century the rise of hospital medicine, especially practiced by the physicians and surgeons, diverged from general practitioners in the community, and all had to compete with quacks and alternative practitioners. Many local medical societies were formed in the early 1830s, as well as the national Provincial Medical and Surgical Association, which later became the British Medical Association. Utilitarian sentiments, and those such as the radical medical MP and campaigning editor of the *Lancet*, Thomas Wakley, promoted the idea of a medical profession based on expertise and service in the public good, as a turn away from the gentlemanly and allegedly corrupt patronage and lay voluntarism of the pre-reform era.[15]

There was some reorganization of the medical profession in the early years of the century. For example, the Apothecaries Act 1815 extended the power of the Society of Apothecaries to regulate members from London to the whole country, although at the cost of some subservience to the Royal College of Physicians.[16] In parallel, the standard of medical education for apothecaries, the emerging general practitioners, was raised. It included training in practical chemistry.[17] That left the medical chemists and druggists behind, last in line behind physicians, surgeons, and apothecaries, and they formed the Pharmaceutical Society of London in 1841 to protect their interests.[18] The Royal Colleges guarded their professional areas assiduously, successfully preventing the formation of a college of general practitioners in 1845.

Doctors, however, had a mixed reputation throughout the century, particularly among the lower classes. This is exemplified in the passage of the Anatomy Act 1832, following a scandal over grave robbing, which allowed the bodies of people who died in public institutions, if unclaimed by relatives, to be used for dissection and then buried. The act required anyone practicing anatomy to be licensed, and created posts of inspector of anatomy as a state responsibility, reporting to the Home Office.[19] Other posts established around this time—such as medical inspector of prisons, asylum visitor, factory act surgeon, and Poor Law doctor—signify the arrival of medical men in state functions. Against the backdrop of internecine squabbles between the different medical specialties, between provincial and metropolitan perspectives, and between contagionists and anti-contagionists, the influence of the medical community steadily grew.

GENERAL REGISTER OFFICE

One organization that became central to developing understanding of mortality and disease in the nineteenth century was the General Register Office, established in 1837 on the initiative of Lord John Russell. Its purpose was to monitor births, marriages, and deaths, reporting annually to Parliament and complementing the decennial census, which had been established in 1801.[20] Although it was a typical British compromise, with registration of births not compulsory following opposition in the House of Lords on religious grounds, it enabled much improved mortality analysis to be made. That work was undertaken in conjunction with the medical community, and included causes of mortality by occupation. Information from birth and marriage data was less comprehensive. Alterations were made in 1874 under Benjamin Disraeli's government, transferring the responsibility for registration of births from the registrar to the parents and effectively completing the registration system. But more substantial changes took place only in 1938, given concerns about fertility and the possibility of a declining population.[21]

Edwin Chadwick recommended Charles Babbage as the first registrar-general,[22] which would have been an intriguing appointment given his uncompromising character. Chadwick had urged the government to make the registration districts coincident with the Poor Law unions, and to ensure that registrars recorded the causes of death. In the event, the first registrar-general was Thomas Henry Lister, Lord John Russell's brother-in-law. He died in 1842 and was succeeded by Major George Graham, brother of Home Secretary James Graham. Graham served until 1879, working closely with the man most associated with the activities of the General Register Office, William Farr, who was the first superintendent of statistics.[23] Chadwick had more success here, as he had recommended Farr to Lister. Farr served in post even longer than Graham, resigning in 1880 after failing to be appointed to Graham's position. Lister was crucial to the appointment of Farr as a medical statistician. When he died, leaving the General Register Office in a disorganized state financially, Graham took determined charge and proved to be an able leader and administrator.[24]

Farr is a central figure in the development of the state's use of statistics, and he came to it from a medical perspective.[25] This approach, encouraged by Chadwick and others, drove the collection and interpretation of statistics, and shaped the focus of sanitary efforts to counter disease.[26] This medical emphasis was also evident in the way occupational statistics were designed and collected by Farr. A key component of the strategy was the publication of local death rates, to stimulate and inform local action.[27] Only later in the century was there pressure to develop occupational statistics to inform labor and employment policies, against a background of labor unrest and industrial decline.[28] Farr influenced debates on public health matters for decades. Born in 1803, the first of five children of a farm laborer, he studied medicine, supported by the patronage of a rich bachelor, became a licentiate of the Society of Apothecaries, and then took up medical journalism. He wrote articles on statistics, which doubtless influenced his appointment to the General Register Office in 1837.

Recognizing that prevention was better than cure, Farr used the annual reports from the General Register Office to raise issues in statistics and medicine, particularly in epidemiology, and to demonstrate the waste of human life caused by preventable conditions in British cities. In addition, he produced special reports—for example, on the cholera epidemics of 1848 and 1866. His classification of diseases for use in reporting deaths, first produced in 1839, was developed and applied for the rest of the century. Farr showed how environmental risks to health could be identified, and developed a theory of disease based on his statistical analysis. He used mathematical analysis to show the trajectories of epidemics of smallpox and cattle plague, and that mortality from cholera varied inversely with closeness of population. That confirmed his zymotic theory of disease, derived from the work of Justus Liebig, which saw the causes of epidemic, endemic, and infectious diseases in the products of organic decomposition and infection through breathing.[29] George Airy also took an interest in this work by collecting meteorological data, sent weekly to the registrar-general's office, to investigate possible correlations between weather patterns and the tables of epidemic mortalities.[30]

Farr gave evidence to many committees. Outside the arena of health they included the select committee on insurance associations from 1852 to 1853, and the select committee on income and property tax in 1861. But public health was a focus. Farr worked closely with Florence Nightingale on sanitary reform in hospitals and the army, including making a report for the army sanitary commission from 1857 to 1858.[31] Related roles included serving on the committee of army medical statistics in 1861, and as an actuary consultant for the Indian sanitary commission of 1859 to 1863. He became an influential figure in public health through studying occupational and geographical factors affecting mortality and morbidity.[32]

CHOLERA RETURNS

In the mid-1840s, a third cholera pandemic emerged out of India. With London excluded from the public health legislation then in train, Lord John Russell had appointed Chadwick to head up the Metropolitan Sanitary Commission (see chapter 11). The likely arrival of cholera spurred them on. Working closely with Thomas Southwood Smith and Richard Owen, Chadwick's commission swiftly produced two reports.[33] A third report, following an outbreak of fever at Westminster School, enabled Chadwick to put on record a case study that, in his view, backed up his theories.[34] Chadwick was a convinced anti-contagionist, going back to his experience of the previous cholera epidemic. At that time, in 1832, Neil Arnott, James Kay, and Southwood Smith had found filth to be the cause of fever in the East End. Chadwick had used those findings in his influential *Report on the Sanitary Condition of the Labouring Population* (1842).

When cholera was reported in Britain in 1846, the government passed the temporary Nuisances Removal and Diseases Act 1846, which allowed the removal of nuisances prejudicial to health, as determined by two medical practitioners. At the same time, Chadwick urged action on the basis of the reports of the Metropolitan Sanitary Commission. Given its terms of reference, shaped by Chadwick himself, its conclusions were hardly surprising. The commission had been instructed to examine "better House, Street, and Land Drainage, Street Cleansing, and Paving; the collection and removal of Soil and refuse, and the better supply of Water, for domestic use, for flushing Sewers and Drains, and cleansing Streets."[35]

The first report dealt with evidence given by medical men from London on the causes of cholera, and statistical data from Farr at the General Register Office. Farr's subsequent *Report on the Mortality of Cholera in England, 1848–49* was aimed at discovering the statistical laws underlying the disease. It was here that he showed a correlation between decreased mortality and increased elevation above the River Thames, which was consistent with an idea that the stagnant soil and water along the river was producing a miasmatic atmosphere.[36] He did not, at this point, accept John Snow's theory of waterborne transmission.

The report of the Metropolitan Sanitary Commission concluded that cholera arose from impure air in humid conditions, generally around watercourses, and would appear in the same places as typhus or typhoid fevers.[37] It was not spread from person to person. Lack of access to pure air, rather than poverty and lack of food, was the main cause that could be addressed. The solutions were cleansing and ventilation, with the report detailing the design of drains and sewers, drawing on evidence from engineers, architects, and surveyors. The commission proposed an integrated system of water supply, cleansing, and drainage. Given that little had been done since 1832, the commissioners predicted a similar disaster if cholera returned.[38]

The second report, with additional evidence from experience abroad, reinforced the conclusions of the first. It also took into account the views of several chemists, including Lyon Playfair, now at the Royal School of Mines, Thomas Graham, and Robert Angus Smith.[39] Playfair and Graham agreed that chlorine had disinfectant properties in relation to typhoid, by destroying "decaying emanations," but not for cholera. Angus Smith offered an analysis of the effects of ventilation, water, drainage through soil, heat, cold, and chemical means of "disinfection." He was clear that organic matter in the air was the source of the problem.[40] There was disagreement among the chemists about whether various disinfectant salts, such as nitrate of lead, could purify the air of "noxious gases," and hence combat epidemic disease.[41]

The report suggested that cholera was not contagious because no "vitiated excretions" were emitted by sufferers to poison the air.[42] Diarrhea was generally the first sign, which it was thought could be converted into cholera or indeed typhus and should be treated as soon as it appeared, as a means of preventing the emergence of cholera itself. Playfair, who volunteered to visit several towns for the General Board of Health during the subsequent epidemic, put this idea into practice and claimed that it largely prevented cholera.[43] That may have been a consequence of the prescription of constipating drugs that prevented people from using the privy and spreading disease. Had the role of the water supply been recognized at this point, much more could have been done.

But Pym opposed Chadwick's approach, recommending quarantine and cholera hospitals. Chadwick's leverage with Lord Morpeth resulted in Lord Lansdowne, president of the Privy Council, agreeing to transfer responsibility for cholera to the General Board of Health. That was effected in the "Cholera Bill," which became the Nuisances Removal and Diseases Prevention Act 1848, although it neither consolidated local cleansing responsibilities nor gave powers to enforce regulations, until the Privy Council finally did so at the height of the epidemic.[44]

This act was passed because the 1846 version was about to expire, but it no longer required medical certification of the nuisance as a health hazard, although such expertise was often involved locally.[45] Sanitary regulations in abundance were directed at the Poor Law guardians and local medical officers, as the General Board of Health sought to combat the epidemic. Serious medical opinion, from the Royal College of Physicians and the Royal College of Surgeons, disputed these approaches. But Chadwick and his colleagues, of whom only Southwood Smith was medical and like Chadwick did not believe in contagion, sidelined those who disagreed and carried on. Ironically, Chadwick's policy of flushing as much filth as possible into the Thames contributed to the ferocity with which cholera attacked parts of London. Southwood Smith fell ill, and even Chadwick contracted suspected cholera. But gradually in 1849, as it had before, the epidemic abated.

As it faded, at least for a while, Chadwick and Southwood Smith tackled the quarantine system too. The General Board of Health produced two reports in 1849 and 1852, in which the sanitary approach was considered unarguable. For Chadwick and his colleagues, scarlet fever, influenza, plague, yellow fever, and cholera all shared a general resemblance, generating fever in the same localities and with its intensity increased or diminished by the same sanitary and social conditions.[46] Even if it were supposed that any of these diseases were contagious, the board argued that quarantine had been shown not to be effective.[47] The fevers derived from filth and overcrowding, so sanitary approaches should be taken to combat them, including on ships. Quarantine establishments should be abolished. The board issued instructions to that effect, against Pym's strong opposition. Although quarantine regulations remained, governments were asked to look to the hygiene of their ships and ports, and to require inspection and certification of ships for cleanliness by a port medical officer. Pym refused to accept the recommendations, declaring that it was impossible to provide every ship with a bill of health.[48]

ENTER JOHN SIMON

What the cholera episode did was to bring to notice one man who would have a major impact on public health policy for the next twenty years: John Simon.[49] Of French extraction, Simon studied surgery under Joseph Green, professor of surgery at King's College London. He carried out some original research in physiology and pathology, becoming a fellow of the Royal Society in 1845, and was appointed lecturer in pathology at St Thomas's Hospital in 1847. In 1848 he became the medical officer of health for the City of London, only the second such post in the country after William Duncan's appointment at Liverpool in 1847. It is clear that the threat of cholera was the impetus for this appointment.[50] Despite his relative youth, his father's City connections and the support of significant figures, including Richard Owen and Benjamin Brodie, surgeon to Queen Victoria and Prince Albert, along with his detachment from the controversial Chadwick, secured the position for him.

Over time, it was public service rather than private practice that took over Simon's life. Although he abandoned private surgical practice in 1855, he still operated for the poor and kept his post as a surgeon at St Thomas's Hospital until 1876. Liberal if not radical in outlook, he came to public health policy with experience both of practical medicine and scientific research. Unlike Chadwick, Simon believed in the role of medicine in promoting public health, through understanding the causes, nature, and trajectory of diseases. But as he started out in 1848, he believed in Chadwick's approaches and in his miasmic theory of disease. He considered clearing away filth the best means of prevention. It worked to an extent, although flushing it into the Thames piled up problems downstream. No one knew that cholera was both waterborne and fly-borne, and there was no known cure.

Simon advised the City Commission of Sewers that the sanitary approaches advised by the General Board of Health would be "as effective as vaccination against smallpox," but he met considerable resistance.[51] The cholera of 1848–1849 came in two waves, with the most devastating in the second, when thousands died in London.

Despite significant local opposition, and not until dozens were dying every week, Simon persuaded the health committee of the City of London Corporation to employ inspectors to carry out house-to-house visits. He instigated a range of sanitary measures, including trying to shut down foul burial grounds and slaughterhouses. His crusading approach, often intemperate at this stage, brought him to national notice, particularly through the supportive pages of the *Times*. This was cemented when he produced his first annual report in November 1849. Backed up by detailed statistics, and in compelling language, Simon set out an agenda for sanitation to counter the social degradation of the crammed-together poor that he documented in shocking terms for his readers. It drew on Chadwick's views but came from the City of London's own employee, not a centralizing bureaucrat, giving it huge impact.

The report urged proper house drainage, unrestricted water supply, the expulsion of slaughterhouses from residential areas, the closure of burial grounds in the city, slum clearance, and rebuilding. All this was to be backed up through inspection and reporting by Poor Law doctors, district medical officers, and the police. The report was a sensation, repeated in similar vein the next year. For the next few years, moderating his attitude and style to increase support, Simon and his colleagues made steady advances. Although progress against vested interests was slow, Simon continued to come to public notice outside the walls of the City of London.

It was an outbreak of cholera in late 1853 in Newcastle that brought Simon's first formal involvement with national government. His reputation was such that he was asked to act as a commissioner with Joseph Hume, son of the MP, and the engineer John Bateman, to report on this outbreak.[52] The commissioners attributed its cause to atmospheric impurity, and its transmission to unfiltered water from the River Tyne. They exonerated the inspectors of the General Board of Health and heavily criticized the Corporation of Newcastle for the insanitary conditions in the city.[53]

Cholera struck London again in 1854, but Simon's measures were now in place and the city came off relatively lightly. In hindsight, that is probably due to the removal of filth preventing fly-borne transmission, and because the water company filtered its water. Simon did not claim that his measures alone had definitively reduced the disease, as he admitted that much was still unknown about its cause and transmission, but the episode reinforced his public reputation.

THE MEDICAL COUNCIL OF THE
GENERAL BOARD OF HEALTH

At this point, Chadwick was ejected from the General Board of Health and Benjamin Hall took charge as president, as described the previous chapter.[54] The position of president was now a ministerial one, and Hall's reign lasted just a year until he was moved by Viscount Palmerston to the Board of Works. The sanitary enthusiast William Cowper, Viscount Palmerston's stepson, succeeded him.

Despite his personal opposition to Chadwick and his engineering methods, Hall had taken sanitation seriously as the national picture became apparent to him. Not only did he continue to support the broad thrust of previous sanitary policy, but he also looked to preventive medicine and the medical community in a way that Chadwick had not. He appointed two medical inspectors, John Sutherland and Gavin Milroy, who reported on the epidemic. Then, persuaded by Simon, Farr, and pressure from the medical community, he created a Medical Council of thirteen medical practitioners and people with scientific expertise to advise him on the epidemic. The members were nominated by the Royal College of Physicians, the Royal College of Surgeons, the Society of Apothecaries, and by Hall himself. Nine were fellows of the Royal Society. They included Simon, Farr, Owen, and James Clark, all nominated by Hall, as well as John Paris, president of the Royal College of Physicians. Arnott was also appointed by Hall, but not Southwood Smith, whose days of service, like Chadwick's, were over.[55]

The Medical Council had three committees, one of which was the Committee for Scientific Inquiries, consisting of Arnott, Simon, Farr, Owen, and the physician William Baly. This committee instigated a research effort, the first substantive example of a government department promoting and funding scientific research for health purposes. The scope included statistics, observation of the air and water of the metropolis, and possible medical treatments. The committee received and published evidence from James Glaisher on the meteorology of London, and from Robert Thomson, George Rainey, and Arthur Hassall on chemical and microscopical observations of the air and water supply.[56] Its report is a fascinating treatise on contemporary beliefs about cholera, but it is hedged around with recognition of the uncertainty of knowledge. The most that could be said, and not borne out by later evidence, was that the influences on cholera "belong less to the water than to the air."[57]

The report made reference to Snow's identification of the Broad Street pump as the source of transmission of cholera in Soho throughout 1854, when he had noticed high mortality among the pumps' users during the 1854 epidemic. He had persuaded the authorities to remove the pump handle, helping to end the outbreak. But the committee argued that the disease had arisen there "not in its containing choleraic excrements, but simply in the fact of its impure waters having participated in the atmospheric infection of the district."[58] In official circles, the miasmatist theory still prevailed even as its influence was waning. Sewage water was seen by the majority of doctors, public health officials, and clergy as one of the many predisposing factors for cholera.[59]

Hall then proposed that Simon should take up the new post of medical officer to the General Board of Health, an appointment confirmed by Cowper when he became president. This move was welcomed by the medical community. For example, in 1856 the reforming medical practitioner Henry Rumsey had argued in his *Essays on State Medicine* that public health should be controlled by medical experts—not, referring to the General Board of Health—by "two lawyers and a barrister."[60] He proposed a central office, led by a minister and advised by a medical council. At the local level, there would be full-time officers, taking charge of all aspects of health, including disease prevention. The state would investigate medical issues, administer health care, and support medical education. Rumsey became a leader in the public health department of the Social Science Association,[61] and coined the term *state medicine* as consisting of investigation, legislation, and administration aimed at improving public health.[62]

Simon became the government's first permanent medical officer, although it was not until 1919 that the chief medical officer of the new Ministry of Health was given the pay and status of a permanent secretary and direct access to the minister.[63] Like Chadwick, Simon always imagined his role to be more than that of an adviser. He sought to shape policy and legislation nationally in the manner in which he had done locally for the City of London. He also shaped the role of local medical officers of health, by suggesting experience of scientific medicine rather than sanitary engineering as the key criterion for appointment. The new Association of Metropolitan Medical Officers promptly elected him to their presidency.

JOHN SIMON AT THE GENERAL BOARD OF HEALTH

Simon's legislative influence was felt early at the General Board of Health. Not only was Hall's decision to appoint medical officers of health in all metropolitan districts a consequence of the example set by Simon in the City of London, but Simon framed parts of the Nuisances Removal and Disease Prevention Act 1855, which formalized sanitary processes for places other than London. This act also added certain kinds of industrial pollution and residential overcrowding to health hazards, again requiring medical certification, although it was ineffective in those regards.[64] Simon's cholera report on the City of London and the commission's report on the Newcastle outbreak influenced Hall to take action.[65] His approach was underpinned by a Christian socialist belief that it was the duty of a Christian society to protect the poor against sanitary evils through legislation and government action.[66]

Of lasting significance, analogous to that of Chadwick's *Report on the Sanitary Condition of the Labouring Population of Great Britain* (1842), was a document Simon issued in the last weeks of existence of the General Board of Health. *Papers Relating to the Sanitary State of the People of England* was published in 1858, with a fifty-page foreword by Simon.[67] The report was written by Edward Greenhow, lecturer in public health at St Thomas's Hospital. Simon was responsible both for the creation of that post, the first in public health, and for Greenhow's appointment. Using unpublished data from the General Register Office, to the chagrin of Farr, whose thunder was being stolen, Greenhow carried out an extensive statistical analysis of death rates from different diseases in different places. He was later employed by Simon to carry out inquiries into diphtheria in 1859 and into pulmonary disease among operatives in various industries in 1860 and 1861.

The *Sanitary Papers* mark a determination by Simon to proceed on the basis of expert scientific analysis, ushering in an era of "state medicine." He was not a centralizer in the mode of Chadwick, but viewed his role as investigating and establishing the facts fully and publicly in a manner that would inform responsible local authorities.[68]

STATE MEDICINE

At this point, an unexpected change of government reshaped public health policy. Viscount Palmerston's government, formed in 1855 after the resignation of the Earl of Aberdeen's coalition during the Crimean War, had survived a general election in 1857 after Viscount Palmerston had been censured over his conduct of the Second Opium War. But in early 1858, following Felice Orsini's attempt on the life of Napoleon III in Paris, the government fell after being defeated on a measure to reduce the charge of conspiracy to murder abroad from the class of misdemeanor to felony. The Conservatives came to power in February 1858 in a short-lived government under the fourteenth Earl of Derby.

THE PRIVY COUNCIL TAKES OVER

What happened next nearly did not happen at all, but it left Simon in a unique and powerful position as an adviser and civil servant. Neither Lord Derby nor the new home secretary, Spencer Walpole, was much interested in sanitary reform, nor were they in favor of centralization. Charles Adderley became vice president of the Privy Council and president of the General Board of Health, and determined to hand over sanitary responsibility to local administrations. The General Board of Health would be abolished, with remaining central functions for town improvements and health given to separate bodies. The first went to the new Local Government Act Office, part of the Home Office, which in the event retained significant central powers. The second, under the Public Health Act 1858, transferred powers to respond to epidemics to the Privy Council along with Simon's post as medical officer, but just for a year.[69]

The Marquess of Salisbury, slum landlord and opponent of sanitary approaches, was now president of the Privy Council. He determined to abolish Simon's post, until persuaded by Prince Albert not to do so.[70] At this point, in June 1859, the Conservative government fell and Viscount Palmerston returned, with Earl Granville as president of the Privy Council and Robert Lowe as vice president. Lowe, with Cowper's help, secured the passage of the Public Health Act 1859, which retained Simon's post on a permanent basis. That outcome was influenced also by lobbying by the Social Science Association, of which Simon had been a founding member.[71]

The significant clauses, shaped by Simon himself, transferred the National Vaccine Establishment to the Privy Council, and gave him the power to report on any matter concerning public health. It was this latter power that was so significant and unprecedented. Simon, on his own authority, could investigate and report on any question relating to public health.[72] Not only was he committed to advancing a scientific and research-based approach to public health, but he could in principle investigate and report on anything he wanted. The contrast between his position and that of, for example, Frederick Abel at the War Office, James Kay-Shuttleworth in education, or Charles Trevelyan at the Treasury, is stark in terms of his power to act individually. Farr, in a less influential role, had however also taken to himself the ability to investigate issues on his own initiative. One further advantage, especially by comparison with Chadwick, was the respect in which Simon was almost universally held by the major newspapers and by parliamentarians. There were some exceptions. Nightingale was hostile, exemplified by their disputes over the siting of St Thomas's and Netley Hospitals.[73]

Simon had a real challenge to influence public health policy across government. Responsibility was fragmented in England and Wales, more so than in Scotland and Ireland. The Privy Council had responsibility for quarantine, and developed separate veterinary and medical departments. Indeed, the council was the place where proto-departments such as education, agriculture, and health could develop, before they were sufficiently large to require separate existences. The Home Office oversaw health-related aspects of factories, burial grounds, and registration. The Poor Law Board oversaw workhouses, union medical officers, and vaccinators. After 1860 it took further responsibility when the local nuisance authorities were transferred from the vestries to the Boards of Guardians. The Poor Law Board appointed its own medical officer in 1865. The Board of Trade had some oversight for water companies and controlled alkali manufactories after 1863. The army and the navy had their own arrangements. Against this, Simon needed substantial persuasive powers, in a body in which the president and vice president could devote limited time to any one policy area. He was fortunate that his friend, the sanitary campaigner and writer Arthur Helps, became clerk to the council in 1860. In 1861 Simon was authorized to sign all orders, regulations, and directives, such was the trust in his judgment.

Simon took a wide view of public health matters. Quite apart from immediate questions of disease, his interests and advocacy extended to areas such as regulation of the prescribing of drugs, the location and design of hospitals, in which he came up against the robust views of Nightingale, and the condition of housing for the poor.[74]

JOHN SIMON'S FIRST RESEARCH AGENDA

The report on cholera for the Committee for Scientific Inquiries in 1855 had suggested that more research was necessary. Simon established a further statistical investigation, drawing also on Snow's work and on Farr's epidemiological study of the 1854 outbreak in London as affected by the water supply. Simon had published his findings in 1856, concluding that "the population drinking dirty water accordingly appears to have suffered 3½ times as much mortality as the population drinking other water."[75] The danger of fecal contamination of water during cholera epidemics was now clear to him, even if Farr did not fully accept it until later.

In 1857 Snow added evidence that the number of deaths from cholera among customers of the Southwark Water Company was six times higher than among customers of the Lambeth Waterworks Company. He attributed this to the fact that the Lambeth Waterworks Company drew its water from above Teddington Lock, where there was limited danger from sewage contamination, while the customers of the Southwark Company received water from the polluted tidal stretch of the river.[76] William Budd drew a similar conclusion about typhoid in Bristol. Snow died in 1858 and did not live to see his theory receive full acceptance.[77]

From 1859 onward, Simon issued annual reports from his position at the Privy Council.[78] They were punchy, based on extensive scientific and statistical evidence, and widely read and reported. Much of the research was carried out by a team of young enthusiastic medical officers, headed by Greenhow and by Edward Seaton, who was appointed the first vaccination inspector under the General Board of Health in 1858. Seaton would eventually succeed Simon as medical officer in 1876. Others included George Buchanan, John Bristowe, and John Burdon Sanderson, who would rise to be Regius professor of medicine at Oxford University and receive a baronetcy. Over the next few years there were reports on diseases, including diphtheria by Greenhow, Burdon Sanderson, and the physician William Gull; on diarrheal diseases by Greenhow; on vaccine quality and the quality of vaccinators by Seaton; on respiratory and lung diseases in industry by Greenhow; on typhus by Buchanan; on possible contagion from imported rags by Bristowe; on an outbreak of yellow fever in Swansea by Buchanan; on suspected plague, which turned out to be cerebrospinal meningitis, by Burdon Sanderson; on conditions in the arsenic, lead, phosphorus and dressmaking trades; and on infant mortality related to the industrial and agricultural employment of women. This built a formidable body of evidence to inform future policy and practical actions.[79]

Much evidence was collected by investigating local outbreaks of disease,[80] a process that made it clear that local government routinely failed to take preventive action on nuisances, and sometime even failed to do so during a crisis. Simon managed to marshal statistical data collected by Buchanan, which demonstrated reduced mortality in towns that had adopted and acted on the public health acts. Armed with this evidence, and with another cholera epidemic emerging, he seized the opportunity to introduce more compulsion into the legislation.

THE SANITARY ACT 1866

In his annual report of 1866, Simon called for new laws to help limit the spread of common diseases such as typhoid, scarlet fever, smallpox, and diphtheria, involving increased powers for local authorities which could be enforced.[81] He had earlier pointed out that the nuisance laws gave no proper powers over finance, water supply, and overcrowding, and he argued that in any case, "the inaction of local authorities has been an absolutely inexcusable neglect of duty."[82]

Cholera arrived again in the summer of 1865 and was at its worst in 1866, coinciding with the cattle plague. It was the last major outbreak, taking around fourteen thousand lives, compared to fifty-four thousand in 1848–1849 and twenty-four thousand in 1853–1854. Improved sanitation and epidemiological knowledge may have played their part, even if washing sewage into the Thames was not in hindsight a sensible policy. The Nuisances Removal and Diseases Prevention Act was put in force for the whole country, ordering local authorities to appoint medical attendants, remove nuisances, and supply medical aid, but Simon could only issue regulations and hope they were implemented. He did manage to persuade the Treasury to support some research, when the incoming Conservative president of the Privy Council, the Duke of Buckingham, persuaded Disraeli to release funds. The conclusion Simon drew was that cholera and typhoid would continue to kill people unless filth was removed and water purified. At this point his model of disease was still zymotic, with filth serving as a medium for propagation.

This outbreak convinced Farr that cholera was waterborne, and that the conclusions of the 1854 committee of inquiry were in error. Farr noticed that the 1866 epidemic only affected a small area of Whitechapel, which was not yet connected to Joseph Bazalgette's drainage system. This indicated that the East London Water Company's reservoirs drawing from the River Lea had been contaminated. In his huge report on the epidemic he remarked that "only a very robust scientific witness would have dared then to drink a glass of the waters of the Lea at Old Ford after filtration."[83]

It was this epidemiological approach that convinced Farr, who had also commissioned Edward Frankland to undertake chemical analysis. But Frankland was of the view that "chemical analysis, like every other mode of investigation, is powerless to detect the presence of matter like the choleraic poison amongst the organic impurities of water, for this poison may be present in quantity fatal to the consumer though far too minute to be detected by the most delicate chemical research."[84] Since cholera came from outside the country and, like the cattle plague, arrived by sea, further preventive measures were introduced. Given the British commitment to unfettered trade, the strict quarantine measures implemented by some countries were not adopted. Instead, the government created Port Sanitary Authorities in 1872, an initiative strongly advocated by Henry Letheby, medical officer of health for the City of London. These authorities could ensure that vessels were reported to Customs for detention, could isolate diseased individuals, and require disinfection. With the support of the Medical Department they were arguably important in keeping any significant outbreak of cholera from Britain during the rest of the century.[85]

The consequence of Simon's exertions was the Sanitary Act 1866, the most significant since 1848. Henry Bruce, the Whig vice president of the Privy Council in June 1866, presented the bill to Parliament, although it was common knowledge that Simon was its architect. Almost immediately Lord John Russell's government fell and the Earl of Derby, whose previous administration had nearly abolished Simon's post, came to power. Fortunately, the Duke of Buckingham, president of the Privy Council, supported Simon, and the government allowed Bruce to continue.

The bill became law with speed, under the threat of cholera. The new act introduced much of what Simon had been arguing although, as a measure passed in haste, it did not relieve all the confusion about local and central responsibilities. It made sanitary facilities available to all health authorities in the country, although to Simon's dismay that included the rural vestries, which had been designated as sewer authorities, and which he thought would not be effective. This act worked in conjunction with the Sewage Utilization Act 1865 (see chapter 11). Local authorities could now take action against overcrowding, although that was difficult in practice. Definitions of "nuisances"—including filth, poor ventilation, and overcrowding—covered all factories and workplaces. All nuisance authorities could deal with industrial smoke. They could compulsorily disinfect houses, remove people to hospital, and bodies to mortuaries. By law they were obliged to inspect their districts and exercise their powers. Duty and an element of compulsion had been added to voluntary powers, with defaults determined by the home secretary. The act not only extended the powers of local authorities but also the power of the center.[86] It applied in part at least to Ireland and to Scotland, both of which received further public health legislation over the next few years in a similar vein.

In practice, the machinery of the act did not deliver the intended consequences on a consistent national basis. The medical community, particularly through the British Medical Association and the Social Science Association, who formed a joint committee in 1868, lobbied hard for greater compulsion in sanitary matters. It also argued for the rationalization of local government, the appointment of medical offers of health everywhere, and better central information. This joint approach was instrumental in the establishment of the Royal Sanitary Commission in 1869, described below.[87]

JOHN SIMON'S SECOND RESEARCH AGENDA

Simon had orchestrated a wealth of research in the field, but it had not borne great practical fruit. He argued that if the state could fund research into the cattle plague, it could surely do the same for human disease. From 1867 to 1871 he instigated an increasing amount of laboratory research, recognizing that significant progress would not be made in public health until more was known about the underlying biology.

The initial emphasis was on chemical aspects of pathology and the aetiology of contagion. The first was carried out by Johann Thudichum, who had studied medicine at the Giessen University and worked in Liebig's laboratory. He had already reported on parasites in meat and would develop a substantial program over the next twenty years. For the second area, Simon himself and Burdon Sanderson initiated work in 1867 on the inoculability of the tubercule. By 1871 Burdon Sanderson had determined that each specific contagium was a living, self-multiplying organic form, as Louis Pasteur had done for putrefactive and fermentation processes. That helped establish the germ theory of contagious disease, though it would not be until 1882 that Robert Koch identified the tubercule bacillus itself. Simon gradually accepted the theory, although by this time he had left government service.[88]

Other inquiries into public health matters continued in parallel. One in particular used empirical social research to provide evidence as to whether the Contagious Disease Acts of 1864 and 1866 should be extended from garrison towns to the wider population. These controversial acts empowered the military and naval authorities in eleven towns to compulsorily superintend the health condition of prostitutes. William Wagstaffe, a surgeon at St Thomas's Hospital, showed that the proportion of people with venereal disease was far lower than the advocates claimed and Simon used this to argue against the necessity for extension. The campaigner Josephine Butler reprinted his work in 1871 as part of a campaign to have the laws repealed, which was finally done in 1886.[89]

But in the general area of sanitary policy, Simon's views on compulsion were hardening. He wanted water companies to be forced to pay damages if they were negligent with the quality of their water, and he wanted permissive legislation on local authorities made compulsory. General pressure on such issues led the Conservative government to establish the Royal Sanitary Commission in 1868. Meanwhile, Simon was under pressure himself. The Duke of Marlborough had replaced the Duke of Buckingham at the Privy Council and deprived Simon of his executive powers in February 1868. Politics intervened again when the Liberals took power in December 1868 under William Gladstone. With Earl de Grey and Ripon as a supportive president, the friendly William Forster as the vice president, and with Bruce at the Home Office and Lowe at the Treasury, Simon had more support. He was able to increase the size of his department, appointing two new general health inspectors in Buchanan and John Radcliffe alongside the six existing inspectors of vaccination. He also retrieved his executive powers as he looked to build a genuine department of health.

To work toward this aim Simon persuaded Graham, the registrar-general, to provide mortality figures broken down by district and disease, which could serve as a basis for inspection visits. Although he succeeded here, he failed to persuade the Poor Law Board to provide sickness statistics. But he had a further success in establishing that the medical department should first investigate any complaints under the Sanitary Act 1866, passing them to the Local Government Act Office, which was still responsible for overseeing sanitary engineering locally, following the demise of the General Board of Heath. Whereas Chadwick had put the engineers first and the doctors a distant second, Simon had reversed the position.

Local inspection was now increasingly driven by the data, rather than by appeals and complaints. Despite the lack of coercive power from the center, Simon's team had an impact because they were highly qualified and credible, met together to plan and share ideas, and were persuasive in person and through their public reporting. They still lacked vital knowledge of disease, although improved techniques of water analysis were giving them more detailed, if contentious, information.

THE ROYAL SANITARY COMMISSION

On coming into office as home secretary, Henry Bruce reappointed the royal commission, arguably the most significant in public health since 1843. Lowe, now the chancellor of the exchequer, objected on the basis that government should not devolve responsibility for this matter to a royal commission, which he thought would tend to weaken the authority of executive departments as well as being expensive.[90] But Bruce prevailed, although he limited the scope to England and Wales, excluding Scotland, Ireland, and London.

The commission was chaired by Adderley, who had been vice president of the General Board of Health in its last days, then vice president of the Privy Council in Lord Derby's short-lived government. In addition to parliamentarians and the Poor Law inspector John Lambert, there were two engineers and five doctors. The medical contingent were Henry Acland, the Oxford professor of medicine; Thomas Watson, who had been president of the Royal College of Physicians; James Paget; William Stokes; and Robert Christison, a Scottish toxicologist. Christison, the only one of the five who was not a fellow of the Royal Society, is notable for being strongly opposed to women becoming doctors. He was instrumental in the campaign to prevent Sophia Jex-Blake and others graduating from the University of Edinburgh. James Stansfeld, the radical politician who would play the major role in implementing the recommendations of the commission, was of the opposite view. Not until the Medical Act 1876 were British medical authorities allowed to licence qualified applicants whatever their gender.

The commission heard nearly one hundred witnesses, including Rumsey, Farr and, on several occasions, Simon. Although the medical and engineering expertise was extensively called upon, the focus of the commission was on the complexity of existing legislation and the machinery of administration, for which they offered 38 pages of observations and 105 pages of proposed revisions to statutes.[91] Simon had originally suggested consolidation of public health matters under the Home Office but later emphasized his closer connection with the Poor Law Board with, in hindsight, dire consequences for him. The commission proposed consolidation of sanitary law on the basis of its suggested revisions and recommended that its application should be uniform, universal, and obligatory. It effectively advocated merger of the medical department of the Privy Council and the Poor Law Board, but it was ambiguous about how inspection for medical, engineering, and Poor Law matters would operate.

Simon had already persuaded the Treasury to grant him more resources, and when the royal commission's report arrived in January 1871, with the recommendation that all central health functions should be under a single minister, he proposed wide-reaching changes to the medical office and another expansion of staff to provide a sufficient national function. More resources were again granted. When Simon made the further case for increasing the vaccine inspectorate on technical evidence, Ralph Lingen, the permanent secretary at the Treasury, protested, "I do not know who is to check the assertions of experts when the government has once undertaken a class of duties which none but such persons understand."[92] That tension between political decision making and specialist scientific advice grew more evident during the century.

It is not easy to assess the impact of public health policies in this period, but it can be argued that the approaches introduced by Chadwick and Simon led to detectable improvements. Although the mortality rate was stable from 1851 to 1871, that can be viewed as an achievement given an increase in urban population of more than 50 percent.[93] In addition, mortality in most European countries was higher. But major and sustained advances against disease had to await the new bacteriological and the viral knowledge, and new vaccines.

THE LOCAL GOVERNMENT BOARD

Stansfeld, an admirer of Bentham and who had become president of the Poor Law Board in March 1871, saw through the Local Government Board Act 1871. This implemented some of the approach of the commission's report by abolishing the Poor Law Board and creating the Local Government Board, which took over Poor Law administration as well as the public health work of the medical department of the Privy Council, the Local Government Act Office and its sanitary responsibilities, and supervision of the General Register Office.[94] Despite his ostensible desire to see public health under one minister, Simon argued that the Privy Council should retain responsibility for scientific research and regulation of the medical profession, as these were not local matters. He kept his position as medical officer at the Privy Council in addition to his public health role at the Local Government Board.

Stansfeld became president of the new Local Government Board, now a cabinet post, and brought forward a further bill in 1872. This became the Public Health Act 1872, which consolidated local administration of sanitary law. But it left crucial organizational matters unresolved, including the appointment of local medical offers. The organizational arrangements would be determined by the Local Government Board rather than Parliament. In the event, that sealed Simon's fate. Gladstone appointed Lambert, who had extensive Poor Law experience, to the post of permanent secretary. In effect, it became a takeover by the ethos of the Poor Law Board, aided and abetted from the sidelines by Chadwick and Nightingale, the sanitarians who believed in practical sanitary engineering over what they saw as theoretical medicine. Lambert, who thought medical experts should be subordinate to the lay secretariat, sought to unify the administration, making Robert Rawlinson the chief engineer and removing Simon's powers to act independently, as far as the Local Government Board was concerned.[95]

Simon was unable to establish a good relationship with Stansfeld, who decided that the Poor Law inspectors, none of whom was medically qualified, would be charged with the sanitary supervision of more than 1,500 local authorities and the creation of the network of local medical officers. Indeed, until the end of the Local Government Board in 1919, supervision of the Poor Law doctors was carried out without any involvement of the board's medical officer.

So although vaccination activities, now combined under Simon, florished, the wider local approach to disease prevention fell back, not least because Stansfeld showed an unwillingness to use the central coercive powers that Simon and other medical groups thought essential. The medical community was up in arms about Simon's treatment, but it had no political effect. Although Playfair raised the issue in the House of Commons, there was no great enthusiasm from Parliament to become involved. Indeed, the relative indifference of Parliament that had allowed Simon to build such an empire also allowed Stansfeld and Lambert to dismantle as they wished.[96]

In February 1874 Gladstone's government departed and Disraeli came to office. George Sclater-Booth, a former parliamentary secretary to the Poor Law Board, was the new president of the Local Government Board. He was not in the cabinet, indicating the relatively low priority Disraeli gave to public health legislation, despite claiming sanitary reform as a Conservative contribution three years earlier.[97] Simon still had influence. From his Privy Council office he continued to sponsor laboratory research and to investigate local disease outbreaks, although his inquiries were reduced in number.[98] He and his colleagues contributed to Home Office legislation on canal boat inspection, and possibly helped in the drafting of measures on river pollution and food adulteration passed in 1876.[99]

Simon's most significant influence was on the Public Health Act 1875, which was based on the report from the Royal Sanitary Commission and taken through Parliament by Sclater-Booth. This major consolidating act, with a similar measure passed for London in 1891, remained in force until 1936 and brought together all the powers to do with public health and the physical environment. To this was added the Artisans' and Labourers' Dwellings Act 1875, strongly influenced by Letheby, which made the appointment of medical officers and sanitary inspectors by local authorities mandatory.[100]

This national emphasis was complemented by the introduction of local acts in some towns from the 1870s, requiring notification of certain infectious diseases such as smallpox, cholera, diphtheria, scarlet fever, and typhoid. Central government legislation caught up after a smallpox epidemic in Sheffield in 1887 and 1888, when Charles Ritchie, as president of the Local Government Board, saw through the Infectious Diseases (Notification) Act 1889.[101] It was compulsory in London and optional for sanitary authorities elsewhere, and became mandatory everywhere in 1899.

Public health provision could not advance substantively further until the scientific discoveries in bacteriology of the 1880s and 1890s had been assimilated.[102] Although the local powers were still permissive under the 1875 act, they cemented the ability of local authorities to control water supplies and sewerage, regulate cellars and lodging houses, and regulate building. Every public health authority was required to have a medical officer and a sanitary inspector, to ensure that the laws on food, housing, water, paving, lighting, and hygiene, or nuisance, were implemented. The Local Government Act 1888 created county medical officers. Nevertheless, the continued confusion of local institutions held back systemic progress, as did the focus of the Local Government Board on the Poor Law, to the detriment of a preventive approach to public health. Local government complexity was a general obstacle to coherent and effective social policy throughout the mid-Victorian period.[103] The Local Government Board was eventually disbanded in 1919, when its functions were distributed between different departments.

Simon also retrieved his ability to make independent reports. Over the next three years he issued eight reports, but it is here that he came to grief. With a loss of sense of proportion, under considerable emotional stress in his private life, and making a cardinal error for a civil servant, Simon's reports roundly criticized his own board. He compounded the error by giving Sclater-Booth an ultimatum that he would resign unless he had full right of access to the minister, control of the board's sanitary administration, and an increase in medical staff. Sclater-Booth rejected his demands and he resigned in May 1876.[104]

The medical community was aghast and scandalized, but to no avail. Politicians are not in the habit of giving up their power, as Chadwick had discovered before. The post of medical officer at the Privy Council lapsed, and Seaton succeeded to the lesser post of medical officer to the Local Government Board. His brief reign lasted until 1879, and he was succeeded by Buchanan, who held the post until 1892. Not until 1919 and the creation of the Ministry of Health did the chief medical officer for England, in the form of George Newman, report directly to the relevant minister. The scientific research work continued nevertheless, leading in 1913 to the formation of the Medical Research Committee, which became the Medical Research Council under the Privy Council in 1920.

REFLECTIONS ON THE MEDICAL PROFESSION

General practitioners argued repeatedly for representation in running the profession, for a unified qualification to enable them to practice in all areas of medicine, and for a single register of qualified doctors. Politicians also recognised the need to assure some level of competence for medical roles increasingly required by the state, in the emerging sanitary initiatives and Poor Law medical services. The Medical Act 1858, for which Simon was a key figure in its drafting, went some way toward meeting the desires of general practitioners, although the elite institutions retained significant control.[105] The act gave statutory recognition to a unified profession, with a single register and with powers of self-regulation under the General Council of Medical Education and Registration, or General Medical Council. But general practitioners were excluded from the governing body until 1886, when the Medical Act Amendment Act finally standardized qualifications.[106] The 1858 act did not outlaw homeopaths, herbalists, or naturopaths, although they were not allowed to pass themselves off as a doctor unless medically qualified. Doctors could qualify through a hospital medical school, a medical corporation, or a university medical faculty to get on the public register. Only properly qualified men could hold a range of public posts such as in the Poor Law medical service, the prison service, or the public health service. In this way the politicians sought to protect the public, while allowing individuals to operate in some aspects of health outside the formal structures of the medical professions. New associations emerged. The Poor Law Medical Officers Association was established in 1868 and the Association of Medical Officers of Health in 1869. Medical practitioners gained positions also in inspectorial roles in the Home Office, for anatomy in 1832, for burial grounds from the 1850s, and for cruelty to animals and vivisection in 1876. Those who achieved these positions had shared backgrounds in the same hospitals and medical schools in London and Edinburgh, and most were fellows of medical royal colleges.[107]

These developments enabled the medical profession more easily to embed itself within public administration, advancing both its own prestige and its ability to improve public health. That was no easy task in mid-century when medical knowledge, before the advent of effective epidemiology and of bacteriology, could demonstrate little impact on treating, let alone eradicating, so many diseases. That it was achieved is due in large measure to the activities of influential groups including the British Medical Association and the Social Science Association. They helped develop an environment in which a larger group of medical men could gain influence on public policy, alongside the gentlemanly elite whose educational background and social connections gave them continuing access.[108] As an indicator of political recognition, thirty-six of them were knighted between 1850 and 1883 and sixteen granted baronetcies.[109] Even so, not until 1897 was the first medical man raised to the peerage, in the form of Joseph Lister.

While many doctors had a problem with low social status, that was even more the case for veterinarians. They dealt primarily with horses and increasingly with other livestock such as cattle, in competition with self-professed animal healers who often had considerable practical experience. The Royal College of Veterinary Surgeons was not established until 1844, although educational and training institutions had been set up earlier, such as the Veterinary College of London in 1791.[110] The prime opportunity for veterinarians in public policy came in the 1860s with the advent of the cattle plague.[111] They gradually gained influence, through advocating and supporting policies of eradication of animal diseases by legislative measures, including slaughter.[112] In 1878 local authorities were required to appoint at least one qualified veterinary inspector, which replicated the requirement for qualified medical officers of health introduced in 1875. The Veterinary Surgeons Act 1881 prohibited unqualified people from using the title "veterinary surgeon," formalizing the profession. But as with the medical community, any influence of laboratory-based research on practice and policy did not come until late in the century.[113]

For the most part, as with scientific and engineering specialists, medical men exerted their influence from outside Parliament. Although a few people who had qualified medically became MPs, such as Joseph Hume and Thomas Wakley early in the century, it was barely possible to combine a medical practice with a parliamentary career. A private income was necessary for entering Parliament, and the demands and characters of the two activities conflicted.[114] Indeed, for much of the latter part of the century, Playfair, the chemist who had abandoned his medical degree after contracting severe eczema in the dissecting room, acted as a significant spokesperson for medical interests in the House of Commons. From around 1885 a dozen or so medical figures became MPs, and several were vocal in debates about vaccination.[115] But they were never a big group.

Doctors had to retain the trust of the public to enable them to practice commercially and to receive government sanction. That meant conveying a humanitarian ethos to counter strong objections and occasional violent protest over questions such as animal vivisection, anatomical dissection with its taint of stealing corpses, and charges of exploiting the poor in hospitals for research purposes.[116] Lacking government support, and with specialization resisted by much of the medical profession until late in the century, the medical research community was slow to develop.[117] Medical science, as contrasted with medical practice, also had to steer a careful line with regard to experimentation, since public opposition to some practices was a powerful force, exemplified by responses to the Anatomy Act 1832. The Cruelty to Animals Act 1876, prompted by similar concerns, limited research on animals to licensed researchers in registered premises. And it was not straightforward to convert knowledge from medical and biological sciences into clinical or other practices to alleviate diseases. As with the natural sciences and engineering, medical practice became specialized in the nineteenth century. This specialization happened later in Britain than in many other countries, becoming evident from the mid-1880s.[118] The elite physicians and surgeons retained their power throughout the century, while the profession as a whole became more unified. While the few doctors who became MPs were mostly from the gentlemanly elite, general practitioners and medical officers were also able to exert influence through local activities, professional associations, and appearances before national committees and commissions. By the end of the century, medical men were thoroughly integrated into state functions.

These two chapters on public health illustrate vividly the importance of tensions between central and local government in constraining the impact of scientific, engineering, and medical advice. That is against a background of competition for influence by experts espousing different medical theories, during a time when understanding of the causes of diseases and means of prevention was changing. Politicians had to navigate this difficult terrain, while also considering factors such as impact on trade and the economy, interfering as little as possible with private property and industry, and upholding individual freedoms and responsibilities.

PART VII

REVENUE AND STANDARDS

13

CHEMICAL ANALYSIS, EXCISE, CUSTOMS, AND INLAND REVENUE

CHEMICAL ANALYSIS WAS VITAL TO THE TREASURY. IT HELPED GOVERN-
ments determine the appropriate level of taxation for many products and
counter fraud. Its importance extended to the Board of Excise, responsible for
duties on substances produced in the country, and to the Board of Customs,
responsible for duties on imported goods.[1]

The political emphasis placed on chemistry in the first half of the nineteenth century is notable in that most analysis was undertaken to protect
revenue, not to protect public health or the environment. Only from mid-
century onward was much attention given to chemical adulteration of foodstuffs that damaged public health as opposed to revenue, or to questions of
pollution. For the first few decades of the century, analysis was undertaken
on an ad hoc basis if deemed necessary at all. A case in point is the evidence
Thomas Brande gave to a select committee on the adulteration of imported
clover and trefoil seeds in 1821. The seeds were doctored with sulfurous compounds and by dyeing. Brande gave the opinion that such treatment would
either be severely damaging or fatal to the seeds.[2] A bill was introduced to
prevent the practice but was not passed. It was a further half century before
the first Adulteration of Seeds Act was passed in 1869.

In time the need for in-house chemical expertise became apparent, which
led to the establishment of the chemical laboratory of the Board of Excise in
1842. The Board of Excise then merged with the Board of Stamps and Taxes in

1849 to become the Board of Inland Revenue. A chemical laboratory for the Board of Customs followed in 1860, and the two laboratories merged in 1894 to create the Government Laboratory. Many substances came under the purview of these boards, but perhaps the most contentious of the issues requiring chemical analysis was the malt tax.

MALT TAX

The malt tax was political dynamite. Over a ten-year period from the mid-1830s to mid-1840s it produced nearly £5 million for the Exchequer, which was a third of the excise revenue and one tenth of the entire ordinary government revenue.[3] The House of Lords, repository for the major landowners, stimulated a request to the chemists as a result of its searching select committee examination into the "burdens on land" in 1846. This was an inquiry that took in taxation, tithes, and all manner of other costs on land of concern to the landowners. What brought in the chemists was the barley crop and its conversion into malt for brewing and distilling.

The challenge of setting appropriate charges for the malt tax is exemplified by a request as early as 1804 by William Huskisson, secretary to the Treasury, to the Board of Excise in Scotland.[4] Huskisson wanted a comparison between Scottish and English barley to inform the setting of duties. This request followed a recommendation by a parliamentary committee for a reduction in the duty on malt made from Scottish barley, which was deemed to be inferior and hence less valuable than the English supply.

The commissioners for the Board of Excise in Scotland gave the task to three people at the University of Edinburgh: Thomas Hope, professor of chemistry; Andrew Coventry, professor of agriculture; and Thomas Thomson, lecturer in chemistry. Their examination was systematic. They carried it out in bulk to replicate real conditions, and used three different breweries to minimize any errors.

This approach implies that the chemists, and those who sought the evidence, were not confident to draw conclusions from laboratory experiments about what would happen in the outside world. The research took two years, and the report was so enormous that it was sent in a box by mail coach to the Treasury, being too large for the post. There was nevertheless disagreement between the analysts. Thomson declared that Scottish malt was 14 percent inferior to English, while Hope and Coventry maintained that it was only 8 percent inferior. Hope gave an explanation for the discrepancy in a note, leaving the Board of Excise and the Treasury to decide between them.[5] This sort of detailed analysis was infrequent, as was analysis for detecting fraud which might deprive the excise of revenue. Indeed, it is clear from evidence to a select committee in 1818 that experienced excise officers would rely on taste rather than chemical analysis to detect problems such as the adulteration of beer.[6]

When a bill was brought forward in 1845 to allow the use of malt duty-free for feeding cattle, John Wood, chairman of the Board of Excise, approached Lyon Playfair, Thomas Graham, and Thomson, now professor of chemistry at the University of Glasgow, to advise. Graham found no evidence that malt was fed to cattle anyway. Thomson and his nephew in Glasgow reported a charming set of experiments to compare the relative value of barley and malt when used as feeds, with respect to the quality of the cows' milk. They took two cows, which they named Brown Cow and White Cow, "of which the White Cow was the handsomest," and monitored them for several months.[7] Soon finding that the cows were too different to be compared using separate feeds, they gave the same feeding regimen to both to measure differences depending on the type of feed. Their conclusion was that barley was better than malt for producing milk, which fitted with a chemical analysis of the two substances.[8] A further experiment purported to show that barley was also better than malt for fattening bullocks. Doubtless to Wood's relief, Playfair argued on chemical grounds that barley lost part of its nutritive value in conversion to malt, and that the benefit of allowing it to be fed to cattle duty-free was "not so great as to warrant Government endangering Part of the Revenue by granting such a Boon."[9]

THE CHEMICAL ANALYSTS

When the chemical laboratory of the Board of Excise was established in 1842, the first chemist appointed was George Phillips.[10] He served until 1874, to be replaced by James Bell until the formation of the unified Government Laboratory in 1894. Phillips was an excise officer, self-taught in chemistry. Prior to 1842 both the Boards of Excise and of Customs had employed consultant chemists for occasional analyses. The most visible of these was Andrew Ure, a rancorous figure who did not get on with Phillips. Ure trained as an army surgeon and then became professor of natural philosophy in 1804 at Anderson's Institution in Glasgow, in succession to George Birkbeck. He in turn was succeeded by Graham in 1830, before Graham left for London to be professor of chemistry at University College London in 1837. The network of leading chemists involved with government was a tight one.

After moving to London, Ure became a consulting chemist, specializing in mineral waters and ores, but providing evidence for many other clients. He was a firm advocate of free trade and the factory system,[11] and joined a growing group of industrial and consulting chemists and analysts. They started to apply chemical knowledge to industrial processes, and many carved out careers as expert witnesses in legal cases and as contracting analysts to government.[12]

Ure became a consultant for the Board of Customs in the early 1830s and also carried out analyses for the Board of Excise and other government bodies. For example, in 1832 and 1833 he was contracted by the Board of Trade to determine the yield of sugar by refining raw material from the West Indies, so the board could set levels of bounties for the export market.[13] He had earlier given evidence to a select committee on the use of molasses in breweries and distilleries. This was instigated as a result of a severe financial depression in the West Indies and to explore whether the importing of molasses as a substitute for malt or corn would affect the brewing and distilling industries in Great Britain. Ure was brought in to determine the weight of molasses equivalent to a weight of malt, so that levels of price interventions could be set.

This select committee examination gives a good indication of whom the committee members trusted. Several brewers and distillers offered their calculations of equivalent values, but the committee seem to have put most weight on Ure's "very elaborate experiments."[14] They also called Thomson, whose capability Ure had savagely attacked in the 1820s, which had led to a lasting personal feud. The committee referred back to Thomson's experiments on malt from Scottish and English barley. With him, they were careful to explore the limits of accuracy in determining the amount of spirit that could be produced from a given ferment. According to Thomson, those limits depended on the volume of the sample taken and on the expertise of the excise officers. Thomson was of the view that few if any were competent enough to make the analysis accurately, unlike the "man of science."[15] It was important, for the integrity of the system and the fair charging of duty, that all parties could trust the decisions of the excise officers. That depended on trustworthy analysis, although it was not the only factor. In the event, the committee determined that the introduction of molasses would be "at the expense of the Landed Interest in this country," damage brewers and distillers and reduce the revenue, while barely benefiting the West Indies.[16] Chemistry came second to more powerful political factors.

Phillips did not reach the eminence of fellowship of the Royal Society, and indeed was required by the Board of Excise in 1845 to matriculate at University College London with other excise officers, although it seems that he did not attend any chemistry classes.[17] The link with University College London continued, with excise officers studying under Graham and then under Alexander Williamson, who succeeded Graham in 1855. Phillips built up a laboratory that carried out increasingly numerous and detailed analyses over the next three decades. From around 100 analyses in 1842 his laboratory was undertaking 9,500 by 1859, when six assistants were employed.[18] He had developed his reputation by detecting the adulteration of tobacco, a subject to which Parliament turned in 1844.

TOBACCO, TEA, AND COFFEE

It was the importance of tobacco revenue that had led to the creation of the chemical laboratory of the Board of Excise.[19] In the early part of the nineteenth century, the government had taken a typical laissez-faire approach to tobacco after suppression of smuggling had resulted in producers adulterating it instead to increase profits. While the government passed an act in 1840 that forbade the mixing of leaves with tobacco, it did not explicitly prohibit the addition of other substances, such as confectionery, salt, and sand. That had the effect of reducing the revenue after producers leapt through the loophole, so the Pure Tobacco Act 1842 was passed in an attempt to limit additions. Enforcement required chemistry, and a laboratory was established with Phillips engaged on tobacco analysis. Using a microscope and chemical techniques he soon detected adulterants, including sugar, numerous different leaves, salts, dyes, and potassium nitrate. Outside help was brought in, and Graham analysed around one hundred samples in the fifteen months after the act was passed.[20]

This work was challenged by Joseph Hume, who chaired a select committee on the tobacco trade in 1844. Hume was anxious to see a reduction in the import duty on tobacco. He asked George Phillips, Richard Phillips, and Graham to analyze adulterated samples, without providing pure tobacco as a reference. They could not detect all the adulterants, although their work held up quite well. Ure was asked separately, and gave an analysis with less clear results.[21] John Lindley, professor of botany at University College London, gave George Phillips's evidence additional credibility by agreeing that adulteration could often be detected using the microscope.[22] The committee was persuaded by this evidence that much adulteration could be detected. They came to no formal recommendation but did express the view that a reduction of one shilling per pound weight in duty, while having a limited effect on smuggling, would lead to a loss of £1.2 million to the revenue, a huge sum.[23]

This failure to reduce the duty did not please the manufacturers, and they took issue with the position of Phillips and Graham that no sugar was found naturally in tobacco. So they employed Ure, Brande, John Cooper, and William Herepath to disprove the claim. These chemists found that sugar did occur naturally at up to 2 percent.[24] Graham admitted the fact, and work continued for many years to establish the natural variability of sugar in tobacco. Adulteration gradually decreased as a consequence of the work of the laboratory, falling from a detection rate of 69 percent of samples in 1844 to 16 percent in 1860. Parliament revisited the legislation several times in the following decades to try to limit adulteration, which had largely ceased by the 1890s.

So many different substances were subject to duties in the early nineteenth century that the laboratory developed a wide expertise. In the case of soap, Richard Phillips and Thomas Thomson were brought in to help. Until the duty was abolished in 1853, pure soap had a higher duty than impure soap. Phillips and Thomson visited soap manufacturers in 1839 and recommended how duties might be charged fairly on different soaps.[25] As to other substances, pepper could be mixed with ground rice, mustard seeds, sago, starches, and rapeseeds. It was still being adulterated in the 1880s.[26] Tea was widely adulterated, both as imported from China or by wholesalers. The Board of Customs established a separate tea laboratory in 1875 to counter the practice, coincident with the passing of the Sale of Food and Drugs Act 1875. Coffee offered further opportunities. After 1840 it was legal to add chicory, and considerable effort was expended to distinguish the two in mixtures. In 1851 Graham and his colleagues, including Bell, carried out chemical analysis. The chairman of the Board of Inland Revenue later commissioned William Carpenter, physiologist and professor of medical jurisprudence at University College London, and Alfred Swaine Taylor, professor of medical jurisprudence and chemistry at Guy's Hospital, to report on coffee mixed with chicory, which at this point was prohibited by a Treasury minute of Lord Derby's government. They recommended that chicory could be added, but only if so labeled.[27]

Manufacturers continued to experiment with different formulations and adulterations to reduce the duty paid, to the extent that the government gradually increased the duty on chicory to match that of coffee. Taxation rather than science was the solution in this case.

WIDER USE OF THE CHEMICAL LABORATORY

The expertise developed by the laboratory was made available to other government departments as its reputation grew in the later part of the century. These included the Admiralty, for analysis of juices to prevent scurvy; the India Office; the General Post Office; the Stationery Office; the Home Office; and the Commission of Works. Water analysis was a further focus, with the establishment of a dedicated laboratory in 1879. The analyses extended to bacteriological work in the 1880s, and by 1899 the laboratory had taken over the analysis of London water supplies when Edward Frankland died.[28]

BEER, WINES, AND SPIRITS

Duties on beer, wines, and spirits occupied both the Board of Excise and the Board of Customs, and were vital to the Exchequer. The charging of duty on spirits by proof required the use of a hydrometer, which had been specified as Clarke's hydrometer in an act of 1787.[29] This proved unreliable, and a committee of the Royal Society concluded in 1790 that measuring specific gravity would be preferable. Given the importance of excise duties to the Exchequer, and inconsistencies in measurement, Nicholas Vansittart, the joint secretary of the Treasury, established a committee in 1802 to examine the problem. William Wollaston worked for fifteen years on the challenge, in particular to test hydrometers, as a paid consultant to the Board of Excise and the Treasury.[30]

After this government-sponsored examination, a hydrometer by Bartholomew Sikes was declared the best. The Sikes hydrometer became the legal basis for measurement in 1816 and spirit duties were simplified in 1825.[31] To this was added examination of liquids prior to fermentation or distillation using a saccharometer invented by Robert Bate. This allowed an estimate to be made of the amount of spirit that would later be obtained. The Sikes hydrometer also drew criticism, so the Board of Excise, wishing to examine a new hydrometer and to establish an accurate standard for measuring alcohol, wrote to the Royal Society for advice.[32] Convened initially in 1832, the Royal Society's Excise Committee drew up specifications over several years for the construction of instruments and tables for ascertaining the strength of spirits. It was a large committee, including Davies Gilbert, Brande, and Michael Faraday. Faraday examined Bate's new hydrometer and his tables, and found them accurate.

After their failure in the 1830s, the West Indian sugar planters agitated again in 1846 for sugar and molasses to be allowed in brewing and distilling. Phillips and a colleague worked for several months to determine the proof yields, and the Board of Excise came to the view that their use would cause no loss to the revenue, provided that the duty on molasses was increased.[33] The results were extended by George Fownes, professor of practical chemistry at University College London.[34]

In 1847 Parliament authorized the use of sugar in breweries and distilleries and molasses in distilleries. Phillips and his colleague therefore needed to find ways that could be carried out by excise officers of measuring the amounts of sugar and malt originally used in ferments. That was a challenge, and in 1852 the Board of Inland Revenue had recourse to Graham, Wilhelm Hofmann, and Theophilus Redwood, professor of chemistry to the Pharmaceutical Society, as an external check. Their method was given statutory authority in 1856, thus embodying an official method of analysis in the law. In recognition of his impact on the revenue, Phillips was awarded £500 in 1862, nearly as much as his annual salary.[35] Malt duty was abolished in 1880, with tax raised on beer instead to simplify revenue collection. Bate's saccharometer was still used to measure the gravity of worts from which the beer was made.

Other complexities arose from the rules on duty. All imported spirits were liable to duty before 1855. That led to complaints from manufacturers of lubricants, who argued that it caused British products to be uncompetitive with those of foreign manufacturers who had paid no duty on the raw material.

They asked for the ability to use spirits duty-free for industrial purposes, and Phillips suggested that adding methanol, which would make the spirit unpleasant to drink, would reduce any possible revenue loss. Graham, Hofmann and Redwood were again asked to advise. They suggested that adding about 10 percent of methanol would make the substance undrinkable. Frederick Abel wrote to recognize the improvement he had seen as a result in the management of workmen making explosives. They tended to drink the alcohol used to moisten the explosives, and the smell of methylated spirit now enabled the culprits to be identified easily and dealt with.[36]

It was the interest of the Board of Customs in wine that resulted in the establishment of further laboratories separate from the Board of Inland Revenue. When an 1860 act set duties based on alcoholic or proof strength, the Board of Customs also set up a chemical laboratory, of which James Johnstone was the first official. In cases of doubt, samples were sent to the Inland Revenue laboratory.[37] The work extended to beer, with demand for analysis fluctuating as the government tinkered with different regimes of duties.

The adulteration of beer was a further problem, which particularly exercised Bell as successor to Phillips from 1874. Some of this interest concerned questions of public health, such as the detection of "grains of paradise," which was used to make beer appear stronger than it was, and potentially poisonous. The majority related to adulteration that might affect revenue, such as the addition of sugar or salt, dilution by water, or the illegal addition of spirit. The analysis required an understanding of the natural variability of many substances in beer, which was a focus of Bell's work following the Sale of Food and Drugs Act 1875.[38]

Concern about the adulteration of food and drink crossed several distinct governmental responsibilities. The primary interests related to taxation and public health. Although the chemical laboratory of the Board of Inland Revenue was directly concerned with protecting revenue, its expertise was relevant to public health problems too. There were therefore interactions between the laboratory and, at different times, the interests of the medical department of the Privy Council, the Board of Health, the Local Government Board, and the Home Office.

FOOD ADULTERATION

The growth of food adulteration, little known in the eighteenth century, sprang from several causes.[39] In the expanding urban areas, the separation of producer from consumer led to a loss of personal connection and hence of local trust and sanction. The situation was exacerbated by free competition and the temptation to replace highly taxed ingredients such as malt by cheaper ones. This resulted in food that was often unhealthy or downright dangerous. The chemist Frederick Accum published *A Treatise on Adulterations of Food and Culinary Poisons*, an analysis of adulterated foodstuffs, in 1820.[40] It included delectations such as Gloucester cheese colored with red lead, tea made of thorn leaves, and coffee adulterated with "pease and beans."

Some adulterants caused sufficient concern to require individual legislation. For example, the Sale of Arsenic Regulation Act 1851 was passed after pressure from the medical profession about unregulated sales of arsenic, which were leading to deliberate or accidental poisonings. Green blancmange, colored with copper arsenate, was one vehicle of death at a public dinner in 1848.[41] Concerted action had to await a campaign by Thomas Wakley, editor of the *Lancet*. In the early 1850s he commissioned the physician Arthur Hassall, known as a practical microscopist, and Henry Letheby, from 1855 the medical officer of health and public analyst for the City of London, to undertake analyses. They found widespread adulteration, ranging from diluting milk to the presence of poisonous coloring matters in sweets, which may have damaged the health of many children.[42] The surgeon John Postgate found similar problems in Birmingham. Wakley's promotion of the findings and the wider press interest developed pressure for action.

While John Simon, as medical officer of the Board of Health, asked Hassall to investigate the adulteration of bread in 1857,[43] Postgate had already encouraged William Scholefield, radical MP for Birmingham, to ask for a select committee, whose remit covered food, drink, and drugs.[44] Simon, then in his role as medical officer of health for the City of London, gave evidence, as did Postgate, Hassall, and Letheby.[45] They were joined by chemists, including Robert Thomson, professor of chemistry at St Thomas's Hospital; Robert Warington, chemical operator to the Apothecaries Company; and Redwood. George Phillips described the work and training of his analysts.

The range of adulterated products described by the analysts included bread mixed with plaster of Paris and alum, acid drops with oils containing prussic acid, cayenne with red lead, and gin with sulphuric acid.[46] That the problems were endemic is illustrated by the navy's decision to establish its own mustard manufactory to supply it with unadulterated mustard, as evidenced by Richard Gay, superintendent of the factory.[47]

The committee found extensive examples of adulteration for pecuniary fraud and of adulteration that caused injury to public health. It recognized that detection was a problem, although it believed that the Board of Inland Revenue, with some sixty to seventy analytical chemists across the country, had stopped or diminished adulteration of many substances. Simon had argued that responsibility should lie with Poor Law medical officers, as this was a public health matter, not an Excise issue, and that these people should be trained in analysis. The committee agreed that the responsibility should lie

with local boards of health or nuisance authorities, who might appoint officers, with some central support from the General Board of Health. This was
just after Edwin Chadwick had departed from the original Board of Health
in 1854. The new one lasted only until 1858, when its responsibilities for local
government were transferred to the Local Government Act Office within the
Home Office.

Badgered by Postgate, Scholefield tried to bring in a bill. But without government support it was defeated by food interests, until the killing of twenty
people in Bradford by peppermint lozenges adulterated with arsenic led him
to try again in 1859. The Food and Drugs Act 1860 was the result. The act
made it illegal to sell food knowing that it was impure and allowed local
authorities to appoint analysts. Letheby was appointed for the City of London. Few others followed. If appointments were made, the appointees were
often trained in medicine rather than chemistry, and it was difficult prove
that a vendor knew that the goods were adulterated. The act was a failure in
practice. The *Lancet* continued to agitate, as did the *Chemical News*, founded by
William Crookes in 1859, and the Social Science Association. Public opinion
gradually built pressure that led to the Adulteration of Food and Drugs Act
1872. This act now obliged local authorities to appoint analysts when required
by central government in the shape of the new Local Government Board. It
also provided for appropriate people to collect samples—such as inspectors of
nuisances, weights and measures, or markets—and to prosecute adulterating
tradesmen, who could no longer employ the defense that they were unaware.

MEAT AND MILK FROM DISEASED ANIMALS

Although not a case of food adulteration, the sale of meat and milk from
diseased animals was a parallel concern. At one level it was tied up with the
question of the control of animal diseases under the prevailing strategy of
slaughter.[48] It was difficult to prevent products from infected animals reaching the public, although evidence of the danger to human health was disputed.
For example, Simon asked John Gamgee, principal of Edinburgh Veterinary
College, to investigate the problem in 1862, but came to the conclusion that
the main concern was putrid meat. Gamgee's revelations may nevertheless
have led to a private member's bill, which resulted in the 1863 act to amend
the Nuisances Removal and Diseases Prevention Act 1855 for England. This
extended the powers of local authorities to seize diseased meat and food.[49] A
year later, having commissioned Johann Thudichum to investigate parasites
in meat, Simon was able to reveal that London as a whole had no protection
against parasite-infected meat.[50] The subsequent Nuisances Removal (No. 1)
Act 1866 recognized that diseased meat could cause sickness and needed to be
removed from sale.

Opinions continued to differ among the experts. Giving evidence to a select committee on contagious disease in animals in 1873, Gamgee was of the view that diseased cattle should not be put on sale for food at any point. In contrast, George Brown, professor at the Royal Veterinary College, London, thought that food from animals infected with foot-and-mouth disease would be safe, and James Simonds, principal of the Royal Veterinary College, thought that meat from animals infected with pleuropneumonia was safe.[51]

FOOD ADULTERATION BECOMES PARTY POLITICAL

In the case of food adulteration, science now became part of the problem. Analytical science was still in its infancy, and the analysts were not generally leading exponents of chemistry. Evidence from different analysts could conflict, the validity of their evidence in court could be questioned, few standards were specified, and there was no clear definition of adulteration. The issue became party political during the 1874 general election. Under pressure from traders who believed they had been unfairly prosecuted, the Conservatives announced that they would repeal the act, while Liberals admitted that it needed amendment.

After winning the election, the Conservatives established a select committee under the agriculturalist Clare Read, containing several trade representatives. It went as they might have expected, if not planned. Grocers complained about false convictions and analytical chemists disagreed with each other. Many local analysts gave evidence, as did more significant figures such as Bell, now principal of the Inland Revenue laboratory; James Wanklyn, public analyst for Buckinghamshire; Hassall; and Augustus Voelcker. Voelcker, chemist to the Royal Agricultural Society, was scathing about the competence of many food analysts, who had been in his view "the greatest enemies to the Food Act."[52] This he thought was a consequence of many analysts being medical men who had no analytical experience. He wanted a higher class of men, better paid and funded centrally by the Local Government Board like the sanitary inspectors. He also wanted adulteration laws extended to animal feed. Voelcker and others pointed to the difficulty of defining and measuring what constituted adulteration in many cases.

The committee concluded that at least people were generally cheated rather than poisoned by adulteration, but agreed that the 1860 and 1872 acts needed repealing and replacing.[53] Harmful adulterations had become rarer since the 1850s after public outcries. Overall, the committee's findings, influenced by its business focus, pointed to the need to allow a more flexible definition of *adulteration* to accommodate natural variability and consumer preferences. The committee suggested ways of making legal processes fairer to the trader, such as an appeal to analysts at the chemical laboratory of the Board of Inland Revenue in Somerset House in cases of dispute.

George Sclater-Booth, president of the Local Government Board and insti-gator of the select committee, brought forward a bill with weak provisions that were challenged in public and within Parliament. With pressure from the Liberal MPs Charles Cameron, who had trained in medicine, and Playfair, the resulting Sale of Food and Drugs Act 1875 continued to require all local authorities to appoint public analysts when ordered by the government.[54] More significantly, they could prosecute without the need to prove that trad-ers were guilty of deceit. The act moved the emphasis to whether the food or drug itself was injurious to health, rather than whether individual ingredients might be harmful. The "chemical officers" at Somerset House were appointed as independent referees to whom appeals could be made in legal cases. This was designed to placate the business community, although it was criticized by the public analysts who regarded the excise officers as having insufficient expertise. The Society of Public Analysts had been formed in 1874, and helped develop standards.[55]

Responsibility for collecting samples was extended to medical officers of health, whose local appointment had been compulsory since the Public Health Act 1872, and to police constables. Although the 1875 act has been seen as a success, reliable sample collection was a problem as was reliable local analysis. As a result, there may have far more adulteration than was recorded in the official figures.[56] The act was strengthened in 1879, and by 1880 almost all authorities had appointed analysts. Enforcement analysis was made compul-sory in 1899, although a few local authorities still ignored it. Only in 1900 did the Local Government Board specify clearly the requirements needed for the post of public analyst.[57] In the early twentieth century, government went a stage further, introducing minimum standards for many foods.

The work of the Inland Revenue laboratory had now evolved from an almost complete focus on revenue protection to the protection of public and of animal health. Its value externally is illustrated by a departmental com-mittee of the Board of Agriculture that was established in 1892 to investigate the adulteration of artificial manures, fertilizers, and feeding stuffs. The main expert member was Bell. The committee recommended better labeling, stan-dardized sampling processes, and the initiation of criminal prosecutions sub-ject to the opinions of authorized analysts.[58] The Fertilisers and Feeding Stuffs Act 1893 tightened regulation.

This episode also contributed to the amalgamation of the laboratories of the Inland Revenue and Customs as the Government Laboratory in 1894.[59] There had been tension between the two laboratories, with Bell seeing his as the senior partner. Now, the new act required an independent referee to adjudicate between a public analyst and the industry when needed. A Treasury committee reviewed the laboratory provision and, with Bell due to retire, recommended retaining the Customs laboratory but that a single person be appointed as temporary inspector general of revenue laboratories. Thus came into being the Government Laboratory under a new chief, Thomas Thorpe, professor of chemistry at the Royal College of Science in South Kensington, where many analysts now trained.[60] He had studied, among others, under Robert Bunsen and Henry Roscoe. This continued until 1911, when the Department of the Government Chemist was created. Like the medical men, the chemists were now an integral part of state bureaucracy.

The experts advising on chemical analysis in relation to taxation found themselves dealing with central government departments on questions of optimising the revenue. However, and particularly in relation to foodstuffs, wider issues of public health also became salient. These brought to the fore familiar tensions between central and local government, and between compulsion and freedom of the individual, in addition to the rights and responsibilities of private industry.

14

WEIGHTS, MEASURES, AND COINAGE

ON OCTOBER 16, 1834, THE HOUSES OF PARLIAMENT WENT UP IN FLAMES. Few of those watching the conflagration would have been aware of the danger to an elegant brass yardstick kept securely in the building. It was irreparably damaged, and with it went the legal standard British measure of length, the imperial unit. The bar had been made by John Bird in 1760 following an eighteenth-century inquiry. Its length was established in relation to the length of a pendulum that beat, or oscillated, every second at the latitude of London.

The yard had been standardized in the Weights and Measures Act 1824. It formed the basis of a system for defining other measures and weights across Britain and its empire, using a mechanism that in principle could be set up anywhere, as opposed to transporting a physical object of reference. It also offered a unit derived from nature that was believed to be constant and thus universally reproducible. The French had taken a similar approach to the definition of the meter. It was calculated as one ten-millionth of the distance from the North Pole to the equator, passing through Paris.

Although the construction of a unit of measurement may seem to be a scientific process, its establishment in law requires politicians. The men of science found themselves in a political as much as a scientific arena, subject to political tensions between central and local powers, and between options of pragmatic gradual development and revolutionary change. To add to the political complexity, the science was not straightforward. Not only did the practitioners of science have to address sources of difference between different observers and a lack of reproducibility but also to deal with variability in the length of a pendulum beating every second at different places on the globe, due to gravitational effects of the shape of the earth, which was not precisely known. That a metrological solution was adopted at all gives an intriguing insight into the relationship between science and policy at this point.[1] That it did not last is equally telling.

THE WEIGHTS AND MEASURES ACT 1824

The Weights and Measures Act of 1824 had a long history.[2] Britain had a centuries-old tradition of using nonstandard weights and measures, differing from county to county, and even within counties. This local variation was accepted through custom and practice. It could offer flexibility in response to trading and economic conditions—for example, to help relieve local poverty at times of distress, when adjustments could be made to benefit the poor. But there were also long-standing concerns that the poor were being cheated in markets by different local measures that did not conform to official standards.

The question of fraud became politically salient, heightened in the early nineteenth century by violent disputes over the Corn Laws. A relatively new member of Parliament, George Clerk, supported by the radical Samuel Whitbread and by Davies Gilbert, asked in 1814 that a select committee make recommendations about standardization.[3] Clerk was a Tory, and became a friend of Robert Peel. In 1819, proposed by Gilbert, William Wollaston, and John Barrow, among others, he was elected a fellow of the Royal Society, being "a Gentleman well versed in Mathematics and attached to Science in general."[4] In the same year he was appointed a lord of the Admiralty.

In 1814 Clerk was an up-and-coming man. He chaired the select committee on weights and measures that he had proposed.[5] It made a series of recommendations about licences for making weights, their official stamping, and penalties for using false weights and measures. It also proclaimed in its opening statement that "the great causes of the inaccuracies which have prevailed are the want of a fixed standard in nature, with which the standards of measure might at all times be easily compared."[6] These standards could be derived from a pendulum for length, from which the volume could then be calculated. Those volumes could be converted into weights through the use of water, given its constant specific gravity at a known temperature. The committee argued that the standards so obtained would simplify the system and be easy to use. It ended by declaring that not to introduce such a system would continue to offer openings for fraud.

The committee, which included Gilbert,[7] interviewed Wollaston and John Playfair, professor of natural philosophy at the University of Edinburgh, using a sequence of questions that neatly and inexorably led to the committee's recommendations. It is difficult to decide whether this was policy-led science or science-led policy, perhaps an element of both, but the report reads in a compelling manner. Both Playfair and Wollaston recommended the pendulum as providing the best natural standard for length, although Playfair stated that the result had not been achieved with all the accuracy that might be obtained. They clearly preferred it to the revolutionary French choice of part of an arc of meridian that defined the meter.

Following this inquiry, Whitbread introduced a bill in 1815 that later became law. It contained some provision for a set of common measures to be held locally for inspectors, but did not contain any reference to a metrological standard. So, Clerk tried to introduce such a standard the same year, with the more radical proposals based on the pendulum and a revised system of weights. But Parliament was not prepared to go so fast.

Meanwhile, the Royal Society had established a Pendulum Committee in 1816 to determine the length of the pendulum that beat every second and to compare British and French standards.[8] This committee was instigated as a consequence of the select committee's report and of a request by the home secretary, Viscount Sidmouth, to the council of the Royal Society.[9]

The Pendulum Committee was chaired by Joseph Banks and included Thomas Young, Henry Kater, and Wollaston, in addition to Gilbert, John Pond, and others. Young worked with Pond, astronomer royal, and Kater to establish how to measure accurately a portion of the length of a pendulum that beat every second. Pond's report of experiments by Kater was sent to Viscount Sidmouth and published by the House of Commons in 1818.[10] Kater, employing some technical innovations, gave a determination of the length of Bird's parliamentary standard and of the French meter. Nevertheless, he considered his experiments incomplete. Pond also indicated that other experiments were proceeding. At no point in the report is there any indication of whether the results might be reliable and accurate enough for practical purposes. The men of science treated it as a purely scientific problem.

They were called on formally again the next year when a royal commission was appointed "to consider the subject of weights and measures." It aimed to explore how practical a more uniform system might be. Apart from Clerk, the commission consisted entirely of scientific experts, although Gilbert was also an MP. Gilbert, who drafted the subsequent legislation with Clerk, was joined by Banks, who chaired the commission, Wollaston, and Kater, with Young as secretary. It is a sign of the firm belief of men such as Gilbert and Banks that science provided the best means to tackle these practical problems.

Their first report in 1819 argued again for the adoption of a natural system based on the pendulum, in order to recover the length of any standard should it be lost or damaged, and for measures of capacity to be based on weighing water. They recommended altering the existing system as little as possible, so as not to disadvantage people engaged in everyday business, and they rejected the metric system as too complicated a change from established British methods.[11] One of the few advocates of the metric system was the third Earl Stanhope, who had been an enthusiast for the French Revolution and had opposed the war with France.[12]

The second report in 1820 confirmed their support for the adoption of Bird's bar as the standard yard, or 36 inches, by comparing it with the length of a pendulum vibrating every second at the latitude of London at 62 degrees Fahrenheit. The length was declared to be 39.13929 inches, very close but not identical to the length of the French meter.[13] That enabled the yard to be established in proportion. The final report in 1821 proposed that Bird's bar be called "the authentic legal Standard of the British Empire," and the gallon containing ten pounds avoirdupois of water be the "Imperial Gallon."[14] Kater had done some further experiments to finalize calculations of quantities by volume.

In practice, these were not radical changes. It was a pragmatic, gradualist approach, in the British tradition rather than that of the revolutionary French. King Louis XVI had invited Britain to discuss an international agreement, to which the government had not responded, and the French had then developed the metric system, which was officially adopted in 1799. Young, as secretary to the commission, was of ideal temperament to draft the reports.[15] He had made this clear at the outset, writing, "We are impressed with a sense of the great difficulty of effecting any radical changes, to so considerable an extent as might in some respects be desirable; and we, therefore, wish to proceed with great caution."[16] He was a sound and soothing pair of hands. The select committee reconvened to add its own short report two months later.[17] It emphatically ruled out moving to a decimal system, which it claimed had caused major problems in France, and endorsed the simplifications recommended by the royal commission. The members added that they imagined that the standards might be adopted in the United States, which would "tend to facilitate the commercial intercourse" with those erstwhile colonies.

It was not yet over. The House of Lords convened a select committee to examine a petition against the proposals from the Glasgow chamber of commerce, which supported a metric system.[18] Their report came to no firm conclusions and left the position open. Gilbert and Clerk were both called as witnesses, as was Patrick Kelly.[19] Kelly was a metrologist, consulted by both Houses of Parliament on matters of coinage and currency. Gilbert admitted that they had not interviewed any tradesmen as "it is manifest that in establishing one uniform set of measures, there must be more or less of inconvenience experienced by those who have been in the habit of using such as are about to be abolished."[20] The men of science knew best. Kelly, however, argued that the emperor of science's clothes were threadbare. He stated that "nature seems to refuse invariable standards; for, as Science advances, difficulties are found to multiply, or at least they become perceptible and some appear insuperable." He regarded the diversity of measures of capacity as "more an imaginary than a real evil," as in practice people understood the differences and set prices accordingly.[21]

This testimony did not stop the passing of the Weights and Measures Act 1824, which was introduced by Clerk and seconded by Gilbert. It incorporated the simplifications developed primarily by Wollaston, and Bird's standard yard, whose length was established by the pendulum method. A review of how the system was working was undertaken in early 1834,[22] and resulted in the 1835 act that formalized the stone and hundredweight.

DESTRUCTION OF THE STANDARDS

Then came the fire of October 1834. Bird's standard was destroyed. Work on pendulums had continued under the Board of Longitude and, following its abolition in 1828, under the auspices of the Resident Committee of Scientific Advice to the Admiralty, consisting of Young, Michael Faraday, and Edward Sabine. But Young died in 1829, and pendulum work was then coordinated by Sabine, with the astronomer Francis Baily also taking a hand and undertaking a wide range of experiments.[23] Baily was given responsibility for the Royal Society's Pendulum Committee, which now included Sabine, George Airy, and William Whewell. Pond was the only surviving member from the original committee. Baily's experiments raised further questions about the accuracy of previous measurements, and Young and Sabine had pointed out other problems. Airy was therefore in a good position to help when Thomas Spring Rice, chancellor of the exchequer in Viscount Melbourne's Whig government, wrote to him in May 1838, asking him to chair a commission to determine "a Standard Weight and Measure, to replace those which were destroyed by the burning of the Houses of Parliament."[24]

The eight commissioners, reduced to seven after Gilbert's death in 1839, were almost entirely men of science, and only one was not a fellow of the Royal Society. Airy was joined by Baily, John Herschel, John William Lubbock, George Peacock, and Richard Sheepshanks, mostly mathematicians and astronomers, with John Shaw Lefevre (not a man of science but a fellow of the Royal Society) and John Drinkwater Bethune, counsel to the Home Office. This group of men took a rational scientific approach by recommending that the government consider decimal coinage and a partial decimal approach to weights and measures, including land measures.[25] Decimal coinage would have to wait 130 years, until 1971, excepting perhaps the introduction of the florin.

The possibility of a metric system of weights and measures would have to wait too, although not so long. It had been a live issue for many years, although ruled out in the early part of the century for being too revolutionary and too French.

Airy's commission examined the state of the standards recovered from the ruins of Parliament. The standard yard was "so far injured, that it is impossible to ascertain from it, with the most moderate accuracy, the statutable length of one yard."[26] To add to the woes, the legal standard of one troy pound was missing. It was therefore necessary to take steps to make and legalize new standards of length and weight. The commissioners did not comment on the fact that the country had survived for several years without them, and indeed would continue to do so until 1855. They took additional evidence and issued a report in 1841 with detailed practical proposals for creating parliamentary and secondary standards.

Most problematic for existing legislation was the evidence they quoted to show that both the measurement of length by a pendulum and of weight by a volume of water were unreliable. They argued that these could not form the basis of reestablishment of the standard, contrary to the provisions of the 1824 act. They concluded that "with reasonable precautions, it will always be possible to provide for the accurate restoration of standards by means of material copies which have been carefully compared with them, more securely than by reference to any experiments referring to natural constants."[27] Thus artisanal pragmatism, based on accurate measurement, replaced the idea of scientific precision derived from natural constants.

Fortunately, the Royal Astronomical Society had constructed its own standard yard in 1832. Indeed, Baily had used Bird's yard in its construction. It could therefore be used instead, along with others held by the Royal Society and the Ordnance Survey, although it was not until the Weights and Measures Act 1855 that this formally happened, making the new copies the "restored Standards."[28] The delay was caused by problems of constructing the instruments, based on Baily's work on determining length and William Hallowes Miller's work on determining weight. Following the commission's report in 1841, a further commission had been established, also under Airy, to superintend the construction of the new standards. To the old commission were added the Marquess of Northampton, the Earl of Rosse, and Lord Wrottesley, who served in sequence from 1838 to 1858 as presidents of the Royal Society. This commission finally reported in 1854, although Herschel was not able to sign its report as he was by then a "confidential Officer of the Government,"[29] as master of the mint. At the end of all this the physical yardstick, not the pendulum, became the basis of the standard for length. The accurate shape of the earth remained unestablished.

THE ROYAL MINT

The nation's coinage was inevitably bound up with issues of weights and measures. There were particular standards of weights for gold and silver, and the currency of pounds, shillings, and pence was, like British weights and measures, not a decimal system.

The Royal Mint and its predecessors have made Britain's coins for more than a millennium. Control of the country's coinage and trust in its authenticity are essential matters of good government. Given the metallic basis of coinage, scientific measurement has been important to ensure the quality of the coins of the realm and to counter fraud. Chemistry was vital for analysis, such as the annual Trial of the Pyx, the formal process for certifying the quality of the coin.[30] Likewise, technical engineering expertise has been essential for the manufacture of coins.

At the start of the nineteenth century the Royal Mint was headed by the master of the mint, who reported to the Treasury. It was an administrative role and a political appointment, and the master has rarely been a man of science. There are a few notable exceptions. The position was held by Isaac Newton from 1700 to 1727.[31] A century later the last two masters of the mint were also men of science, Herschel from 1850 to 1855, and Thomas Graham from 1855 until his death in 1869. After Graham's death the Royal Mint was absorbed into the Treasury and the role of master of the mint subsumed into that of chancellor of the exchequer.

The Privy Council Committee on Coin had been established in 1787 to oversee the currency and the Royal Mint.[32] It was reconstituted in 1798 when it contained, as a symbol of its political importance, the entire cabinet and the speaker of the House of Commons, as well as Banks. The committee soon determined to renovate the Royal Mint and took expert advice. John Rennie, engineer of the Birmingham firm of Matthew Boulton and James Watt, which made coins and medals, inspected the machinery and made a highly critical report. Rennie recommended introducing steam presses. Boulton and Watt steam engines were duly installed in the renovated Royal Mint, which was moved from the Tower of London to a new building just outside and completed by 1810. In parallel, Henry Cavendish and Charles Hatchett were asked to examine wear on coins. They advised that pure gold and silver were unsuitable, and that the amalgams in use were an effective compromise between hardness and color. Later, in 1823, Thomas Brande was asked to carry out an enquiry into the best blend of constituents for the steel used in dies. As with Brande's appointment to the Royal Institution, this employment seems to have been instigated by Hatchett. Brande's research improved the life of the dies. He was appointed superintendent of machinery in 1825, retaining his post at the Royal Institution.[33]

The Royal Mint, now in its new buildings, was reorganized in 1815 by William Wellesley Pole, the new master of the mint. He abolished various sinecures and put many posts on a salaried basis rather than their reliance on fees, which were being increasingly regarded as opportunities for corruption. Government and parliamentary enquiries into the Royal Mint in subsequent decades concentrated on its management and labor practices rather than technical issues.[34] These enquiries culminated in a report by commissioners in 1849, who argued that the situation could be remedied "in no other way than by an entire reconstruction of the system on uniform principles adapted to present circumstances, and to the general progress of improvement in manufacturing science."[35] They offered a plan for reorganizing the Royal Mint.[36]

Lord John Russell's liberal government appointed Herschel as master of the mint in 1850, after a select committee inquiry into the position of members of both Houses of Parliament who were also Crown officials. The post was thereby taken away from a politician and given to a man of science. All staff now came under Herschel's management. Brande's role was widened, with a salary increase to £900, and he resigned his post at the Royal Institution in 1852. Herschel also appointed William Allen Miller, professor of chemistry at King's College London, and Graham, professor of chemistry at University College London, as consultant assayers, thereby furthering the patronage of chemistry by government.[37]

But Herschel's reign was an unhappy one, given the organizational upheaval and pressures of new coinage issues.[38] He resigned, citing ill health, in 1855 and was succeeded by Graham, whose place as assayer was taken by Wilhelm Hofmann. Graham was known to government, in addition to his role in the Royal Mint, through service on the ventilation of the new Houses of Parliament, the casting of guns, and the analysis of water supplies. One innovation that Graham introduced was the use of bronze instead of pure copper for coinage. But Graham's administration, like Herschel's, was not a happy one. He found the resistance of the institution to change a great struggle over many years.[39] When he died in 1869, the Treasury official Charles Fremantle, who had become deputy master, submitted a formal report to the Treasury.[40] The result was the abolition of the separate post of master and its incorporation into the responsibilities of the chancellor of the exchequer. The deputy became, in effect, the chief executive and started to shape the Royal Mint into its modern form.

The two men of science had not been a great success in management terms. The Treasury view was that great scientific attainments "were not essential to the successful administration of the Department which depended rather upon active and intelligent control and the application of well-trained experience in matters of business."[41] The Royal Mint did still need scientific and technical advice, perhaps the driving force for the appointment of Herschel and Graham. Brande had died in 1866, and William Roberts-Austen, who had been Graham's private secretary, was appointed as chemist in place of the assayer to the Royal Mint. Thus ended the experiment of appointing people with scientific expertise to the leadership of the Royal Mint. Fremantle remained in charge until 1894.

DECIMAL COINAGE

While the influence of external scientific advice on the Royal Mint was limited, mathematical and scientific experts did try to influence public policy on coinage. The question of decimal coinage was a live issue throughout the nineteenth century, given its presence on the continent. Nor was it a new idea even then. As far back as 1682 the economist William Petty had argued the advantages of decimal arithmetic for accounts, suggesting that farthings be made five to the penny rather than four.[42] The French had replaced the livre with the franc in 1803, divided into one hundred centimes. Soon after the Napoleonic Wars, in 1816, John Croker raised the question of decimal coinage in a debate on the Earl of Liverpool's Coinage Act. His suggestion was not supported. John Wrottesley, father of the later president of the Royal Society, proposed an inquiry into decimal coinage in 1824 but was opposed by the Royal Mint. Ideas of rationalization were in the air, which may have contributed to the fact that the currencies of England and Ireland were made uniform in 1826.

Airy's commission had recommended moving to the decimal system for currency and for weights and measures in 1841, but nothing came of it. Nevertheless, following a motion by John Bowring in 1847 in the House of Commons, the florin was introduced in 1849 as a sop to decimalization, being one tenth of a pound. Airy's commission reiterated its view in its report of 1854.[43] In March 1853, before they made their final report, the commissioners wrote to the chancellor of the exchequer, William Gladstone, to urge consideration of decimal coinage given that a large coinage of copper was being contemplated. Gladstone replied that the issue was too important for any rapid decision, and that the government would support the establishment of a select committee to examine the question.

The leading light of this committee was William Brown, MP, who traded with the United States and had been struck by the ease and convenience of the introduction of the decimal system there. Evidence from scientific perspectives was given by Augustus De Morgan, Herschel, as incumbent master of the mint, Airy, and Lieutenant General Charles Pasley.[44] De Morgan was professor of mathematics at London University from 1828 to 1831, and at University College London from 1836 to 1866. He was secretary of the Astronomical Society for many years, but never became a fellow of the Royal Society, which he considered too open to social influences to be efficient. He was a visible advocate of decimalization. Pasley, a royal engineer, had published a book about decimal coinage in 1834. All argued for the simplicity of the decimal system in calculations, and they suggested means of preparing the public for the new system. Herschel in particular proposed a substantial period of public education to prepare the lower classes for such a change.

Brown's committee deliberately sought out both "scientific opinion" and evidence of the "practical inconveniences" of the existing system of coinage and means of remedying them. All the witnesses selected were supportive of decimal coinage, including Herschel as master of the mint, though the committee did unsuccessfully invite opponents. The witnesses differed only in how they would address the practical consequences of implementation, and presented evidence that it would lead to greater accuracy, simplify accounts, and reduce the labor of calculation, by up to four fifths according to De Morgan. They pointed out how widespread the use of decimals already was, both abroad and even at home. They noted that a decimal system of coinage and of weights and measures was in use across Europe, and that even "Chinese boys" could readily make decimal calculations. The colony of Canada was introducing a decimal currency. At home, commissioners were told that the Royal Mint already used a decimal system of weights for purchasing bullion and that many teachers used it for calculations in their classes.

The committee considered that two main barriers to the adoption of decimal coinage existed: resistance from the public used to the traditional system and contractual issues if the existing currency were made illegal. The committee members were satisfied by evidence that the public would accept changes based, for example, on the ease with which decimal coinage had been introduced in the United States, and considered the legal problems readily soluble. They did not propose abolishing the pound sterling, recognizing the "political and moral association connected with it," as De Morgan expressed in his evidence.[45] There were some challenges with respect to setting individual costs in a decimal currency, such as the penny postal rate, but they were not considered insuperable. The report ended with a resounding endorsement of the introduction of decimal currency, recommending the pound and mil scheme in which a mil was a thousandth of a pound, close to the value of a farthing.

There was extensive subsequent support in professional circles. For example, the Third Annual Conference of the Representatives of Institutions and the Council of the Society of Arts, held in 1854, unanimously assented to the resolution that "this Conference earnestly desires an early adoption of the decimalization of weights, measures, coins, and accounts."[46] Two institutions were formed around this time to advocate for decimalization. Brown was instrumental in establishing the Decimal Association in 1854, which included about 230 members of the House of Commons. It was followed by the International Association, a British branch of the International Decimal Association that had been established after the 1855 Exposition Universelle in Paris.

Many declarations and petitions were made for legislation. Brown successfully moved motions in the House of Commons in support of decimal coinage, although not for its actual introduction. So with the issue still politically live, and Gladstone adamant about the need for widespread public support if decimal coinage were to be introduced, a royal commission took evidence from 1856. The three commissioners were Lord Monteagle, Lord Overstone, and John Hubbard. Lord Monteagle had been chancellor of the exchequer from 1835 to 1839 in Viscount Melbourne's government as Thomas Spring Rice, becoming Lord Monteagle in 1839. He was a strong supporter of decimalization. Overstone was the banker Samuel Jones Loyd, architect of Peel's Bank Charter Act 1844 and raised to the peerage in 1850. Hubbard was a financier and Tory politician. The commissioners sought evidence from opponents of decimalization in general or of the pound and mil scheme, and submitted a report in 1857 to inform their further deliberations and public debate.[47]

The commissioners found no compelling evidence for change based on international experience, pointed out that weights and measures had not been decimalized, and argued that both the pound and mil scheme (which would keep the pound) and the penny scheme (which would abolish it) had drawbacks compared to the status quo. A persuasive argument was that of public acceptability. They concluded that "these difficulties are partly of a moral character, arising from the violent disturbance of established usage and habits, especially amongst the uneducated classes, which are the least qualified to comprehend, and the least disposed to acquiesce in, such disturbance of their customary course of acting and thinking; and partly of a mechanical character, arising from the non-interchangeability of the old and the new coins."[48]

Lord Overstone's draft report, presented with the commissioners' conclusions, reveals the different interests that the commissioners considered. Paragraphs summarize "public feeling," the "scientific view," the "educational view," the "commercial view," and the "official view." Lord Overstone noted that those with scientific expertise were not of one mind. Airy, De Morgan, and Herschel were in favor of decimalization, even if it did not lead to a metric system of weights and measures, which some also advocated. But the commissioners found other scientific experts who disagreed, even if they admitted that "the weight of scientific authority is in favour of the projected change."[49]

The same was the case for other interest groups. It was hardly likely that the commissioners would come to any other political conclusion than to kick the issue into the long grass.

In circumstances where the scientific experts were predominant, such as Airy's two commissions, they were able to advocate for the decimalization of weights, measures, and coinage, with apparent support from witnesses. But the real issues were political, with serious concerns about public acceptability, exemplified by Gladstone's position. Even with a united front the scientific experts would have had a major challenge. As soon as they could be shown to be in the least bit divided, their influence was more limited.

REVISITING THE METRIC SYSTEM

It was the International Exhibition of 1862 in London that proved to be a catalyst for reexamining the metric system. Sponsored by the Society of Arts, Manufactures and Commerce, the exhibition brought together twenty-eight thousand exhibitors from thirty-six countries, on the site now occupied by the Natural History Museum. It was a worthy successor to the Great Exhibition of 1851. The six million visitors were able to see leading British inventions, including the electric telegraph, submarine cables, and the first plastic, Parkesine.

But it was the lack of alignment of the British system of weights and measures with the metric system on the continent that exercised many men of science and industry. The issue had been highlighted in the Great Exhibition of 1851, when jurors were perplexed by the proliferation of different standards, and raised again at the first meeting of the International Statistical Congress in Brussels in 1853. The Society of Arts had petitioned the Treasury for a common system, and the international jury at the Exposition Universelle in 1855 had recommended adopting a universal system. Then at the London meeting of the International Statistical Congress in 1860, Prince Albert had argued for harmonisation. This meeting prepared a report for the next congress in 1863 in Berlin, which adopted a resolution in favor of assimilating the coin of different countries.

The political lead in Britain was taken by William Ewart, a reforming Liberal politician, who had carried a bill to abolish hanging in chains in 1834, and a bill for free public libraries in 1850. He also instigated the blue plaque scheme to recognize the link between a famous person or event and a building. Ewart chaired the Select Committee on Weights and Measures in 1862, which was held during the exhibition so that foreign experts could be readily examined. Its political focus was on trade, and the committee interviewed about forty witnesses, including merchants and working men.[50] In addition to the foreign experts, the British witnesses included many men of science and industry, such as Airy, De Morgan, William Hallowes Miller, William Farr, William Fairbairn, and the industrialist William Siemens.

Having pointed out that British practices incorporated ten different systems of weights and measures, and the French only one, the committee was keen to explore whether the metric system was difficult to use. It soon became apparent that it was already in widespread use for practical reasons, in scientific research for ease of international communication and in manufacturing for better accuracy. For example, the assistant superintendent at the Royal Gun Factory in Woolwich revealed that decimal measures were used in making Armstrong guns, prime symbols of British military power,[51] and Graham, now master of the mint, explained the use of the decimal system at the Royal Mint. In response to the question "Have you ever had occasion to observe whether English people have great difficulty in acquiring the metrical system in France?," Graham answered, "I think not; in fact, I was rather struck with the facility with which English ladies made use of it in keeping their accounts."[52]

This large bundle of supportive evidence orchestrated by Ewart led to the eleven recommendations of the committee. The first was that the metric system should be made legal, although not compulsory until the public was generally in favor. The second recommended the creation of a Department of Weights and Measures within the Board of Trade to verify standards, superintend inspectors, and report annually to Parliament. These key recommendations were followed up to an extent. The Weights and Measures (Metric System) Bill 1864 became the act that implemented the first recommendation,[53] although in a watered-down form that merely legalized the metric system in contracts. It left traders possessing metric weights still technically liable to arrest under the 1835 act. The second followed in 1866, when the Standards of Weights, Measures and Coinage Act 1866 created a department of the Board of Trade called the Standard Weights and Measures Department, which was responsible for the primary and subsidiary standards.[54]

This change of responsibility for the standards, from the Exchequer to the Board of Trade, led to Airy being asked in May 1867 to chair yet another royal commission to report on the condition of the standards. The commissioners issued four reports, covering in particular the metric system, inspection, and abolition of the troy weight, until their work was completed in 1870.[55] Airy was joined by men of science including Sabine, who was at this point president of the Royal Society, Graham, William Hallowes Miller, and the Earl of Rosse. On the controversial topic of the metric system, the commissioners noted that no shopkeepers or ordinary tradesmen had been consulted in earlier inquires and that they had expressed no obvious desire for a change. They concluded that it would be highly probable that any attempt to change, in a country "where the people are more accustomed to self-government than in other European countries," would "totally fail."[56] But they did support permissive introduction alongside the imperial system, and reiterated support for a decimal system of coinage.

REVISITING THE CURRENCY

In parallel with his work on the Standards Commission, Airy now found himself on another royal commission, this time on international coinage.

The first International Monetary Conference had been held in Paris in 1867, prior to which the French government had invited the British to join the new currency arrangement between France, Belgium, Italy, and Switzerland. Other countries were pledging to join, including the United States. Britain sent two commissioners to represent the country, with strict instructions not to commit to anything. They were Graham, as master of the mint, and Charles Rivers Wilson, a senior Treasury civil servant.

In February 1868, a few days before Benjamin Disraeli succeeded the unwell Lord Derby as prime minister in a minority Conservative government, a royal commission was appointed to consider the report from the two commissioners to the conference and to make recommendations. Unlike the royal commission on decimal coinage with its trio, or duo, of commissioners, this one had fourteen.[57] They were chaired by the Liberal politician Viscount Halifax, and about half were members of Parliament, with Wilson as secretary. Three commissioners were scientific experts—the ubiquitous Airy, the banker and naturalist John Lubbock, and Graham himself. But they were hardly there for any scientific expertise. Graham only asked one question, and questions from Airy and Lubbock were concerned, as were those of the other commissioners, with varied matters of administrative detail.

The International Monetary Conference had recommended the adoption of a single gold standard, and that each country should strike a gold coin of twenty-five francs, able to serve as an international coin. This offered a focus for discussing the desirability of a common international system. The commissioners were aware that they would need at least to look constructive, given Britain's international reputation for advocating free trade. The witnesses examined by the commissioners were generally supportive of the move to an international currency agreement, as it would give trade advantages.

But the commissioners saw many disadvantages. Not least, that without a common system of weights and measures, and of a money accounting system, neither of which were within their terms of reference, advantages of currency uniformity would be limited. Overall, they thought the practical disadvantages outweighed the advantages. But their main worry was that the striking of a gold sovereign of twenty-five francs would devalue the sovereign—worth twenty-five francs, twenty cents—and hence the pound. That might seem a small amount, but the commissioners were concerned that "the inconvenience, vexation, and seeming injustice would be apparent to a large portion of the population."[58] Again, fear of negative public reaction was advanced as a barrier to action. In the true tradition of British exceptionalism, the commissioners instead considered that the French and others might adopt the value of the pound, given its effective position as an international currency. They rejected the idea that Britain should adopt a gold coin to the value of twenty-five francs, but did recognize the potential for a uniform international system of coins and standards. Since compromises would probably be required in all countries, they suggested the need for further intergovernmental conferences to consider the difficulties. Subsequent conferences were held in 1878, 1881, and 1892, but no agreement could be reached.

TOWARD THE TWENTIETH CENTURY

After the 1866 act, modifying legislation on weights and measures followed throughout the rest of the century, with governance and inspection systems developing in tandem. By the time of the Weights and Measures Act 1878, local authority departments were established with responsibility for apparatus for commercial use and to counter fraud.

Right at the end of the century a further select committee was established in 1895, chaired by Henry Roscoe, the chemist and MP, to examine further changes. The committee questioned witnesses representing official, commercial, manufacturing, trade, educational, and professional interests. The professional witnesses included the engineers Frederick Bramwell and Benjamin Baker. In a succinct one-page report, supported by reams of evidence, the committee recommended that the metric system be legalized for all purposes, and made compulsory after two years by an act of Parliament.[59] The committee was all but unanimous in its recommendations. There was substantial outside support, including a delegation of forty-six chambers of commerce who went to lobby Prime Minister Arthur Balfour.[60] The men of science were not the only ones in favor. The subsequent Weights and Measures Act 1897 legalized the use of the metric system of weights and measures as the twentieth century approached. Its use was not made compulsory, although witnesses including Baker, but not Bramwell,[61] had advocated that. Balfour, like Gladstone before him, argued the need for widespread supportive public opinion before such a step were taken.

So, in the British tradition of piecemeal and pragmatic development, the national system of weights and measures evolved gradually throughout the nineteenth century. It moved from a fragmented and often localized system to one in which common standards, definitions, and inspection processes were implemented. Even the metric system, shunned initially in the context of its association with revolutionary France and of British exceptionalism and imperialism, had a foot in the door of the legal system of public weights and measures by the end of the century, although those in the scientific community, the engineers, and many international traders had been using it for decades. Scientific expertise played a major role in defining the measurements and suggesting modes of technical simplification. The skills of systematizing, calculating, and of practical experimentation and measurement underpinned the changes. But politicians were needed to advocate for alterations to the law. Gilbert played that role in the early century, Ewart in mid-century, and Roscoe at the end. The men of science who were most involved were a limited group, including Banks, Gilbert, Wollaston, and Young at the beginning of the period, and Airy, Baily, Miller, Sabine, Herschel, Graham, and Roscoe later on. Political and aristocratic cover was provided by people such as the Marquess of Northampton, the Earl of Rosse, and Lord Wrottesley. But despite the desire of so many in the scientific community to move to the metric system and to decimal coinage, British tradition ensured that such change was not politically feasible. Most of that change awaited the second half of the twentieth century. And we still have the mile, the pound sterling, and a pint of beer.

Conclusion

CONSTRAINTS
ON INFLUENCE

As THE NINETEENTH CENTURY UNFOLDED, SCIENTIFIC, ENGINEERING, and medical practitioners became increasingly involved in state activities, while government grew in size and complexity. This increased involvement evolved in parallel and in interaction with government growth, and was part of the growth itself. On the one hand, it reflects the increasing range and impact of military, social, and political pressures for action on issues to which expert advice could be relevant. On the other, it derives from the growth of scientific, engineering, and medical knowledge and disciplines, allied to the increasing authority of experts in these fields. Myriad contingent circumstances influenced the nature and speed of state actions in response to any particular policy issue. In civil policy areas, particular manifestations of liberal culture set major constraints. These were the sanctity of private property, a laissez-faire approach to capitalist private industry, emphasis on individual freedom and responsibility, and the importance of local government.

The preceding chapters revealed the penetration of scientific advice in government departments and in Parliament, and showed it as subservient to politics.[1] That reflects the fact that scientific advice is provided in a political context, while advisers vie with each other and with politicians for influence. Expert advice informed policies and their implementation, but did not wholly determine them, since many other issues were salient to politicians.

Definitions and meanings of measures such as the tonnage of a ship, the standard unit of length, or the level of the malt tax, were ultimately decided by politicians. So, effectively, were the numbers of annual deaths in coal mines, on the railways, or from explosives that were deemed acceptable. Choices of ships for the navy, guns for the army, or the sites and designs of hospitals were political decisions that were influenced by technical considerations, but as one set of factors among others.

When it comes to the implementation of policy and to executive action, as opposed to policy creation and legislation, there may be more scope for the influence of expert advice when technical considerations are important. Even so, when executive decisions are made by politicians or civil servants, they take place in a political context. Regulatory actions can be seen to illustrate this reality in many of the previous chapters. It is possible to blur the line between delegated executive authority and political decision-making. But when politicians argued with the decisions or advice of their civil servants or scientific advisers, the evidence here showed that it was politicians who tended to prevail. Britain never became a technocracy even if, by mid-century, technology was becoming key to state power within a liberal culture.[2]

Scientific, engineering, and medical advisers were widely involved in both military and civil domains. In the military arena—encompassing the Admiralty, Board of Ordnance, War Office, navy, and the army—their advice was primarily directed toward policy implementation and technical development. Such advice could come from people such as George Airy and Frederick Abel, who were civil servants employed under the Admiralty and the War Office, respectively, and from many other civilians, such as Michael Faraday. It also came increasingly from advisory committees.

In civil policy, scientific advisers were involved across matters including infrastructure and transport systems, lighthouses, manufacturing, mining, energy systems, agriculture, fisheries, and public health. Here, more than in the military arena, they were engaged in policy development as well as implementation. That is most evident in the work of the "statesmen of science" or "statesmen in disguise" such as Lyon Playfair, Edwin Chadwick, and John Simon but also in the ways by which inspectors were able to inform the nature of subsequent legislation in their areas of responsibility. From the 1830s, social policy, which had been largely the responsibility of local rather than central government, became more centralized, bureaucratic, and positivistic.[3]

Much government legislation established regulatory systems, and individuals advised on both the nature of regulatory requirements and their implementation. Implementation was typically monitored by inspectorates. Those covered in this book may be defined as "enforcement" inspectorates, concerned with securing compliance with legislative requirements, as opposed to "efficiency" inspectorates, such as those for education, policing, and prisons, which oversaw public services.[4] Most inspectorates required technical expertise—for example, for mines, explosives, railways, shipping, gasworks, atmospheric and water pollution, cruelty to animals, and fisheries. The inspectors had varied executive powers, limited by the relevant legislation. They could also influence the use of ministerial powers and sometimes had judicial functions. They tended to work with rather than against private employers, by persuasion rather than coercion. In this respect they could be said to be advisers as well as inspectors.[5] In areas under the purview of the Privy Council, such as aspects of public health, inspectors and other civil servants could influence the nature of Orders in Council as one mode of executive government action. They could also influence legislation—for example, in areas including fisheries, burials, explosives, mines, and gas and electricity.

This mode of acting by persuasion is equally evident in the culture of the Local Government Board. It took a "diplomatic-political" model of leadership, rather than a "technical-bureaucratic" model of control, although the latter eventually prevailed with the creation of the Ministry of Health in 1919. This reflects a culture of liberalism and pluralism rather than collectivism or statism, within a central government fiscal approach that was generally not prepared to give financial support to local services, or to dismantle the rights and privileges of local property holders.[6]

COMMITTEES AND COMMISSIONS

The Admiralty and the War Office made extensive use of committees for technical matters. This approach increased over time in civil departments such as the Home Office and the Board of Trade, as the technical requirements of legislation expanded. In the military sphere, these committees concentrated on the development of ships, weapons, and explosives, and facilities such as hospitals. The constitution of some of the committees has been explored in earlier chapters and is worthy of further study. Many military committees contained no civilian members, and relied on the professional expertise of army and naval officers. Others incorporated civilians, and whether they did or not will have reflected factors such as the views of politicians and senior officers, belief in the self-sufficient professionalism of military staff, and their openness to external perspectives.

Whatever their constitutions, the committees were advisory, under a hierarchy of military command and ultimately of political decision-making. John Herschel may have thought that the scientific experts should have decision-making power in Admiralty committees over spending money, but Thomas Young's view that this was ultimately a political responsibility is the more realistic. Governments could also keep advisers under control by making them members of advisory committees rather than boards. Examples include the Admiralty's Resident Committee of Scientific Advice and in a different field the advisory committee to the short-lived Central Board of Health from 1831 to 1832. Technical advice often raises operational and political considerations beyond the purview of the advisers. For example, the development of new weapons had implications for military tactics, and for command and control in the field, as well as political implications for expenditure. These were not technical issues.

One feature of the politics of the nineteenth century that greatly increased opportunities for the provision of external advice was the growing use of parliamentary select committees and royal commissions. As government grew in size, departmental committees were added to the mix. All could, and often did, make use of external expertise.

In the early decades of the nineteenth century, government became increasingly legislative in addition to executive. Rather than seeing social policy as the concern of Parliament as a whole, with measures introduced by private members, the government started to take a more active and sustained role.[7] This transition can be followed in many of the policy areas discussed previously—when, for example, legislation for the regulation of mines or shipping might be introduced by private members and then taken over by the government. As this process developed, both Parliament and government would seek to explore relevant knowledge and experience. Few select committees or royal commissions were established before 1820, but in the reform period of the 1830s and 1840s there was a veritable explosion of them. Criticism of the inefficiencies of Parliament and the executive in the 1850s then led to the more frequent use of royal commissions to investigate social issues in the 1860s, and hence to legislation based on their findings from the end of the decade.[8] For example, the Contagious Diseases acts were part of that movement of "scientific" legislation that sought, through the 1860s and early 1870s, to remove hazards to public health by using expert commissions of inquiry, followed by piecemeal measures. Further pressure, and inquiries into the working of the legislation, led to amending acts that tightened central control and increased punishments for noncompliance.[9]

Select committees and royal commissions are different beasts.[10] Select committees could be instigated by individual enthusiastic and persuasive MPs, generally with government support or at least acquiescence, and required approval by a vote in Parliament. They did not have technical advisers as they do in the twenty-first century, so it would be a matter of chance if any members were expert in a technical sense.[11] Nevertheless, reading the reports of the committees leaves one in no doubt that the lay members were willing and able to explore substantial technical detail. The onus was on the witnesses to explain themselves clearly and convincingly. Royal commissions, by contrast, are tools of government and appointed by ministers. They would have a measure of political balance to maximize the likely acceptability of any findings and recommendations and contain relevant expertise on the subject in question. A good example of the careful management of the constitution of such a body is the royal commission on vivisection. Its membership was simultaneously representative of all views but biased to obtain a particular outcome. Both select committees and royal commissions generated reams of evidence that brought issues, proposals, and the state of scientific, engineering, and medical knowledge to public view and discussion, and underpinned subsequent parliamentary debate and legislative action. Royal commissions in particular brought a wide range of external expertise into political and public discourse. This effective centralization of knowledge from the 1830s onward was crucial to the centralization of power,[12] and to the impact of advisers, hence my focus on the Parliamentary Papers.

AREAS OF EXPERTISE

This book has dealt with science, engineering, and medicine as distinct but overlapping areas of expertise. The specialists in these fields came from different social and cultural traditions, and from disciplines and professions at different stages of development. The engineers were demonstrably useful. Their canals, roads, bridges, mines, ships, and railways shaped the environment in which the Victorians lived and worked. They generated employment and wealth. Despite many setbacks and criticisms, the visible success of their constructions lent them a natural authority on which Parliament and government drew. On the occasions in which they failed in spectacular style, such as the Dee Bridge and Tay Bridge disasters, there were big inquiries, but they were not damaging to the overall reputation of engineering.

It is arguable that the engineers had an easier route to influence than those of other disciplines. Both in science and medicine the evidence brought to political discussion often rested on conceptions of invisible agents—such as molecules, miasmas, germs, and electricity—that were posited to be responsible for the conditions in question. Engineering knowledge was visibly reliable to the nonexpert, or at least reliable enough, whereas the characterization, detection, and measurement of invisible agents was problematic, not least when it was contested among the scientific and medical specialists themselves.

In this respect, medicine can be seen to sit between engineering and science in its intrinsic ease of influence. People were familiar with doctors and interacted with them in ways that most did not with the emerging men of science. The elite physicians and surgeons moved in the upper social circles, and were respected and credible to Parliament and government. For the men of science, gaining influence required establishing personal relationships and trust. Indeed, it was the person, as much as the credibility of the scientific process, that conferred authority.[13] A few, like Joseph Banks or Herschel, had automatic entry into elite society by virtue of birth and education. Most, such as Faraday, did not. Here, the networks created by the major scientific institutions, including the Royal Society and the British Association, and the Athenaeum, helped those connections to be made and developed.

It has often been argued that members of the scientific community had to scheme and fight for cultural influence during the nineteenth century, especially from mid-century, against the forces of established religion. Their materialist view of the world, it was claimed, encapsulated in Charles Darwin's theory of evolution by natural selection and John Tyndall's iconoclastic addresses on the nature of science and its relationship to religion, tended to lead to atheism, socialism, and similar so-called ills.

There are truths in these arguments, but in terms of practical politics, the case falls apart. Throughout the nineteenth century, the scientific experts were never in danger of failing to have influence on the development or implementation of policy. The issue for them was that it was not always on their terms or as wholehearted as they would have liked. In the 1870s, for example, when governments avoided implementing the recommendations of the Devonshire Commission for a ministry of science and for public funding for scientific research, or constrained research through the Cruelty to Animals Act 1876, the scientific community felt threatened. It sought to paint politicians as failing to appreciate science and the application of scientific processes to politics and administration.[14]

Despite these challenges, political access for the experts was substantial, and politicians generally treated them with respect, right across the century. In the hundreds of select committees, royal commissions, and departmental committees dealing with matters with scientific, engineering, or medical dimensions, multiple experts were almost invariably called to give evidence or to serve on them. Exceptions include some military examples, when senior officers may have considered in-house expertise sufficient and appropriate. If the experts' views were not always as influential as they would have liked, the reasons are due to broader social and political factors that have little to do with their expertise or access. These shaped the inevitable contestation that underlay interactions both between and within political and expert communities.

Most scientists do not think like politicians. A revealing example is William Thomson's claim that the need for a separate Parliament in Ireland was a "scientific absurdity" given the existence of the electric telegraph.[15] Politics stands above science, and politicians are prepared to disregard scientific advice when stronger forces or values are in play, such as the views of their immediate electorate. The questions that matter to politicians are by definition political. Science can inform, but it can also be suborned, and it frequently offers contested and conflicting perspectives, especially at the limits of knowledge. Whom to trust or whom to choose to bolster one's case becomes a matter of reputations, personal connections, beliefs, and values. When scientific advisers were divided in their opinions, as for example on decimalization, their influence was reduced.

Expert disagreement is common. Arguments between experts before select committees and commissions were analogous to arguments between expert witnesses for the prosecution and defense in legal cases. Such disputes raised serious questions both about the validity of expert knowledge and about the proper role of the expert. For many, expert witnessing, when the experts were in the pay of one side or the other and hence tainted by potential bias, was incompatible with science as the disinterested seeking after truth. This was an opinion congruent with a belief that science as a search for knowledge was a higher calling than science as a means of earning a living. William Odling, the epitome of chemical consultants, contested this view. He saw the adversarial system as part of the process of testing scientific knowledge,[16] and he recognized that scientific questions in relation to public policy were subsidiary to legal or political ones.[17] Indeed, it can be argued that consensus on scientific questions that are more than marginally relevant to policy is impossible.[18] That in itself is an argument for an incremental approach to policymaking, which turns out to be the characteristic approach in the nineteenth century, and not just for reasons of science.

The adversarial approach had implications also for the process by which local bills were scrutinized in Parliament—for example, those brought by local authorities seeking to build waterworks. The powerful interests who could pay for experts would have an advantage over those defending the wider public interest. Chadwick used this argument to advocate for independent central experts to advise.[19] It became part of the discussion about expert central regulation in the context of local accountability. Concern about experts paid by particular interests, such as water and electricity companies, increased toward the end of the nineteenth century. Such experts could be seen as fallible, untrustworthy liars, by comparison with authorities who were unpaid and had no financial interests in matters on which they were pronouncing. Indeed, the term *expert* was often used derogatively in the latter half of the century, instigated by media critique of their actions.[20] But it is not just private businesses or individuals that might select particular expert witnesses. Members of parliamentary committees and royal commissions could likewise invite witnesses or members whom they considered would best strengthen their case, whether those individuals were paid or not. Those giving advice had to position themselves as authoritative.

SIGNIFICANT INDIVIDUALS

One consequence of a focus on the history of science in policy, as opposed to the history of the development of scientific knowledge, disciplines and institutions, is that some lesser known names become more significant. While Michael Faraday may excel in both dimensions, people like Thomas Brande, Davies Gilbert, William Odling, George Airy, Goldsworthy Gurney, John Percy, Frederick Abel, Robert Angus Smith, Arthur Hassall, Frank Buckland, Edward Frankland, John Tyndall, William Crookes, William Pole, John Lefroy, Vivian Dering Majendie, and August Dupré rise in historical importance when the impact of science on public policy is considered. Alongside them are the "statesmen," such as Playfair and Simon, who occupied powerful administrative positions in government by virtue of the expertise they could command and project. Some major scientific figures, such as Darwin, are almost invisible. In Darwin's case, that might reflect his unwillingness to become publicly involved with politics, even if he took a personal interest in political affairs.[21] Equally, it may reflect the state of development of the biological sciences, which had less to offer to the solution of pressing political problems than chemistry, engineering, and medicine.

Two substantive reasons for the continuing influence of advisers, even if it was not to their entire satisfaction, can be identified. The first is that so many of them already were, or soon became, respected members of the small elite that governed Britain. Most may not have had automatic entry by birth and education to the political class, but the widespread interest in new knowledge, the cachet of attending a Royal Institution discourse, and the kudos of having the latest knowledge explained by its discoverers at soirées and dinners placed them on the cultural map. They mingled in London's clubs, especially the Athenaeum, and at meetings of the burgeoning scientific, engineering, and medical societies. Their politics tended to be moderate, and they could work with governments of any political hue. Several were elected to Parliament, whether as Whigs, Liberals, or Conservatives, and many members of the House of Lords had strong scientific interests, often serving as presidents of scientific institutions. Many members of both Houses of Parliament were fellows of the Royal Society, members of the British Association, and of other specialist institutions. Knighthoods, baronetcies, and even peerages toward the end of the century, were conferred on many advisers.

The second reason for their continuing influence is that the knowledge that the men of science, engineering, and medicine generated and could communicate offered practical solutions to pressing problems. Even when that advice failed disastrously, such as Humphry Davy's "protectors" for the navy, or proved too complex, as in some of the suggestions for ventilating and lighting the new Houses of Parliament, overall confidence was not lost. The advisers had their uses, too, as scapegoats for deflecting blame and delaying decisions. If science could not offer a definitive solution, why should politicians gainsay that? Political inaction could be justified by the apparent uncertainty of scientific knowledge, and by pointing to disputes among the advisers as to the correct conclusions to be drawn. It was ever thus.

People making and implementing policy need reliable practical advice for the messy real world. That world is less controlled than the laboratory, making the results of laboratory experiments, and the theoretical conclusions and predictions derived from them, challenging to implement in practice. Sometimes this was explicitly acknowledged, as in the method chosen for chemical analysis for the malt tax, and for the analysis of coals by Henry De La Beche and Playfair. Persuasive individuals like Faraday could communicate their advice in the context of a convincing theoretical explanation, while at the same time acknowledging its uncertainties and limits. Even then, as in the case of Faraday's investigation of the Haswell colliery explosion, his analysis could be challenged by those with long practical experience of mining. Everything was negotiable and contestable.

SCIENCE AND PARTY POLITICS

If scientific advice did not lead to immediate radical legislative change, only to halting steps over decades, that reflects the political contests that resulted in incremental approaches to legislation, as much as the complexity of turning such advice into practical policy. Britain avoided social, political, and administrative revolution in this period, as politicians sought to reconcile progress with the maintenance of order. Revolution inspired by scientific knowledge was never likely. Gradual reformation was the consequence, built under shared conceptions of the past, morality, and community,[22] while powerful interests ceded authority as slowly as they could get away with.

Scientific evidence and scientific disputes rarely bore heavily on the big political questions that decided elections, such as franchise reform, Ireland, land rights, taxation, education, poor laws, religion, and foreign policy. On the other hand, science, engineering, and medicine offered practical options for improving social and economic conditions, and for fighting wars and extending the British Empire. In most cases, the implications of the advice did not become party political issues. To take a couple of exceptions, the use of iron rather than wood in naval ship construction became politicized in the 1860s, and food adulteration became party political in the 1870s. Conservatives could oppose centralizing aspects of public health legislation,[23] although legislation on social reform become party political only late in the century.[24] And a party dimension could become significant on occasion in some constituencies, as the disputes over vivisection in Reading attest.[25]

On the whole, governments of different political colors were able in turn to pass modifying and consolidating legislation underpinned by science as the nineteenth century advanced, in areas such as public health, transport, energy generation, and manufacturing. Scientific advice not only shaped legislation in these policy areas but also its implementation through regulatory systems and inspectorates. Even if legislation in the areas addressed in this book was often uncontroversial in terms of party politics, it could be greatly affected by the contingencies of falls of governments, the pressures of more urgent parliamentary business, changes of ministers, and election results. Examples abound. Public health policy, and Simon's role within it, was greatly influenced by unexpected changes of government and then by John Lambert's determination to ensure lay control. Similarly, the introduction of effective lightning conductors on ships was affected by changes of government, alongside the abolition of the Navy Board, and the resistance of John Barrow. Many decisions were contingent on specific events. The death of twenty people by arsenic poisoning from peppermint lozenges led to the Food and Drugs Act 1860.

The manner and speed with which these legislative changes took place depended on beliefs about property rights, individual and corporate responsibilities, the nature of local government, and the legitimate powers of central government. These were not scientific questions. The relatively slow pace of change can be explained if one recognizes the difficulty of turning scientific, engineering, and medical knowledge into social benefit, the essential conservatism of so much of the political class and of those entitled to vote, and the power of vested interests.

In addition, party discipline, never strong in the early part of the century, weakened after the Conservative Party split in 1846 following the repeal of the Corn Laws. Conservatives tended to support paternalistic philanthropy rather than interventionist social policy, but social amelioration was more attractive to them as a means of countering social unrest than franchise reform in mid-century.[26] They were also keen to advocate for local government as guardians of property representation, and to counter any centralizing tendencies of Liberal legislation. The "age of equipoise" around the 1850s saw a string of minority and coalition governments, when MPs had greater individual power than under a strict party system. But after the Reform Act 1867, with its extension of the franchise, governments were more dependent on constituency and party views, which also weakened the power of sectional interests such as railway companies.[27] Private members' bills became less common than earlier in the century as governments became more proactive under party discipline and administration became more professional and extensive. Conservatives as well as Liberals passed consolidating legislation in areas of social policy. Benjamin Disraeli publicly claimed the Conservative contribution to sanitary reform in 1872, even if it was a political tactic to appeal to working-class sentiment for election purposes.[28] His government passed the Artisans' and Labourers' Dwellings Act 1875, with its powers of compulsory purchase and clearance, along with the Public Health Act 1875, although George Sclater-Booth, then president of the Local Government Board, was not given a place in Disraeli's cabinet, indicating the less than wholehearted priority that Disraeli gave to social reform.

FACTORS INFLUENCING THE RECEPTION OF SCIENTIFIC ADVICE

Four overarching factors emerge from the evidence here to set the political context within which scientific advice was constrained in areas of civil policy. These are private property rights, the laissez-faire approach to private industry, the importance of personal freedom and responsibility, and the constitution and role of local government in relation to central government.[29] These four forces, as that is how they manifested themselves, are all aspects of the liberal political culture and state that developed in the nineteenth century, largely regardless of party politics.

Property gave social and political power, nationally and locally. The franchise was based on a property qualification, and the major landowners were a potent force in both Houses of Parliament. Property owners were keen to preserve their rights, but the state was prepared to step in under particular circumstances, so that the limits of private rights became a continuing political issue. Aside from policies associated with science, engineering, and medicine, government showed itself willing to reform aspects of private property that had come to be seen as antithetical to the public interest, for moral and other reasons. Examples include the dispossession without compensation of the borough proprietors by the Reform Act 1832, the dispossession with compensation of West Indian slave owners in 1833, and the confiscation of private property rights and their dedication to public use through the reform of municipal corporations in 1835.[30] The state was also prepared to enforce redistribution of private land to railway companies, as it had earlier with canals and roads, and to bring the lighthouse service into public administration.

Individual property rights could conflict, as illustrated by disputes over fishing rights and pollution in rivers. An example is the impact of the emission of toxic gases from the alkali industry on the surrounding countryside. The inability of common law and the local legal system to deal with major nuisances such as this—even given the deep pockets of a landowner of the status of the Earl of Derby—resulted in elements of central regulation informed by science, which constrained the rights of private property. This failure of the legal system to come to terms with economic, social, and technological change encouraged the formulation of remedial measures dependent on an understanding of science, engineering, and medicine, and administered by a bureaucracy.[31]

The political and economic system was rooted in a belief in the role of private industry and private capital, and the efficiency of the market. The role of government was to keep out of the way as far as possible. That philosophy was challenged by the existence of private monopolies and by other circumstances that could pit public interests against private ones. When public interests, such as the cost and quality of services, and of physical safety and individual health, came up against the private interests of industrialists or other interest groups, any laissez-faire philosophy was tested.[32] The result, depending on a range of circumstances, was inspectorates for factories, mines, alkali industries, railways, explosives, gas supply, fisheries, shipping, anatomy, vivisection, and even burials.[33] The necessity for public health interventions, and thus the growth of government in this domain, was positioned as a consequence of market failure.[34]

Expert advice was important. But the question of whether or not to introduce centralized regulation, and its extent, involved a play-off between conflicting ideas on economic laissez-faire, notions of individual responsibility, local accountability, public safety, responses to public pressure, and cost. The timing of these centralizing interventions by government varied by policy area, depending on factors such as public visibility and the extent of existing voluntary regulation. Centralization of national road infrastructure took place in the 1810s and 1820s. Issues in factories, public health, and railway safety appeared in the 1830s. In the 1840s they extended to urban conditions encompassing water supply and sanitation, and to the mining sector. Not until the 1850s did food adulteration and housing come to significant prominence. But even at the end of the century, as described vividly in Robert Sherard's *The White Slaves of England*, conditions in many working environments were appalling.[35] Regulation was slow and piecemeal. It was always up against the interests of private enterprise, which was vocal and influential in Parliament, within a political culture that saw minimal regulation as an economic necessity. The power of private industry, and perhaps the weakness of scientific knowledge and technological ameliorations, is illustrated by so many examples in preceding chapters. The problem of fisheries is a case in point. By 1825, the government had effectively given up on the idea of legislating to require manufacturing and mining industries to cease polluting waterways in a way that would preserve fisheries.

Neither the nature nor timing of regulatory legislation was inevitable. There are so many possible counterfactuals. Coaches on the roads might have been subject to regulation as early as 1809 if the select committee's recommendations had found support. Steamboats might have been regulated in 1818 if Charles Harvey had remained an MP, in 1831 if reform debate had not convulsed Parliament, in 1836 if shipping interests had been weaker, or in 1839 if the Whig government had been stronger. In any event, a strong government was finally able to pass the Steam Navigation Act in 1846. By contrast, if William Mackinnon had not tried to bring in six bills to abate smoke, any legislative action might have happened even later than it did in 1847.

Even so, not all aspects of employment came under regulatory scrutiny. The two largest fields of employment, agriculture and domestic service, were unaffected by government regulation in the course of the nineteenth century. This book is a study in the contingency of political outcomes involving science, engineering, and medicine. Indeed, effective action against industrial activities damaging to public or private interests needs the alignment of several factors. Taking the case of smoke abatement, the passing of the Clean Air Act 1956, following measures taken during the previous century, required the coalescence of adequate scientific knowledge and practical technology, practical means of surveillance of the law, and measures that were politically practicable.[36] Rarely did those factors coincide sufficiently to generate effective legislation. Legislation on coal mines or shipping provides a lesson in just how slowly regulation could be tightened over an entire century, even in the absence of significant issues of contested party politics.

Public ownership provided another route to securing an acceptable level of public benefit, despite the prevailing belief in the effectiveness and efficiency of private business. But whether that happened with a particular industry or service was a matter of economics and politics rather than of science and technology. Before the nineteenth century, infrastructure was not financed by the state. Taxes were not levied to improve roads and rivers, nor did the government build its own railways or lighthouses. Railways remained in private ownership throughout the century, but by contrast lighthouses did not. Whereas the nature and structure of the railway system encouraged investment in development and improvement, there was no compelling commercial reason for investing in technologies to improve lighthouses. The state eventually intervened.

Many services such as water supply, sewerage, and gas were increasingly taken into public ownership. With gas, for example, that was a consequence of the fiscal situation of local authorities, in addition to rate-payer pressure for efficient services and low tax.[37] Indeed, public ownership and municipal operation was an alternative means of dealing with private monopolies to regulation by price and other constraints.[38] Public or private ownership affected the manner in which both regulatory action and scientific advice were applied. Once lighthouses were taken into public ownership, the remaining issues were primarily scientific and technical, although they were affected by budget constraints and personal politics.

Linked to the idea of laissez-faire was a belief in the importance of institutional and individual responsibility. Many believed that the taking of responsibility by the state ran the risk of removing the incentive for organizations and people to be careful about the consequences of their actions. Greater regulation would reduce the companies' and individuals' own sense of responsibility, so that central measures aimed at increasing public safety would in fact worsen it, an argument applied to the railways.

Ideas of compulsion also went to the root of civil liberties, as illustrated by debates about vaccination policy.[39] While some saw vaccination as a matter of personal choice, others highlighted the wider public interest. Playfair took the latter view, making his position on compulsory vaccination clear in the House of Commons when he declared that "individual disbelief in a remedy which science and experience had confirmed beyond all reasonable doubt was no justification for relieving the conscience of the individual at the expense of society."[40] The case of vaccination illustrates other attitudes of lay politicians toward such medical intervention. Whereas in 1802 support in Parliament was based on mercantilist considerations, by 1807 it was humanitarian arguments that carried weight. Medical evidence was interpreted in those wider contexts. Parliamentarians were also able to argue that it was not necessary to understand the scientific basis for vaccination in order to believe in its effectiveness. They showed themselves prepared to be precautionary when clear scientific evidence was lacking, as in their responses to the possible but unquantifiable impacts of sewage contamination in the River Thames.

One further concern expressed about regulation was the desire not to constrain future scientific and technological development by legislating too quickly. For example, after the Dee Bridge collapse, the commissioners who reported on the application of iron to railway structures could say, "And in conclusion, considering that the attention of engineers has been sufficiently awakened to the necessity of providing a superabundant strength in railway structures, and also considering the great importance of leaving the genius of scientific men unfettered for the development of a subject as yet so novel and so rapidly progressive as the construction of railways, we are of opinion that any legislative enactments with respect to the forms and proportions of the iron structures employed therein would be highly inexpedient."[41] The commissioners in this case were all scientific and engineering specialists, who nevertheless felt able to make this political judgement.

As far as local government is concerned, the entire nineteenth century is a saga of the evolution of local government systems, electoral and administrative, in interaction with the growing central state. This patchwork of organizations, often overlapping, different in rural areas, towns, and the metropolis, had arisen over centuries.[42] It included municipal boroughs and corporations, improvement commissions, among them paving, lighting, and police commissioners, nuisance authorities, sewage commissions, Poor Law boards of guardians, magistrates and courts, and vestries or parishes.[43] Behind all this, throughout the century, was a widespread belief in the delegation of powers over local affairs to community elites, a counterbalance to any centralization, as the central state sought to induce localities to accept common national minimum standards. Members of the various bodies were generally elected by ratepayers, making them representatives of the property interest. Magistrates came from the same social spectrum, and there was a constant tension between local responsibility and central desires for coherence and common standards.[44] If magistrates proved unable or unwilling to implement legislation designed to improve conditions as seen from the center, pressure grew for stricter regulation.

THE ADVISERS AND THEIR NETWORKS

Considering that this book covers an entire century, the influential advisers are not numerous. That reflects both the importance of close personal relationships in establishing networks of influence and the limited number of senior positions available, especially in the sciences, before the growth of universities. The medical community was a large one. Yet few medical people were able to penetrate to positions of political influence, which required social connections. Most of those who advised government were senior members of the medical royal colleges, whether of London or Edinburgh, along with some of the more notable local medical officers of health from mid-century onward.

It was a similar position with engineering. The huge expansion of steam-driven industry and of civil, mechanical, and then electrical engineering in the nineteenth century opened myriad opportunities, yet the number of engineers with national reputations was quite limited. The engineers were busy with their careers, which frequently took them to far flung parts of the country to build their bridges, railways, harbors, waterworks, and sewage systems. Many were based away from the metropolis, where they would generally be called to give evidence or sit on commissions. Nevertheless, the networks of northern engineers, based in centers like Manchester and by the River Tyne, were powerful and provided many advisers and contractors to Parliament and central government, among them George and Robert Stephenson, William Fairbairn, James Nasmyth, Joseph Whitworth, and William Armstrong. The engineers, who well understood commercial opportunities, client relationships, and contractual obligations, were responsive to political needs.

The men of science had a particular challenge. With the exception of some chemists, who could make careers in the new chemical industries and in chemical analysis, and also earn significant incomes in consultancy roles and as expert witnesses in legal cases, the opportunities were fewer. Jobs did not pay well in scientific research and often required a person to hold several roles simultaneously, as was the case for Faraday, Tyndall, and Frankland. Many men of science, especially early in the nineteenth century, were military and naval officers, physicians, aristocrats, bankers, and merchants. A career as a scientific researcher was a rarity, requiring either a private income or the ability to carve out time from lecturing or other duties, at places like the Royal Institution or the few universities and medical schools. Nor was it evident that scientific research and knowledge translated easily and reliably into practical advice.

Nevertheless, widespread interest in the potential of scientific knowledge, belief in the underlying optimism of the Baconian method, and the credibility of the leading protagonists from Banks, Davy, and Faraday to Airy, Playfair, Abel, Tyndall, and Frankland, enabled the scientific advisors to reach positions of influence. The elite chemists managed to gain political influence as they asserted and demonstrated their usefulness, followed later in the century by the physiologists and bacteriologists. The mathematicians and physicists had less of a problem with status. Mathematics was held in high regard at leading universities, giving its luster also to natural philosophy and physics. With astronomy, mathematics was demonstrably useful in helping meet the critical military and civil demands for accurate navigation and surveying.

A tight-knit network of individuals was based around a small number of institutions in which people were employed or met on a regular basis. Scientific societies formed one set of networks. So many of the scientific advisers featured in the previous chapters were fellows of the Royal Society that its significance is obvious. While fellowship may reflect reputation and ability as much as confer it, and result from achievements and personal connections already established prior to election, the regular scientific meetings and those of the governing council offered plentiful opportunities for developing close relationships.

It is clear from reports of committees and commissions that fellowship of the Royal Society could both confer authority and be used to claim it. Members and witnesses of committees and commissions who were fellows of the

Royal Society are almost invariably recognized as such in the printed proceedings by the letters *FRS*. While similar recognition is given to *CE* for civil engineer and to *FRCP* and *FRCS* for the medical royal colleges, the Royal Society generally takes first place.[45] On occasion a witness would seek to buttress his authority by calling on his fellowship. Victor Horsley did this in 1887.[46] He was only thirty at the time and perhaps felt the need to assert his credibility by mentioning his recent fellowship of the Royal Society. He was designated "FRS" in the committee's report.

The opportunities for networking afforded by the Royal Society were soon complemented by an increasing number of specialist scientific societies. The British Association, although only meeting annually on a formal basis, commissioned reports and established committees that operated between the meetings. Some of the networks were informal, such as the X Club, an influential group of nine people that included Tyndall, Thomas Huxley, Joseph Hooker, and Frankland, who met monthly from late 1864.[47] In addition to the desire of its members to promote a naturalistic science as a counterpoint to religion, and to seek state support for science and for scientific and technological education, almost all were involved in state activities.

A second set of networks was formed by the scientific institutions in which the key figures were employed. Most of the advisers came from a limited number of institutions, and many moved between them, or worked for more than one simultaneously. It was a small world. This tight network encompassed military and civil institutions, and some advisers straddled both.

On the military side were the Royal Observatory and Board of Longitude, with the Royal Engineers, Royal Artillery, the Ordnance Survey, and the Royal Military Academy at Woolwich. On the civil side were both private and government institutions, in the shape of the Royal Institution, the Museum of Economic Geology, the Royal School of Mines, and the Royal College of Chemistry. While some constraints operated in the Admiralty, Ordnance, and War Office because of national security, exemplified by the government's acquisition of Armstrong's patents, there was much overlap between military and civil research. For example, the British Association established substantial committees and programs of work on metals for artillery and on guncotton.

One could pick any individual or institution and unravel a series of interconnected relationships, but a few suffice to make the case. Take Brande.[48] Brande met Charles Hatchett, studied under Frederick Accum, and then met Davy. When Davy resigned the professorship of chemistry at the Royal Institution in 1812, Brande replaced him and soon became superintendent of the house. He also married Hatchett's younger daughter. Hatchett may have introduced him at the Royal Mint, where his association started in 1823. In 1852 he left the Royal Institution when he was required to work full-time at the Royal

Mint as superintendent of the coining and die department, under Herschel as master. Herschel was then succeeded by Thomas Graham. Brande's successor as superintendent of the house at the Royal Institution was Faraday, who had likewise established an early relationship with Davy.

Almost anyone significant in science lectured at the Royal Institution, which also enabled interactions with the elite of society who attended the lectures. Faraday was instrumental in appointing Tyndall as professor of natural philosophy in 1853. He then involved Tyndall in the royal commission on lighting picture galleries by gas and effectively handed to Tyndall his position as scientific adviser to the Board of Trade and Trinity House. Frankland was professor of chemistry at the Royal Institution from 1863 to 1868, and Odling succeeded Faraday as Fullerian professor of chemistry in 1868. Frankland had worked under Playfair, taught with Tyndall, and was succeeded in Manchester by Henry Roscoe. In Faraday's reign also, Huxley and Richard Owen were Fullerian professors of physiology and comparative anatomy. Later in the century in that post came Horsley, Michael Foster, John Gamgee, and Edwin Lankester, who all interacted with government.

Another linked group coalesced around the Museum of Economic Geology and the Royal School of Mines.[49] Henry De La Beche founded the Museum of Economic Geology, which was opened to the public in 1841 with Richard Phillips as the first curator. Staff soon included Andrew Ramsay, John Phillips, Warington Smyth, Playfair, and Hooker. The institution was under the control of the Department of Woods, Forests and Works, and became the Museum of Practical Geology in Jermyn Street in 1851, with the Government School of Mines and Science Applied to the Arts, when it was formally opened by Prince Albert. Percy joined at this point.

The Royal College of Chemistry, under Wilhelm Hofmann, opened in 1845 with Abel as one of the first students and with Brande as one of its prime movers. It merged into the Government School in 1853, and the whole organization was taken into the Department of Science and Art under the Board of Trade, although it then moved in 1856 to the Privy Council on the creation of the Education Department of the Privy Council. Hofmann became professor of chemistry of the unified institution in 1853 and Playfair moved into an administrative position in the Department of Science and Art before becoming a member of Parliament. Huxley and George Stokes joined the School of Mines in 1854, and Roderick Murchison took over as director after De La Beche died in 1855. The Government School became the Royal College of Mines in 1863, when Hofmann left for Germany and Frankland became professor of chemistry. Tyndall had already succeeded Stokes as professor of physics in 1859. Key scientific advisers were close to each other and to government by virtue of their positions. One could pick others, such as Graham, and trace connections from Glasgow to London, where staff at University College London became providers of much expertise to government.

There were similar and overlapping groups centred on Admiralty institutions and on the Royal Military Academy at Woolwich and the War Office. At Woolwich, Faraday was professor of chemistry from 1829 to 1852, followed by Abel, whom he had recommended, until 1888. Peter Barlow served as professor of mathematics, as did Francis Bashforth. Percy lectured in metallurgy. Others with roles from the military side included Lefroy, Andrew Noble, and Majendie. Dupré and James Dewar were drawn into advisory work on explosives.

Dupré gives another illustration of this interconnectedness of people and places, as he studied with Justus Liebig, was assistant to Odling at Guy's Hospital medical school, acted as a consultant to the medical department of the Local Government Board and the Metropolitan Board of Works, advised both the Home Office and War Office on explosives, and served as public analyst for Westminster from 1873 to 1901.

In the 1870s, William Froude told the Devonshire Commission that scientific advice was not requested from the scientific community as a "constituted assemblage" of practitioners, but sporadically from "one or two individuals" of reputation who were familiar to politicians. He named Thomson as one of the key individuals the Admiralty consulted.[50] That was true throughout the century, with advice generally sought from individuals, rather than corporately from institutions. Apart from a burst of government requests to the Royal Society to establish committees in the 1820s and early 1830s, advice was mostly obtained from individual fellows. Such people would often be recommended by the Royal Society but they would act independently, either singly or as members of committees and commissions.

EMPLOYMENT OF ADVISERS

Two groups can be distinguished in employment terms, the full-time employees of the state, and the advisers and consultants. The state employees, or civil servants, themselves divide into two types. One encompasses significant managerial, policy, and research responsibilities. In this category may be listed Airy, Chadwick, Simon, Farr, Playfair, Lefroy, and Abel. Herschel and Graham might qualify too, in their role as master of the mint, although this was not a full-time position. The other type, representing many more individuals, is formed by the inspectorates dependent on significant scientific, engineering, or medical expertise. Most of these were full-time positions. Notable figures here, indicating the diversity of expertise and background of inspectors, include Majendie, Robert Angus Smith, Joseph Dickinson, and Frank Buckland. Such full-time employees of the state, or nearly full-time in Angus

Smith's case, were complemented by part-time consultants and inspectors. These were either salaried or paid for specific pieces of work. Salaried part-time advisers include Faraday and Tyndall at the Board of Trade, with Huxley and George Busk paid as part-time inspectors respectively for fisheries and for experiments using animals. Much more numerous were the advisers hired to produce reports on an individual basis, or appointed to committees and commissions. The latter might be paid or not, depending on the social status and employment position of the relevant individual.

ROLES OF ADVISERS

One way of looking at advisors is to examine how they envisaged their role as experts. A modern classification, which is useful to explore the different motivations, distinguishes four types.[51] The first, the "pure scientist," seeks to add to the knowledge that might be available to decision-makers, but without any thought for its potential utility. This attitude resonates with the fundamental belief in the Baconian method of many of those engaged in scientific research in the nineteenth century, including Faraday and Tyndall. Nevertheless, it did not prevent them from offering advice based on new knowledge when the opportunity arose.

The second is the "science arbiter," who responds to scientific questions posed by politicians and decision-makers. This is a common role adopted by most of the advisers throughout the nineteenth century, although some were prepared to stray at times into making political or normative judgements about the wider implications of their advice.

The third is the "issue advocate," who takes a more overt stance of aligning with a particular interest group or political agenda. Examples in previous chapters range from chemical analysts working for particular water companies or local authorities to the overtly political advisers like Chadwick and Simon.

The final type is the "honest broker," who seeks to integrate scientific knowledge with wider concerns to explore alternative possible courses of action. Like the science arbiter, this type of approach may be undertaken in committees or commissions. Frankland has been accused of "colouring facts" and seeking to make chemical analysis a basis for social action in his approach to water analysis and identification of unsafe water supplies.[52] That would make him an issue advocate. An alternative view argues that his methodology and the communication of his results was professional and open, making him more of an honest broker.[53]

Bureaucracy and liberal democracy evolved together in the nineteenth century, which left professional expertise in a paradoxical position. The independence of professional experts was recognized, sanctioning their ability to regulate society as inspectors. Yet these interventions were a potential threat to personal liberties, making their expertise at the same time both nonpolitical and political.[54]

One should remember that scientific advisers will themselves have political views. While they may try hard to stick to the "facts" of the case, they may also deliberately or unconsciously advance their views when giving advice. One medical witness, Charles Routh, was clear in giving evidence to the royal commission on the Contagious Diseases Act in 1871, that a doctor's political stance could color his views on whether venereal disease was contagious or not. He told the commission, "The fact has been stated by Sir Thomas Watson himself in a very forcible manner in his practice of physic, that persons who hold conservative opinions as a rule are contagionists, whereas persons who hold opposite opinions are non-contagionists. I might quote that gentleman's views to show that some persons for instance will not hold that the plague is contagious and the like, and therefore evidence which is conclusive to one man is not conclusive to another."[55] He continued, making the point clearly, "I merely draw a conclusion that a person of strong conservative views would from the peculiarity of his frame of mind interpret the same facts in a different way from one with liberal views, and that the frame of an observer's mind would have a great deal to do with the manner in which he viewed the question of contagion and non-contagion."[56] Liberals would tend to argue for sanitary measures that enabled people to live in greater freedom while taking personal responsibility. They would be less supportive of quarantine, seeing it as a restriction on personal freedom. There are many examples in previous chapters of advisers giving views on social policy that depend on political as much as on scientific outlooks. These include Playfair on approaches to sanitation and Dickinson on mine regulation.

Much of the advice given was nevertheless of a technical rather than a political nature, and separate from social policy. Lieutenant Colonel William Hope, who had won a Victoria Cross in Crimea and invented a form of shrapnel shell for rifled guns, gave a classic description of idealized scientific advice of this type, in his response in 1887 to a royal commission. His report had been requested after the bursting of the forty-three-ton breech-loading gun on HMS *Collingwood*, a disaster that consigned this weapon to the scrap heap.[57]

Hope argued that it was essential to bring in the expert men of science alongside the men of practice, given the complexity and technical nature of ballistics and gun construction, and the errors bestrewing the War Office's official textbook on the construction of ordnance, which he said had not been checked by experts. Solving the problems required "great patience and deep thought," he argued, from men such as "Sir Wm. Thomson, Professor Tyndall, Professor Cayley, one of your own number, Dr. Percy, the first living authority on steel, Professor A. W. Williamson, Mr. Crookes, the talented discoverer of Thallium, and if to such men it is desired to add some of not quite such academic habits of mind, the House of Lords could furnish two men of science of scarcely less eminence, in Lord Rayleigh, and the Earl of Crawford and Balcarres, while the House of Commons could furnish Sir Henry Roscoe and Sir Lyon Playfair." He continued, "It is difficult for those who have not had the advantage of working on committees with men of this stamp, accustomed all their lives to original research, to realize how difficulties of the most complex and puzzling kind disappear, as if by magic, when discussed by a few such men, the knowledge and experience of one supplementing that of another." Hope stressed the importance of bringing in the "very first men of science in the country" to settle questions that were necessarily beyond the competence of regimental officers, "however meritorious." By implication, he was claiming that having civilian members of the Ordnance Committee, in Frederick Bramwell and William Barlow, both engineers, was not enough. His analysis quoted Thomson, Peter Tait, and James Clerk Maxwell in theoretical support of his findings, and ended with a ringing call for the "immediate appointment of a Committee of the very first mathematicians, physicists, chemists and metallurgists in the country to evolve the British heavy gun of the future. They should be empowered to call in, for consultation, any other men, whether of science or practice, whose advice they may wish for, or whose experience and practice they may desire to examine into: and they should be authorised to spend *any* sums of money they may consider necessary in experiments." It was the infuriated and somewhat naive tirade of a man who had indeed predicted the failure of the gun, but highlights the fact that, even late in the century, the best external expertise was not integrated as a matter of course into the development of such a critical article as artillery.

SCIENCE AND BUREAUCRACY

The idea of control by a central bureaucracy met resistance throughout the nineteenth century. Much Victorian legislation was permissive rather than imperative—for example, in public health. There was a prevailing negative view of the role of the state, and an emphasis on the importance of local government. State action, which was tellingly construed as intervention, was seen in terms of more or less interference. The state acted, and hence central government grew, only when attempts at voluntary, local, or individual action was determined to have failed.[58] Treasury policy reflected a lack of enthusiasm for central expenditure, exacerbated by retrenchment of public expenditure in the 1850s and 1860s.[59] It required considerable personal influence to extract money from the Treasury to support scientific work. The alkali inspectorate found it hard to gain approval for funding comparable to that of the factory and mines inspectorates, a consequence of the parsimony both of the Treasury and the Local Government Board. This was despite the fact that the inspectorate generated revenue for the state from 1881 and a net surplus from 1892.[60]

There was no master plan or even desire in the nineteenth century to create a centralized bureaucracy and effectively a welfare state. Indeed, there was a prevalent assumption that social stratification and poverty were part of the natural order. Policy in each area responded to its own characteristics and contingent events, even if the consequence was a large growth in government by the end of the century. Politicians sought a balance between central and local powers, and between private and public interests. The approach taken was to establish centralized knowledge, and to offer uniform procedures, information, and advice to inform local actions. So there was central legislation with supervision and inspection, but not central administration and delivery. For example, the Royal Sanitary Commission declared in 1871 that "we would leave direction only in the Central power. It must steer clear of the rock on which the General Board of Health was wrecked; for so completely is self government the habit and quality of Englishmen, that the country would resent any Central authority undertaking the duties of the local executive."[61]

The growth and professionalization of the civil service is a notable feature of the nineteenth century. But, like reform of the army, it took several decades to work through. Until the later decades of the century, scientific advisers were still navigating a civil service that was small and personal. Recruitment was through personal connection, a hangover from the sinecure posts and profitable offices that were bought and sold in the eighteenth century but largely gone by the 1830s. The Northcote-Trevelyan report of 1854, which set the basis for civil service reform, was a consequence of the desire for a more economical and efficient public service, allied to William Gladstone's belief that competition by examination and access by merit would also raise the moral tone of the civil service, giving a more strenuous ethic in public life.[62] Open competition took until 1870 to penetrate the civil service in any substantial manner, and until the twentieth century to reach all its corners.[63] At the same time many external appointments were made in the new areas of policy and legislation, exemplified by the inspectorates and by senior leading figures such as Chadwick, Simon, Thomas Southwood Smith, Robert Rawlinson, and George Porter, the "statesmen in disguise."[64]

SCIENCE AND KNOWLEDGE

Different groups of people construed and used scientific knowledge in different ways. By the late eighteenth century the prevalent model, expressed in the fellowship and activities of the Royal Society, was of science as a gentlemanly pursuit, practiced as a kind of leisure activity. At the same time, for many landed aristocrats, science became synonymous with improvements in agriculture.

A more explicit utilitarian role for science developed, exemplified by the formation and original activities of the Royal Institution. The founders of the Royal Institution believed that scientific knowledge could be applied to ameliorate the human condition and promote a smoothly functioning social order.[65] Developing science therefore mattered, and people as different as John Stuart Mill and William Whewell believed that doing so would improve morality, politics, and society.[66] While their vision for the ways in which science was best developed were different, such disputes over philosophies did not greatly trouble the politicians. What they sought was reliable knowledge to inform practical decisions.

Nevertheless, the differences between Mill and Whewell represent two approaches to developing scientific knowledge that continued to resonate down the century.[67] Mill's approach, based on a Baconian empiricism, held that science was a means of producing knowledge of material value, by accumulating inductive observations sufficient to make deductive propositions. It was a utilitarian approach for practical ends. Whewell by contrast thought that science should be pursued idealistically, as inductive reasoning led to truth and affirmed religious faith. It was morally improving rather than utilitarian. Whewell also differentiated between "permanent" knowledge that was well-established and should be taught, and "progressive" knowledge that was still in debate. Such knowledge was possibly untrue and therefore potentially dangerous. The motivations of many of the men of science could in practice encompass both approaches. For example, while Tyndall believed passionately that a researcher should be left to his devices to pursue science for its own sake, he was also of the view that such an approach would be most likely to lead to practical applications, even if those could not be envisaged at the outset and the research had not been initiated with that motivation.

In the world of practical politics, such philosophical distinctions and concerns had little place. Politicians wanted practical advice, and the complexity of real-life problems meant that new and disputed knowledge would inevitably be drawn into discussion. Many advisers, of whom Faraday is a prime example, were careful to try to explain the limits of their understanding. When those views differed between advisers, politicians had to draw their own conclusions about whom to trust. Practice rather than theory mattered to the politicians, even if theoretical constructs like molecules and miasmas were often invoked. Not that theoretical concepts were irrelevant. Without Faraday's field theory the Atlantic cable would have been more difficult to develop. Likewise, germ theory had practical consequences well before it received general acceptance following the identification and characterisation of germs themselves.[68]

TOWARD THE TWENTIETH CENTURY

For the whole of the nineteenth century, there was limited state support for fundamental research in science, engineering, or medicine. The government was only prepared to fund research directed at immediate practical ends in the service of the state. In the early part of the century almost all such research was undertaken for military purposes, and particularly under the auspices of the Admiralty. Later, as government intervened in more areas of public policy, individual departments supported specific research. That often required Treasury sanction, which was generally grudging. The most significant areas were in public health, where Chadwick and Simon instigated large and different work programs,[69] and in technical domains under the Home Office and the Board of Trade. Examples of the latter include the development of lighthouses, of gas generation and supply, and of safety in mines. The government grant scheme, administered by the Royal Society, was the one sop to general provision, but it only funded experimental apparatus, on the assumption that individuals would have adequate financial resources on which to live. Not until World War I did either the state or most major industries promote scientific and technological research more broadly.[70] This is despite the fact that the minimal state of the early nineteenth century had grown into a bureaucratic behemoth by comparison at its end, employing a wide range of technical expertise in management, inspection, and advisory roles.

The scientific community continued to seek what it saw as proper government support for science as the nineteenth century ended. Foster expressed this view clearly in an article in 1904. He contrasted the government constraint, requiring the researcher to answer a specific question given that government reasonably demanded results for its money, against the freedom of the independent researcher to follow wherever interest led. The state-aided inquirer, of whom Abel provides a classic example, could not deviate from the task set by his political masters. Foster suggested that the state could follow the example of universities which paid for teaching while enabling research. He proposed government-funded laboratories and workers, attached to universities. That would bring routine work into contact with leading-edge, open research, and benefit both.[71]

The first government research institution of this type was the National Physical Laboratory. It was formed with the support of the prime minister, the Marquess of Salisbury, following a report by a committee of nine established in 1898 under Lord Rayleigh. Their recommendation, which was accepted, was for an institution independent of any government department but in close contact with the Board of Trade, given the board's responsibility for many standards that required accurate measurement.[72] The Chemical Research Department followed at the Royal Arsenal, Woolwich, in 1907, becoming the High Explosives Research Department in 1912, which was the forerunner of the Atomic Weapons Establishment at Aldermaston. The year 1909 saw the formation of the Development Fund for Agriculture and Fisheries Research.

Then, in 1916, in the middle of World War I, and reflecting the position that it is easier for scientists to demonstrate their value in war than it is in peace, the government created the Department of Scientific and Industrial Research.[73] The British Science Guild,[74] which had been formed by the editor of *Nature*, Norman Lockyer, in 1904 as a conservative, imperialist pressure group, sought formal control of science funding by a national council of scientists.[75] However, despite lobbying by the British Science Guild and others, the government retained administrative control of science, while the Royal Society likewise retained its informal influence by controlling nominations to the department's advisory body. Politicians, as they had throughout the nineteenth century, kept control, while the scientific community was able to strengthen its influence. The expert remained on tap but not on top. The British state was not prepared to trade a parliamentary democracy for technocracy, but was keen to retain access to the best scientific advice. As government increasingly invested in fundamental science, the Medical Research Council became the first of the civilian research funding institutions in 1919, followed by the Agricultural Research Council in 1931. The first minister of science, Lord Hailsham, was appointed in 1959, nearly a century after the recommendation of such a post by the Devonshire Commission.[76]

More formalized processes for providing scientific advice to government, as opposed to state funding of scientific research, were introduced over time. In the twenty-first century the government's chief scientific adviser reports to the prime minister, while departmental chief scientists report within their own departments, forming a network that extends across government. Compared to the nineteenth century, it is a difference of scale and organization, but not of kind. That reflects both the increased scope and complexity of science and the development of a larger, more bureaucratic, and more professional civil service.

Advisers can only operate in a political context. During the nineteenth century, although the political tide oscillated between liberals and conservatives, the center ground remained relatively stable. Coalitions were not infrequent, and ministers such as Viscount Palmerston could smoothly span any divide. Matters requiring scientific advice, even if strongly contested, were rarely party political. In the nineteenth century the majority of members of Parliament sought measured reform and steady social improvement while preserving the constitution and state institutions, defending the rights of property, respecting a form of local democracy, and maintaining an ordered, stable society. Or, as the Conservative Earl of Derby put it at the start of his second administration, "The same course must be pursued—constant progress, improving upon the old system, adapting our institutions to the altered purpose they are intended to serve, and by judicious changes meeting the demands of society."[77]

The scientific, engineering, and medical advisers, generally moderate in their politics, whether Liberal or Conservative, and frequenting the same salons, clubs, and associations as the politicians and senior civil servants, could swim with the tide. They gained influence despite the weakness of any technocratic tradition in the central British state, exemplified by the generalist ethos of the civil service, of Parliament, and of ministerial culture. But it was always a struggle, and their influence was constrained by powerful social and political forces.

Notes

INTRODUCTION

1. "Politics is the art of the possible, the attainable—the art of the next best," Otto von Bismarck, 1867.

2. On the concept of "men of science," disciplines, and professionalization, see Barton, "Men of Science." The first women appointed to a royal commission were Lady Frederick Cavendish, "widow"; Sophie Bryant, "Doctor of Science"; and Eleanor Sidgwick, "wife of Henry Sedgwick," for an inquiry into school education in 1894. They are listed at the end after all the men, who included the chemist and MP Sir Henry Roscoe; see PP C7682 (1895), iii; Harrison, "Women Members and Witnesses," 132–43, 216–23.

3. Ross, "Scientist." The word was first suggested by William Whewell in response to remarks from Samuel Taylor Coleridge at the meeting of the British Association for the Advancement of Science in 1833, and first used in print the following year.

4. MacLeod, *Government and Expertise*, 256 n11; Leggett, *Shaping the Royal Navy*; Leggett and Davey, "Expertise and Authority in the Royal Navy"; and Ashworth, "Expertise and Authority in the Royal Navy" explore these through the Admiralty.

5. R. Yeo, *Defining Science*, discusses science and its place in early Victorian culture.

6. Pickstone, "Science in Nineteenth-Century England," explores the changing meaning and structure of "science" over the century.

7. See, for example, Goldman, *Science, Reform, and Politics in Victorian Britain*.

8. Egerton and Pickstone, "United Kingdom," 160–62; Daunton, *State and Market in Victorian Britain*, 1–60.

9. Knight, *Britain against Napoleon*, 389.

10. It is not my intention to deal critically with concepts of the state as such. See, for example, Joyce, *State of Freedom*, and copious references therein. On the importance of taxation and the state, see Ashworth, *Customs and Excise*, 355–57, 370–83. On liberalism and liberal culture, see Joyce, *State of Freedom*; Hawkins, *Victorian Political Culture*, 16–18; Otter, *Victorian Eye*, 10–12; Otter, "Making Liberal Objects"; Parry, *Rise and Fall of the Liberal Government*. Scott, *Seeing like a State*, sees the modernist state, which developed in Britain during the nineteenth century, as based on creating, from a central perspective, a "legible" and manageable population and

environment. The standardization of measurements and systems made centralizing interventions possible in areas such as public health and poor relief as well as supporting the more familiar state functions such as taxation and political surveillance. The broadly liberal culture in Britain, and the democratic participation of local communities, as a counter to centralization, allowed improvement in social conditions based on the potential of science and technology, without leading either to revolution or to authoritarian approaches that would probably have failed.

11. But see Jackson, *Case Studies in Scientific Advice*, no. 10. On the development of approaches to "social science," see E. Yeo, *Contest for Social Science*. On the Social Science Association, see Goldman, *Science, Reform, and Politics in Victorian Britain*; Jackson, *Case Studies in Scientific Advice*, no. 11.

12. Alter, *Reluctant Patron*, 75.

13. Butterworth, "Science and Art Department"; Cardwell, *Organisation of Science in England*; C. Russell, *Science and Social Change*, 235–53; MacLeod, *Creed of Science in Victorian England*; Barton, *X Club*, 292–361.

14. But see Poole and Andrews, *Government of Science in Britain*, and Macleod, *Public Science and Public Policy*. MacLeod, *Government and Expertise*, explores aspects from 1860. Gascoigne, *Science and the State*, takes a longer-term view of relationships between science and the state.

15. Roderick Murchison is an example; see Stafford, *Scientist of Empire*.

16. I am grateful to one of my anonymous reviewers for pointing this out.

17. In terms of party politics, perhaps paradoxically, the most significant Liberal prime minister, William Gladstone, had little time for science and its practitioners, while Conservatives such as Robert Peel and the Marquess of Salisbury showed greater interest. On the concept of "party" and its changing meanings, see Hawkins, *Victorian Political Culture*, 100–111. On the development of the British two-party system, see O'Gorman, *Emergence of the British Two-Party System*. On conservatism, see H. Finer, *Theory and Practice of Modern Government*, 312–17. On liberalism, see H. Finer, *Theory and Practice of Modern Government*, 318–21. On the importance of "scientific liberality" to Whig society, manners, discourse, and politics, see Bord, *Science and Whig Manners*. On Whigs and government, see Mandler, *Aristocratic Government in the Age of Reform*.

18. On laissez-faire, individualism, collectivism, and state intervention in the nineteenth century, see Daunton, *State and Market in Victorian Britain*; Bartrip, "State Intervention in Mid-Nineteenth-Century Britain"; Perkin, "Individualism versus Collectivism"; Fraser, *Evolution of the British Welfare State*, 91–114; A. Taylor, *Laissez-Faire and State Intervention*. On the pragmatism and religious moral underpinning of Victorian political culture, see Hawkins, *Victorian Political Culture*.

19. Clifford, *History of Private Bill Legislation*, analyzes private bills and acts across many policy areas.

20. Parliamentary committees scrutinizing these bills met over short periods,

making outcomes arbitrary and inconsistent; see Bellamy, *Administering Central-Local Relations*, 197–200.

21. For example, on public health, see Keith-Lucas, *English Local Government*, 15–18.

22. Frankel, "Blue Books and the Victorian Reader."

23. Gosnell, "British Royal Commissions of Enquiry," explores royal commissions, although concentrating on the early twentieth century; see also Clokie and Robinson, *Royal Commissions of Inquiry*; Lauriat, "'Examination of Everything'"; Todd, *On Parliamentary Government*, 345–59. There was also some vigorous opposition to royal commissions and the like in the nineteenth century, an aspect of the objections to creeping "centralization"; see Joshua Toulmin Smith, *Government by Commissions Illegal and Pernicious*.

24. Joyce, *State of Freedom*, 191.

25. On the Académie des Sciences, see Crosland, *Science under Control*.

26. Ben-David, *Scientist's Role in Society*, 97.

27. Ringer, *Decline of the German Mandarins*, 14–42.

28. Ben-David, *Scientist's Role in Society*, 120.

29. On the key Physikalisch-Technische Reichsanstalt, see Cahan, *Institute for an Empire*.

30. Baldwin, *Contagion and the State in Europe*. See also Joyce, *State of Freedom*, for a comparative approach to the state.

31. Ben-David, *Scientist's Role in Society*; Ben-David, *Centers of Learning*.

32. On the growth of government, see Dicey, *Lectures on the Relation*, 62–69, the model of administrative and interventionist development in MacDonagh, "Nineteenth-Century Revolution"; a critique of both in Parris, "Nineteenth-Century Revolution in Government," who argues for an influential role for Benthamite ideas throughout the century; and Hart, "Nineteenth-Century Social Reform." MacDonagh, *Early Victorian Government*, assesses the period from 1830 to 1870. MacLeod, "Statesmen Undisguised," reviews significant works on administrative history. For background on the early changes in government and administration during the Napoleonic Wars, see Knight, *Britain against Napoleon*. For a non-Whiggish analysis of the evolution of social policy, see Fraser, *Evolution of the British Welfare State*. On the politics of an industrializing society, see Checkland, *Rise of Industrial Society in England*, 325–81.

33. Kitson Clark, "'Statesmen in Disguise,'" 34.

34. Howarth, *British Association for the Advancement of Science*; MacLeod and Collins, *Parliament of Science*; Morrell and Thackray, *Gentlemen of Science*.

35. MacLeod, "Whigs and Savants."

36. Goldman, *Science, Reform, and Politics in Victorian Britain*; Jackson, *Case Studies in Scientific Advice*, no. 11.

37. See also Burn, *Equipoise*, which offers a vivid account of this period from 1852 to 1867.

38. As delineated in F. Turner, "Public Science in Britain," but see also Hall, "Public Science in Britain."

39. R. Yeo, "Science and Intellectual Authority."

40. R. Yeo, *Defining Science*, 32.

41. On nineteenth-century patronage, see Bourne, *Patronage and Society in Nineteenth-Century England*.

42. PP 1713 (1854); PP 1870 (1855).

43. Kitson Clark, "'Statesmen in Disguise.'"

1. FOUNDATIONS

1. This also reflects the emergence of a distinct scientific community in society in the seventeenth century; see Ben-David, *Scientist's Role in Society*, 45–87.

2. For a more detailed tabulation, see Lyons, *Royal Society*, 341–42.

3. Gascoigne, *Science in the Service of Empire*; Gascoigne, "Royal Society."

4. R. Yeo, "Idol of the Market-Place," explores nineteenth-century interpretations of Francis Bacon. Howes, *Arts and Minds*, 1–28, analyzes the implications of Baconian ideas, and the founding of the Society of Arts.

5. Howes, *Arts and Minds*, 12–19.

6. Lloyd, "Rulers of Opinion," 14. As an example, Elizabeth Ilive, mistress and later wife of George Wyndham, third Earl of Egremont, was awarded a medal in 1798 for a lever she had invented to lift stones.

7. Lloyd, "Rulers of Opinion."

8. "Madder," *Transactions of the Society for the Encouragement of Arts, Manufactures and Commerce* 1 (1783): 9–11.

9. Knight, *Britain against Napoleon*, analyses the development of the machinery of government in this period.

10. Knight, *Britain against Napoleon*, 96.

11. Gascoigne, *Science in the Service of Empire*, 133.

12. Gascoigne, *Science in the Service of Empire*, 14.

13. James, '*Common Purposes of Life*.'

14. Foote, "Science and Its Function," 438–41; Knight, *Humphry Davy*.

15. The main engineering institutions were located a few minutes' walk from Parliament, where they remain. The scientific and medical institutions were not so close. The Royal Society nevertheless was located with the Navy Board in Somerset House, although its residence preceded the Navy Board's arrival, until it moved to Burlington House in 1857.

16. Wheeler, *Athenaeum*.

17. Morrell, "Individualism and the Structure of British Science."

18. *Report of the BAAS* (1851): li.

2. THE ROYAL SOCIETY AND THE BRITISH ASSOCIATION

1. Jackson, *Case Studies in Scientific Advice*, appendix 1.

2. Lyons, *Royal Society*, 342.

3. Lyons, *Royal Society*; Allibone, *Royal Society and Its Dining Clubs*. Hall, *All Scientists Now*, details the development of the Royal Society.

4. For Davy's life, see Knight, *Humphry Davy*.

5. PP 144 (1818).

6. Miller, "Between Hostile Camps;" and Fulford, "Role of Patronage," explore Davy's presidency.

7. Fulford, "Role of Patronage," 470.

8. RS CMP/1/58, 105.

9. Crowther, *Statesmen of Science*, 1.

10. MacLeod, "Of Medals and Men."

11. Jackson, "John Tyndall and the Royal Medal," 159.

12. MacLeod, "Science and the Civil List," 4–11.

13. Parker, *Sir Robert Peel*, 441, 446.

14. For example, in 1827 he attended a meeting of the Board of Ordnance to discuss Trevithick's design for a recoil gun mounting. See Todd, *Beyond the Blaze*, 102.

15. Todd, *Beyond the Blaze*, 184, 199.

16. PP 462 (1817), 4; Todd, *Beyond the Blaze*, 182–84.

17. Todd, *Beyond the Blaze*, 89–91.

18. PP 305 (1824).

19. This Royal Society commission, which included Edward Sabine, Francis Beaufort, and George Rennie, oversaw the surveying of the Thames from Sheerness to London Bridge, examining levels and tidal flows. See RS CMO/11/14–15; RS DM/4/46–48; RS CMB/1/27.

20. For the committee minutes, see RS CMB1/8; RS CMB/1/26. For the report to government, sent to George Harrison, assistant secretary to the Treasury and Fellow of the Royal Society since 1807, see PP 370 (1823).

21. PP 2 (1819).

22. PP 64 (1820).

23. Hall, *All Scientists Now*, 47. On parallel "decline of science" arguments in medicine, especially in relation to a desire for professional status, see Warner, "Idea of Science in English Medicine."

24. Hall, *All Scientists Now*, 43.

25. For example, Joseph Hooker reflected on his presidency from 1873 to 1878 that he found "P.R.S. to be a great power with ministers"; quoted in Barton, "Influential Set of Chaps," 73.

26. Harrison, "Scientific Naturalists," 52–72. One exemplar, the engineer

Frederick Bramwell, remarked that the letters *FRS* stood for "Fees Raised Since," after his election in 1873; quoted in Harrison, "Scientific Naturalists," 51.

27. Jackson, *Case Studies in Scientific Advice*, appendix 2.

28. Charles Babbage, *Reflections on the Decline of Science in England* (London: B. Fellows [etc.], 1830), 8. The remark was hardly true.

29. Joseph Banks had supported the formation of the Linnean and Horticultural societies, but then objected to the independence of the Geological and Astronomical from the Royal Society.

30. For histories of the formation and early years of the BAAS, see Morrell and Thackray, *Gentlemen of Science*; Orange, "Beginnings of the British Association;" Howarth, *British Association for the Advancement of Science*.

31. *Report of the BAAS* (1831): 22.

32. Higgitt and Withers, "Science and Sociability." Lloyd, "Rulers of Opinion," traces the generation of a largely passive female audience for science to the early years of the Royal Institution.

33. *Report of the BAAS* (1842): xxxiv–xxxv. Egerton was the only person raised to the peerage in Peel's resignation honors list in 1846, as Earl of Ellesmere. Egerton had supported him on repeal of the Corn Laws.

34. Jackson, *Ascent of John Tyndall*, 212–13; Turner, "Rainfall, Plague and the Prince of Wales."

35. Morrell and Thackray, *Gentlemen of Science*, 327; Howarth, *British Association for the Advancement of Science*, 211–40.

36. Howarth, *British Association for the Advancement of Science*, 213.

37. Morrell and Thackray, *Gentlemen of Science*, 342.

38. R. Knox, "Observations on the Natural History of the Salmon," *Report of the BAAS* (1832): 595–98.

39. George Harvey, "On the State of Naval Architecture in Great Britain," *Report of the BAAS* (1832): 607–8; and Jeremiah Owen "On Naval Architecture," *Report of the BAAS* (1833): 430–33.

40. G. R Porter, "Suggestions in Favour of the Systematic Collection of the Statistics of Agriculture," *Report of the BAAS* (1839): 116–17.

41. David Anstead, "On Mining Records and the Means by Which Their Preservation May Be Best Ensured," *Report of the BAAS* (1844): 42–3.

42. PP 332 (1829).

43. PP 380 (1832).

44. Morrell and Thackray, *Gentlemen of Science*, 345.

45. *Report of the BAAS* (1854): xlvii; W. Fairbairn, "The Patent Laws: Report of the Committee of the British Association," *Report of the BAAS* (1858): 164–67.

46. PP 3419 (1865). See also Smith, *Board of Trade*, 193–200, on patents and the Board of Trade.

47. "Address by Professor Trail," *Report of the BAAS* (1851): xlii.

48. *Report of the BAAS* (1849): xix–xx.

49. *Report of the BAAS* (1851): xxix, l. This suggests a responsive rather than proactive outlook.

50. Layton, "Lord Wrottesley," gives an extensive study.

51. "Anniversary Address. November 30, 1855," *Proceedings of the Royal Society of London* 7 (1856): 570.

52. *Report of the BAAS* (1851): xxix–xxii.

53. The letter to Palmerston is in Layton, "Lord Wrottesley," 239–42.

54. Prouty, *Transformation of the Board of Trade*, 51–5 describes the Hydrographic and Meteorological Offices.

55. *Report of the BAAS* (1853): xxxii–xxxiii.

56. *Report of the BAAS* (1854): xlii–xliii; Jackson, *Case Studies in Scientific Advice*, no. 16.

57. PP 2234 (1857).

58. On the controversies, and implications for the trustworthiness of scientific advice, see Anderson, *Predicting the Weather*.

59. Morrell and Thackray, *Gentlemen of Science*, 114.

60. Morrell and Thackray, *Gentlemen of Science*, 247.

61. Morrell and Thackray, *Gentlemen of Science*, plate 23.

62. Morrell and Thackray, *Gentlemen of Science*, 116. In 1837 Spring Rice undertook an inquiry into pensions and sounded out Faraday in the process; see James, *Correspondence of Michael Faraday, Volume 2*, 476–78.

63. For an analysis of the government grant, see Harrison, "Scientific Naturalists," 26–48.

64. Hall, *All Scientists Now*, 39–40. For the minutes, see RS CMB 1/28 Donation Fund Committee.

65. *Report of the BAAS* (1862): xliii.

66. The development of this nonstate voluntary activity is described in Egerton and Pickstone, "United Kingdom," 151–58.

67. "Scientific Administration," *Nature* 2 (1870): 449. See chapter 3; the vessel was the turret ship HMS *Captain*, which had sunk a few weeks earlier.

68. Goldman, *Victorians and Numbers*, 3–17; Cullen, *Statistical Movement in Early Victorian Britain*, 1–6.

69. Ashworth, "'System of Terror,'" 153.

70. Cullen, *Statistical Movement in Early Victorian Britain*, 13–14.

71. Brown, *Board of Trade and the Free Trade Movement*, 78–93; Prouty, *Transformation of the Board of Trade*, 8–9; Smith, *Board of Trade*, 209–24.

72. Cullen, *Statistical Movement in Early Victorian Britain*, 19–20.

73. See Goldman, *Victorians and Numbers*, 17–24.

74. Morrell and Thackray, *Gentlemen of Science*, 291–96, describe the formation of the statistical section of the BAAS; see also Goldman, *Victorians and Numbers*,

33–46. For the formation of the Statistical Society of London, see Goldman, *Victorians and Numbers*, 47–56; Hill, "Statistical Society of London;" Goldman, "Origins of British 'Social Science'"; Hilts, *"Aliis Exterendum."* E. Yeo, *Contest for Social Science*, 64–76, deals with local statistical societies; see also Goldman, *Victorians and Numbers*, 57–100.

75. *Report of the BAAS* (1833): xxvii.

76. Hill, "Statistical Society of London," 131.

77. Goldman, *Victorians and Numbers*, 193–4.

78. Goldman, "Statistics and the Science of Society," 428.

79. Quetelet was also an influence on Prince Albert, whom he tutored in mathematics and statistics over many years. Prince Albert's well-received address as president of the International Statistical Conference in London in 1860, at the request of William Farr, was a statement of this view of the science of statistics, and its value in giving the statesman and legislator "a sure guide in his endeavours to promote social development and happiness;" see Shoen, "Prince Albert," 302.

80. Bonar and Macrosty, *Annals of the Royal Statistical Society*, 122.

81. MacLeod and Collins, *Parliament of Science*, 52, 72.

82. *Report of the BAAS* (1833): xxix.

83. Morrell and Thackray, *Gentlemen of Science*, 281–83.

84. Donnelly, "Representations of Applied Science," describes the development of academic chemistry and research against the backdrop of industrial utility. On science and practice, especially in relation to chemistry, see Bud and Roberts, *Science versus Practice.*

85. Russell, *Chemists by Profession*, 135–57; Russell, *Science and Social Change*, 225, 229–30. A century later, having united with the Chemical Society, it would form the Royal Society of Chemistry in 1980.

86. Russell, *Chemists by Profession*, 103–12.

3. ADMIRALTY AND NAVY

1. An example is Sir Charles Napier's call for a select committee in 1845, to examine the Admiralty's strategy for the construction of ships, and especially the introduction of steam. His motion was soundly defeated by 22–93 votes; see *Hansard* 77 (February 13, 1845): 383–441.

2. This was Admiral Sir Joseph Yorke, brother of Charles Philip Yorke, first lord of the Admiralty from 1810 to 1812; Roger Knight, personal communication.

3. I am grateful to Roger Knight for pointing out the double noncoincidence to me.

4. Johnson, "Board of Longitude," gives a brief history. On the historical and conceptual background, see Dunn and Higgitt, *Finding Longitude.*

5. Waring, "Board of Longitude and the Funding of Scientific Work," 56–58.

6. Johnson, "Board of Longitude," 65.

7. For detail on Joseph Banks and the Admiralty, see Gascoigne, *Science in the Service of Empire*, 123–27.

8. See Hall, *All Scientists Now*, 10–11.

9. For an extensive analysis of the relationship between the Royal Society and the Admiralty in this period, and Humphry Davy's role, see James, "Davy in the Dockyard."

10. PP 275 (1817), 73. I am grateful to Rebekah Higgitt for context here from her unpublished lecture, "1818 Longitude Act."

11. Waring, "Thomas Young," 41–3. Robert Seppings later became surveyor of the Royal Navy.

12. A. Lambert, "Science and Sea Power," gives a full discussion of this episode.

13. John Barrow, "Seppings *Improvements in Ship-building* & Young *Remarks on the Employment on Oblique Riders*," *Quarterly Review* 12 (1815): 444–66; Knight, *Britain against Napoleon*, 247.

14. See A. Lambert, "Science and Sea Power," 18.

15. Barrow, *Sketches of Royal Society*, 62.

16. Waring, "Board of Longitude," gives a detailed analysis.

17. Waring, "Thomas Young," 167.

18. Miller, "Royal Society of London," 318.

19. Waring, "Board of Longitude," 59–64.

20. For three examples of lifesaving inventions, see Jackson, *Case Studies in Scientific Advice*, no. 1.

21. Knight, "Introduction of Copper Sheathing," gives the early history.

22. RS CMO/10/24–26. James, "Davy in the Dockyard," gives a detailed analysis.

23. For the minutes, see RS CMB 1/9 Copper Sheathing of Ships Committee (May 15, 1823).

24. For the minutes, see RS CMB 1/5 Coal Tar on Ships Committee March 1822 to June 1822; and the report at RS CMO/10, 4–6.

25. RS CMO/10/42–45. For the minutes, see RS CMB 1/11 Lightning Conductors on Ships Committee June 1823.

26. Jackson, *Case Studies in Scientific Advice*, no. 3.

27. For the minutes, see RS CMB 1/21 Lightning Conductors Committee (December 1827).

28. James, "Michael Faraday and Lighthouses"; James, "Military Context of Chemistry."

29. RS CMO/10/78. For the minutes, see RS CMB 1/13 Optical Glass Committee (May 1824–December 1828).

30. Jackson, "John Tyndall and the Early History of Diamagnetism."

31. Hall, *All Scientists Now*, 43. The Admiralty also sought other involvement when necessary. For example, it funded the organization by the Royal Society of the Thames Levelling Commission in 1830, consisting of Davies Gilbert, Francis

Beaufort, John Pond, George Rennie, and Edward Sabine, among others. This commission managed a detailed survey of the Thames from Sheerness to London Bridge, carried out by John Augustus Lloyd. See RS CMO/11/66–67, RS DM/4/46–64, RS CMB/1/27, and Lloyd's report in *Philosophical Transactions of the Royal Society of London* 121 (1831): 167–97.

32. Waring, "Thomas Young," 185, 189.

33. Waring, "Thomas Young," 113.

34. James, *Correspondence of Michael Faraday, Volume 1*, xxxvi.

35. "Parliamentary Intelligence," *Times*, July 5, 1828, 5.

36. Hall, *All Scientists Now*, 43, although the Admiralty did not given a reason for asking for a chemist.

37. James, *Correspondence of Michael Faraday, Volume 1*, xxxvi and letters 456, 459, 460, 463. Michael Faraday was also asked to advise on Marc Brunel's gaz engine in 1832 but declined since he might not be seen as impartial, having discovered the principle on which it worked. The Admiralty also asked Charles Hatchett, William Allen, and John Rennie; see James, *Correspondence of Michael Faraday, Volume 2*, letter 617; Marsden and Smith, *Engineering Empires*, 68.

38. PP 367 (1835), 13–14.

39. James, *Michael Faraday*, 46.

40. James *Michael Faraday*, 46–8; James, *Correspondence of Michael Faraday, Volume 4*, xl–xli.

41. Reid, *Memoirs and Correspondence of Lyon Playfair*, 159–61.

42. Chapman, "Science and the Public Good," 41–42.

43. George Airy to John Tyndall (June 10, 1872), RI MS JT/1/TYP/8/2664; (June 20, 1872), RI MS JT/1/TYP/1/46.

44. "Address," *Report of the BAAS* (1851): li.

45. Negotiations over the Greenwich Meridian also involved approaches to the Royal Society. Its Prime Meridian Committee advised the Department of Science and Art on international negotiations in 1890; see *Royal Society Council Minutes* 6 (1890): 50.

46. Airy, *Autobiography of Sir George Biddell Airy*, 134–36. For the involvement of George Airy, William Scoresby, the Royal Society, and the Board of Trade in tackling the problem, see Jackson, *Case Studies in Scientific Advice*, no. 2.

47. Airy, *Autobiography of Sir George Biddell Airy*, 169.

48. Perkins, "Airy," discusses George Airy's government commissions.

49. Jackson, *Case Studies in Scientific Advice*, no. 18.

50. Jackson, *Case Studies in Scientific Advice*, no. 21.

51. Chapman, "'Extraneous Government Business,'" 42.

52. PP 394 (1867).

53. Howarth, *British Association*, 225. On the technical and political issues around the development of steam, including the in-house creation of the Admiralty's

Council of Science, see Leggett, *Shaping the Royal Navy*, 59–88. See also A. Lambert, *Steam, Steel and Shellfire*, 14–46; D. Brown, *Before the Ironclad*, 44–60; and A. Lambert, *Battleships in Transition*.

54. PP 915 (1848), 3.

55. Reid, *Memoirs and Correspondence of Lyon Playfair*, 94–5.

56. PP 915 (1848), 5–18.

57. PP 915 (1848), 5.

58. PP 1086 (1849), 3–8.

59. PP 1345 (1851), 3–10.

60. Reid, *Memoirs and Correspondence of Lyon Playfair*, 95; A. Lambert, *Crimean War*, 344–45; D. Brown, *Before the Ironclad*, 135–60.

61. Morriss, *British Naval Technology*; Ashworth, "'System of Terror.'"

62. D. Brown, *Before the Ironclad*, 15–24. On the development of the steam wooden battleship, especially under Captain Sir Baldwin Walker, surveyor of the navy from 1848 to 1861, see A. Lambert, *Battleships in Transition*.

63. *Report of the BAAS* (1835): 107–8.

64. For John Scott Russell's involvements, see Emmerson, *John Scott Russell*, 178.

65. PP 747 (1853), iii. Russell had been instrumental in the development of steamships for the Royal Mail Steam Packet Company; see Crosbie Smith, "'The Great National Undertaking.'"

66. Emmerson, *John Scott Russell*, 35.

67. Emmerson, *John Scott Russell*, 85–86.

68. The Admiralty did not purchase an iron ship until 1845, despite their earlier use as gunboats by the East India Company; see Headrick, *Tools of Empire*, 17–42. For John Scott Russell's influence, see Leggett, *Shaping the Royal Navy*, 89–125. On the development of *Warrior*, see D. Brown, *Before the Ironclad*, 174–87, and on ironclads generally, see Hamilton, *Anglo-French Naval Rivalry*, 64–105.

69. C. Smith, *Coal, Steam and Ships*, 186.

70. Leggett, "Neptune's New Clothes," 76–9. On iron hulls and armor plate, see A. Lambert, *Steam, Steel and Shellfire*, 47–60.

71. Emmerson, *John Scott Russell*, 177.

72. Leggett and Dunn, *Reinventing the Ship*, 89.

73. Emmerson, *John Scott Russell*, 292.

74. In 1877 William Froude was asked with George Rendel and Mr. J. Woolley to advise on whether the *Inflexible*, with unprotected ends, was a safe vessel to take into combat. They found the design to be satisfactory; see PP C1917 (1878). For Froude's approach, see Emmerson, *John Scott Russell*, 291–96.

75. This case became significant in discussions of ministerial responsibility. Hugh Childers, as first lord of the Admiralty, was formally responsible for all that passed there. Although the court martial exonerated Robert Robinson, Childers laid the blame on him. He was not reappointed as comptroller of the navy and third

lord; see S. Finer, "Individual Responsibility of Ministers," 381. On Cowper Coles, see Leggett, *Shaping the Royal Navy*, 126–64.

76. PP C477 (1872); Leggett, *Shaping the Royal Navy*, 165–96.

77. PP C477 (1872), x, xxi–xxvi.

78. PP C477 (1872), xxxvii–xxxviii.

79. Leggett, "Naval Architecture," 74–75.

80. John "Jacky" Fisher had supported William Thomson's compass at a time when it meeting resistance in the Admiralty, leading to their effective later introduction of depth sounding and Thomson's involvement in new ship designs; see Dunn, "'Their Brains Over-Taxed.'"

81. Leggett, "Naval Architecture." For detail of this period, see Leggett, *Shaping the Royal Navy*, 197–270.

82. For a view of the changing culture of the Admiralty with respect to scientific expertise, see Leggett, *Shaping the Royal Navy*; Hamilton, "Three Cultures at the Admiralty."

4. WAR OFFICE, ARMY, AND ORDNANCE

1. Hamer, *British Army*, 2–3.

2. PP C5062-I (1887), 101.

3. Quoted in Hamer, *British Army*, 22.

4. The political responsibility for these choices was explored later in the century by a royal commission. For a new gun, a committee of officers would draw up the specification. It would be designed under the superintendent of the gun factories, and the Ordnance Committee would supervise trials. The secretary of state for War would decide on adoption, and also made the momentous decision on the conversion of artillery from muzzle-loaders to breechloaders. See Hamer, *British Army*, 44–45, and PP 5062 (1887), xv. A further example is the decision on whether to include a thirty-five-ton gun in the estimates in 1870, and if so, whether it should be made by William Armstrong or Joseph Whitworth; see PP 308 (1870).

5. Jackson, *Case Studies in Scientific Advice*, no. 6.

6. Jackson, *Case Studies in Scientific Advice*, no. 7.

7. Jackson, *Case Studies in Scientific Advice*, no. 8.

8. See Skentelbery, *History of the Ordnance Board*, and Forbes, *History of the Army Ordnance Services*, for general histories.

9. See also Ashworth, "Roaming Eye of the State."

10. PP 2396 (1858).

11. The Ordnance Geological Survey had already been established in 1835, after pressure from Charles Lyell, Adam Sedgwick, and William Buckland, under the Board of Ordnance, with Henry De La Beche in charge. Responsibility passed to the Office of Woods, Forests, Land Revenues, Works and Buildings under the Geological Survey Act 1845. See McGregor, "Social Research and Social Policy," 147.

12. Hall, *All Scientists Now*, 13.

13. RS CMO/8/184–86, 190–92. In 1827 the board asked for further advice on lightning conductors for powder magazines from a Royal Society Committee; see RS CMO/10/303–4, 317.

14. RS CMO/8/218–19, 223–26.

15. James, "Military Context of Chemistry," 38.

16. Gladstone, *Michael Faraday*, 72.

17. Macleod, "Royal Society and the Government Grant."

18. James, "Time, Tide and Michael Faraday," 33–34.

19. James, *Michael Faraday*, 58.

20. James, *Correspondence of Michael Faraday, Volume 3*, xxxi; Faraday to Colonel Cockburn, June 20, 1843, letter 1502 in James, *Correspondence of Michael Faraday, Volume 3*.

21. Francis Montagu Smithy, *A Handbook of the Manufacture and Proof of Gunpowder, as Carried on at the Royal Gunpowder Factory, Waltham Abbey* (London: Eyre and Spottiswoode, 1871), 7.

22. James, *Correspondence of Michael Faraday, Volume 4*, xl; George Butler to Faraday, March 3, 1852, and Faraday to George Butler, March 4, 1852, letters 2501 and 2502, in James, *Correspondence of Michael Faraday, Volume 4*.

23. Mauskopf, "From an Instrument of War," and "Long Delayed Dream."

24. *Report of the BAAS* (1862): xlii. The members of the committee included the chemists John Gladstone, William Allen Miller, and Edward Frankland, and the engineers William Fairbairn, Joseph Whitworth, James Nasmyth, John Russell, and William Armstrong. Frederick Abel provided several reports to the committee, and joined formally in 1863. See *Report of the BAAS* (1863), *Reports of the State of Science*, 1–36.

25. *Report of the BAAS* (1865): 264.

26. Howarth, *British Association*, 225–6. Thomas Sopwith and Frederick Abel carried out the experiments on mining at Allenheads; see PP 179 (1869).

27. Mauskopf, "Long Delayed Dream," 5; *Report and Proceedings of the Gun Cotton Committee*, 7.

28. Skentelbery, *History of the Ordnance Board*, 13.

29. Brigadier Hogg, quoted in Skentelbery, *History of the Ordnance Board*, 13.

30. Lyon Playfair and James Graham were consulted by the Select Committee in 1854 "with great advantage"; see Skentelbery, *History of the Ordnance Board*, 23.

31. Skentelbery, *History of the Ordnance Board*, 15–16.

32. Mauskopf, "Long Delayed Dream," 5–7, gives the chronology.

33. PP C586 (1872), 41.

34. PP 243 (1874), 339–42.

35. PP 371 (1872), 17–19.

36. The reports of the War Office committee on lithofracteur are in PP 243

(1874), 329–31, 343–7, 347–50. Two reports on experiments with dynamite at Llanberis are at 350–56 and 356–65.

37. Skentelbery, *History of the Ordnance Board*, 37.

38. Skentelbery, *History of the Ordnance Board*, 36–38. The ten committees, with their dates of establishment, were Explosive Substances (May 1869), Stores and Fitments in Magazines (November 1870), Mounting and Working of Heavy Rifled Guns (June 1872), Range Finders for Field and Coast Batteries (December 1872), Transport and Storage of Gunpowder and Guncotton (November 1874), Siege Carriages (February 1878), Ordnance (July 1879), Machine Guns and Machine Arms (February 1879), Friction Tubes (May 1880), and Rockets (undated).

39. PP C5979 (1890), 102–3.

40. *Report of the BAAS* (1855), *Reports on the State of Science*, 100–108. The committee included William Fairbairn, the Duke of Argyll, Joseph Whitworth, James Nasmyth, and William Macquorn Rankine.

41. Heald, *William Armstrong*, 85–90; Warren, *Armstrong*, 33–42. PP 448 (1862), 156–63, gives a summary of the history of William Armstrong's early involvement, his meeting with the Duke of Newcastle, and his initial report in 1855. Bastable, *Arms and the State*, details Armstrong's impact on British naval power.

42. PP 448 (1862), 8.

43. *Hansard* 152 (March 4, 1859): 1318–20. The report of the Committee on Rifled Cannon is published in PP 448 (1862), 166–71. See also Jonathan Peel's evidence to the select committee on improved ordnance: PP 448 (1862), 111–15.

44. Indeed, Joseph Whitworth's pressures would result in seven committees examining his claims, aided by Colonel Eardley Wilmot, who had been displaced by Armstrong as superintendent of the Royal Gun Factory; see Heald, *William Armstrong*, 115.

45. PP 3605 (1866).

46. The committee and trials are described in Heald, *William Armstrong*, 121–5; Warren, *Armstrong*, 95–101.

47. Russell, *Edward Frankland*, 415–16; Edward Frankland, "On the Influence of Atmospheric Pressure on Some Phenomena of Combustion," *Philosophical Transactions of the Royal Society of London* 151 (1861): 629–53.

48. On Whitworth's heavy guns, see Atkinson, *Joseph Whitworth*, 254–71.

49. PP 459 (1865); PP 3998 (1868).

50. PP 3998 (1868), 10.

51. PP 4003 (1868).

52. Robert Mallet was elected Fellow of the Royal Society in 1864. A seismologist who had started work in his father's iron foundry business, he had invented a forty-two-ton mortar during the Crimean War, although it was too late to be used in action. When it was ready it was found to cost £675 per shot and was rejected by the War Office; see Weintraub, *Uncrowned King*, 319.

53. PP C258 (1871), 8.

54. *Report on Experiments Made with the Bashforth Chronograph to Determine the Resistance of the Air to the Motion of Projectiles 1865–1870* (London: HMSO, 1870), 153.

55. *Report on Experiments Made with the Bashforth Chronograph to Determine the Resistance of the Air to the Motion of Projectiles 1865–1870* (London: HMSO, 1870), 155–61.

56. Heald, *William Armstrong*, 188–89. The fortification and armament of military and home mercantile ports was a further consideration. For example, a committee chaired by Edward Stanhope, the secretary of state for War, provided estimates in 1888 for structural works and armaments in twenty-five ports. Frederick Bramwell was a member of this committee; see PP C5305 (1888).

57. Warren, *Armstrong*, 181.

58. Marc Brunel had also applied ideas of mass production to the manufacture of blocks for the Admiralty; see Marsden and Smith, *Engineering Empires*, 147.

59. Fries, "British Response to the American System," describes the revolution in detail.

60. PP 236 (1854), iii–xi.

61. Heald, *William Armstrong*, 122. For a decidedly pro-Whitworth account of the development of the rifle, see Atkinson, *Joseph Whitworth*, 215–54.

62. Atkinson, *Joseph Whitworth*, 215–8.

63. On the impact of breechloaders, see Headrick, *Tools of Empire*, 96–104.

64. PP 187 (1861).

65. Walker, "Rise and Fall of the Second British Empire," 43.

66. PP 139 (1863), 16.

67. Atkinson, *Joseph Whitworth*, 253.

68. PP 462 (1865).

69. PP 435 (1866), 19–32.

70. PP C299 (1871), 5–6. The committee's decision, and the choice of the scientific advisers, who had been paid £50 each, became an issue in Parliament when the report was published. Edward Cardwell, resisting a motion by Colonel Walter Barttelot and John Hick to establish a select committee to examine the decision, made much of the independence of the committee and its scientific advisers; see *Hansard* 205 (April 28, 1871): 1872–910.

71. Introduction of the Lee-Metford is detailed in Ford, "Revolution in Firepower."

72. PP 543 (1864).

73. *Hansard* 34 (June 21, 1895): 1685–712.

74. Hamer, *British Army*, 161–3, 235.

75. Mauskopf, "Long Delayed Dream," 7–13.

76. For a brief analysis, see Reid, *Tongues of Conscience*, 1–9.

77. Mauskopf, "Long Delayed Dream," 13.

5. AGRICULTURE

1. Wilmot, *"Business of Improvement,"* 46, 77–81. Bord, *Science and Whig Manners,* explores science and Whig agrarian interests.

2. Wilmot, *"Business of Improvement,"* 29.

3. Mitchison, "Old Board of Agriculture."

4. Clarke, *History of the Board of Agriculture.* For a picture of John Sinclair's life and activities, see Fussell, "Impressions of Sir John Sinclair."

5. James, "Agricultural Chymistry."

6. Howes, *Arts and Minds,* 62–8.

7. Mitchison, "Old Board of Agriculture," 42.

8. James, "Agricultural Chymistry."

9. For an account of the actions of the board in relation to potato growing, see Lidwell-Durnin, "Cultivating Famine."

10. Mitchison, "Old Board of Agriculture," 69.

11. James, "Agricultural Chymistry"; Gascoigne, *Science in the Service of Empire.*

12. PP 311 (1817).

13. Mitchison, "Old Board of Agriculture," 65.

14. PP 57 (1813), 7.

15. PP 57 (1813); PP 339 (1814); PP 26 (1814).

16. PP 57 (1813), 3.

17. PP 57 (1813), 30.

18. PP 26 (1814), 63–68, 88.

19. PP 612 (1833).

20. PP 612 (1833), x.

21. Peel, *Private Letters of Sir Robert Peel,* 183.

22. On Justus Liebig's approach, see Brock, *Justus von Liebig,* 145–82.

23. Justus Liebig had asserted that plants obtained nitrogen from ammonia in the air, and his fertilizers made phosphate and potash insoluble, so they were not taken up by the plant. John Lawes and Joseph Gilbert, in experiments at Rotham-sted, disproved Liebig's claim about ammonia; see Brassley, "Agricultural Research in Britain," 468.

24. Ogawa, "Liebig and the Royal Agricultural Society," 145–47.

25. Reid, *Memoirs and Correspondence of Lyon Playfair,* 77.

26. Ogawa, "Liebig and the Royal Agricultural Society," 141.

27. James, *Correspondence of Michael Faraday, Volume 3,* 96–98.

28. Reid, *Memoirs and Correspondence of Lyon Playfair,* 58.

29. Reid, *Memoirs and Correspondence of Lyon Playfair,* 86; Parker, *Sir Robert Peel,* 162.

30. PP 28 (1846). Other commissioners were sent out throughout Ireland; see PP

33 (1846). Arthur Hassall had made earlier suggestions as to how the spread might be controlled; see Gray, *By Candlelight*, 74–77.

31. Parker, *Sir Robert Peel*, 280.

32. *Hansard* 83 (February 9, 1846): 529.

33. Parker, *Sir Robert Peel*, 237–8; Weintraub, *Uncrowned King*, 175.

34. Mårald, "Everything Circulates," 67–68. On zymotic processes and natural theology, see Hamlin, "Providence and Putrefaction."

35. Douglas, *Purity and Danger*, 10–11; Hamlin, "Providence and Putrefaction," 392–400. Angus, "Cesspools, Sewage, and Social Murder," explores the relationship between night soil, sewage and agriculture.

36. PP 308 (1846), 2.

37. PP 474 (1846).

38. PP 1472 (1852), 144.

39. R. Lambert, *Sir John Simon*, 258.

40. PP 2372 (1858), 30–37. During the first year of the inquiry, Henry Austin, as chief superintending inspector of the General Board of Health, had produced a report for the board on deodorizing and utilizing the sewage of towns; see PP 2262 (1857).

41. PP 233 (1857), appendix 1, 1–60. This was a report to Douglas Galton, James Simpson, and Thomas Blackwell, who had been asked to comment on plans for drainage of the metropolis by the Metropolitan Board of Works.

42. PP 2372 (1858), 27.

43. Russell, *Edward Frankland*, 208.

44. PP 2882 (1861), 40–41.

45. PP 160 (1862), iii; Sheail, "Town Wastes," 201–2; Mårald, "Everything Circulates," 69–71. Wilhelm Hofmann, John Way, Edward Frankland, and Augustus Voelcker were all associated with the Royal College of Chemistry, a powerbase of the chemists.

46. PP 469 (1862), iii.

47. PP 487 (1864), v–vii.

48. PP 3472 (1865), 3–4.

49. Sheail, "Town Wastes," 193; Brock, *Justus von Liebig*, 254–72.

50. Sheail, "Town Wastes," 203–5.

51. S. Finer, *Life and Times of Sir Edwin Chadwick*, 300–301.

52. *Hansard* 177 (March 8, 1865): 1358.

53. William Thiselton-Dyer became director from 1885 to 1905 on the retirement of Joseph Hooker.

54. PP 274 (1880), iii–vi.

55. Dearce, "Correspondence of Charles Darwin."

56. Other reports produced by the department include one on rust or mildew in wheat, which included evidence back to the original Board of Agriculture and

more recently from Augustus Voelcker and John Lawes; see PP C7018 (1893). The Intelligence Department developed out of the work of Charles Whitehead.

57. PP C6647 (1892), 8.

58. PP C7138 (1893), vii–viii.

59. PP C3977 (1884).

60. A series of reports on swine fever was also summarised and published in 1886; see PP C4843 (1886). On livestock and disease, see Jackson, *Case Studies in Scientific Advice*, no. 13.

61. PP 5679 (1889), 20–128. Robert Koch had established in 1876 that specific bacteria, which could form resistant spores, were the cause of anthrax.

62. PP C5995 (1890), 137–61.

63. PP C5481 (1888), iii–xxi.

64. Coleman, *Conservatism and the Conservative Party*, 191.

65. Jackson, *Case Studies in Scientific Advice*, no. 13.

66. On Eleanor Ormerod, see Clark, "Eleanor Ormerod," and Sheffield, *Revealing New Worlds*, 139–94. Like Florence Nightingale, she was no women's rights activist. She gained a substantial international reputation, and in 1900 was the first woman to receive an honorary LLD from the University of Edinburgh.

67. Gratwick, *Crop Pests in the UK*.

68. Charles Whitehead describes his involvement with the Agricultural Department of the Privy Council and the Board of Agriculture in Whitehead, *Retrospections*, 75–98.

69. PP C4543 (1885); PP C4680 (1886); PP C4836 (1886).

70. PP C4944 (1886); PP C4986 (1887); Sheffield, *Revealing New Worlds*, 183

71. PP C5217 (1887).

72. PP C5275 (1888). As early as 1873, in evidence to the Select Committee on Wild Birds Protection, suggestions had been made for the establishment of a British government entomologist; see Clark, "Eleanor Ormerod," 435.

73. "Notes," *Nature* 63 (1900): 160–63.

74. RS MS/506/21.

75. PP C8540 (1897), 160. This broad conclusion of the cause of the depression has been supported by more detailed later studies. Fletcher, "Great Depression of English Agriculture," describes the clashes among the commissioners. See also M. Turner, "Output and Prices in UK Agriculture."

76. PP C8540 (1897), 160–72.

77. PP C3309 (1882), 24.

78. PP C3096 (1881), 923–40. See also PP C2778 (1881).

79. PP C3096 (1881), 950.

80. Brassley, "Agricultural Research in Britain," discusses the success or failure of agricultural research in Britain from 1850 to 1914.

81. For a history of its formation, see DeJager, "Pure Science and Practical Interests."

6. FISHERIES

1. At a speech in Blackpool on May 24, 1945.

2. For aspects of the history, see Coull, "Development of Marine Superintendence;" Sutherland, "Scottish Continental Herring Trade."

3. See, for example, the parliamentary committee report on a petition in relation to the import of Swedish herrings, which had caused a loss to the merchants: PP 65 (1805). The government did not provide compensation; see PP 149 (1806).

4. Coull, "Development of Marine Superintendence," 50–51.

5. Desmond, *Devil's Disciple*, 306–7.

6. PP C3106 (1863), 24–29.

7. Buffery, "Changing Landscapes," 125.

8. *Hansard* 11 (March 30, 1824): 33–35.

9. PP 427 (1824).

10. PP 427 (1824), 144–45. Humphry Davy published *Salmonia; Or, Days of Fly Fishing*, which ran to many editions, in 1828.

11. PP 173 (1825), 3–4.

12. PP 393 (1825), 3.

13. PP 393 (1825), 61–62.

14. PP 393 (1836).

15. PP 393 (1836), vii.

16. PP 393 (1836), 254–61, 268–78, 278–83.

17. PP 393 (1836), 274–75.

18. MacLeod, "Government and Resource Conservation," 117n18.

19. PP 456 (1860), xi–xii.

20. PP 2768 (1861), i.

21. PP 2768 (1861), iii.

22. PP 2768 (1861), xxiii–xxxvi.

23. On the inspectorate, see Pellew, *Home Office*, 142–43.

24. MacLeod, "Government and Resource Conservation," 123.

25. Bompass, *Life of Frank Buckland*, 182.

26. For Frank Buckland's life, see Bompass, *Life of Frank Buckland*. See also Lightman, *Evolutionary Naturalism in Victorian Britain*, vol. 11, 1–27, which explores Buckland's popularization of science, and his antievolutionism in the context of his natural theology, differentiating him from Thomas Huxley.

27. Bompass, *Life of Frank Buckland*, 211.

28. PP 361 (1869), 77.

29. PP 368 (1870), iii–vi.

30. PP 676 (1833), 3.

31. PP 3596 (1866), cvi–cvii.

32. PP 428 (1875).

33. Bartrip, "Food for the Body," 288–89.

34. PP C1695 (1877); PP C1979 (1878).

35. PP C1819 (1877); Bartrip, "Food for the Body," 289.

36. Bartrip, "Food for the Body," 289–99.

37. MacLeod, "Government and Resource Conservation," 135–7.

38. An excerpt from William Harcourt's letter is quoted in MacLeod, "Government and Resource Conservation," 138.

39. Huxley, *Life and Letters*, 2:22.

40. Thomas Huxley published this as "A Contribution to the Pathology of the Epidemic Known as the Salmon Disease," *Proceedings of the Royal Society of London* 33 (1882): 381–89. He argued that it was important to remove diseased fish, although he noted that "in practice, it may not be worthwhile to adopt that treatment." An earlier report by Frank Buckland, Spencer Walpole, and Archibald Young, commissioner of Scotch Salmon Fisheries, had investigated fungal disease in salmon, with advice from several naturalists. George Rolleston proposed sweeping the bottom of rivers with hemp impregnated with coal tar; see PP C2660 (1880), 113–15.

41. PP C4328 (1885).

42. PP 271 (1885), 188–203.

43. MacLeod, "Government and Resource Conservation," 143–44. Parts of it still exist at the Scottish Fisheries Museum in Anstruther.

44. PP 3596 (1866), lxxiv–lxxx.

45. PP C3741 (1883), xviii–xx.

46. Southward and Roberts, "100 Years of Marine Research," 465. This paper gives a history of one hundred years of marine research at Plymouth.

47. Southward and Roberts, "100 Years of Marine Research," 474.

48. MacLeod, "Government and Resource Conservation," 141n135.

49. MacLeod, "Government and Resource Conservation," 147.

50. Nevertheless, naturalists did pursue research, as exemplified by the extensive report on the life history of salmon brought together in 1898 at the request of the inspector of salmon fisheries for Scotland by Noel Paton, superintendent of the Royal College of Physicians of Edinburgh; see PP C8787 (1898).

7. TRANSPORT INFRASTRUCTURE AND ENGINEERING

1. Buchanan, *Engineers*, gives a history of the development of the engineering profession; Buchanan, "Engineers and Government," its connections with government; and Buchanan, "Rise of Scientific Engineering," covers scientific engineering education.

2. Jackson, *Case Studies in Scientific Advice*, no. 21.

3. Great-great-great-grandfather of the author.

4. On the connection between the Royal Society Club and the Smeatonian Society for Civil Engineers, see Allibone, *Royal Society and Its Dining Clubs*, 81–85.

5. Buchanan, *Engineers*, 60.

6. Buchanan, *Engineers*, 63.

7. Buchanan, "Engineers and Government," 43–44.

8. Thomas Telford's life is documented in contemporary terms in Smiles, *Life of Thomas Telford*, and later in Rolt, *Thomas Telford*.

9. Guldi, *Roads to Power*, explores this "infrastructure state."

10. Guldi, *Roads to Power*, 52.

11. Hamlin, *Public Health and Social Justice*, 265–66.

12. PP 212 (1806).

13. PP 321 (1806).

14. PP 321 (1806), 30.

15. PP 240 (1811), 5–6.

16. PP 179 (1809).

17. PP 238 (1809), 40.

18. PP 271 (1809), 81–82.

19. PP 271 (1809), 46.

20. PP 301 (1820), 6. This committee noted the improvements already made on the road between Shrewsbury and Holyhead by appointing first a parliamentary commission and then fifteen commissioners in place of local trusts, who employed a "scientific surveyor." A further commission, extending the responsibility to bridges, harbors, and the roads between London and Holyhead, and Howth to Dublin, was appointed in 1824, including Davies Gilbert as one of the commissioners; see PP 305 (1824). An act to consolidate turnpike trusts north of London was passed in 1822.

21. PP 547 (1836).

22. On turnpike trusts, see Bogart, "Turnpike Roads of England and Wales."

23. PP 256 (1840), 5–14.

24. PP 324 (1831); Todd, *Beyond the Blaze*, 110.

25. Marsden and Smith, *Engineering Empires*, 132–34.

26. Swann, "Engineers of English Port Improvements," lists dozens of engineers involved in port improvements up to 1830. Almost none had a military background.

27. PP 166 (1810), 2; PP 203 (1810), 13–15, 19–22; PP 462 (1816), 5–8. The work of the elder John Rennie was crucial to the development of Plymouth Harbour at the end of the Napoleonic Wars; see Knight, *Britain against Napoleon*, 382–85.

28. PP 370 (1835), 121–23; PP 398 (1836), 139–44.

29. PP 290 (1884), xiv.

30. Airy, *Autobiography of Sir George Biddell Airy*, 176–77.

31. PP 334 (1836), iii.

32. PP 567 (1836), vii

33. PP 368 (1840).

34. PP 549 (1843), v.

35. PP 611 (1845), 39–41.

36. PP 611 (1845), 215–16.

37. PP 611 (1845), 11–12.

38. PP 821 (1847).

39. PP 411 (1847), 1–2.

40. PP 411 (1847), 8–29.

41. PP 411 (1847), 40.

42. PP 390 (1851).

43. PP 262 (1857); PP 344 (1858).

44. PP 2474 (1859), 7.

45. *Hansard* 157 (March 23, 1860): 1152.

46. *Hansard* 180 (June 13, 1865): 165–78.

47. PP 255 (1883); PP 290 (1884), x–xi.

48. PP 290 (1884), xiii.

49. *Hansard* 303 (March 12, 1886): 643–97.

50. *Hansard* 76 (July 16, 1844): 973–78.

51. PP 665 (1845), 124, 133.

52. Airy, *of Sir George Biddell Airy*, 169; PP 665 (1845), 194–95.

53. PP 665 (1845), xiv.

54. PP 692 (1846); PP 874 (1847); PP 943 (1848).

55. PP 692 (1846), v.

56. Simmons, *Railway in England and Wales*, and Parris, *Government and the Railways*, give extensive histories of government involvement in the railways, and H. Smith, *Board of Trade*, 124–46. On railway legislation, see Clifford, *History of Private Bill Legislation*, 43–203.

57. *Hansard* 46 (April 11, 1839): 1314–15.

58. PP 517 (1839), xiii.

59. PP 517 (1839), 210–11.

60. Lubenow, *Politics of Government Growth*, 107–36, discusses the railways in the context of government growth.

61. *Hansard* 51 (January 21, 1840): 419–21.

62. PP 50 (1840), 4.

63. PP 299 (1840), 6–7.

64. PP 437 (1840), 7. Charles Wheatstone's evidence is in PP 474 (1840), 1–5.

65. PP 437 (1840), 8.

66. PP 287 (1841), 2–3. For the start of the Board of Trade's responsibilities, and the acts of 1840, 1842, and 1844, see Simmons, *Railway in England and Wales*, 33–38.

67. John Frederick Smith and Peter Barlow recommended the west coast route, which the government accepted, but Parliament passed acts in 1845 authorizing

both. They also recommended, in a separate report to the Treasury, that Holyhead should be the destination for the route to Ireland; see Simmons, *Railway in England and Wales*, 31.

68. PP 354 (1841), 44.

69. PP 354 (1841), v.

70. The key recommendations, which formed the basis of legislation, are in PP 166 (1844), 5–7, with detailed analysis and evidence in PP 318 (1844). McLean, "Origin and Strange History of Regulation," 4–17, and Bailey, "1844 Railway Act," analyzes William Gladstone's approach to regulation.

71. For a discussion of regulation of railways, roads, bridges, and canals in this period in the light of later experience, and the formation of the Railways and Canals Commission in 1873, see Stern, "Regulation and Contracts for Utility Services," 200–205.

72. H. Smith, *Board of Trade*, 211.

73. On atmospheric railways, see Jackson, *Case Studies in Scientific Advice*, no. 19.

74. Parris, *Government and the Railways*, 68.

75. *Hansard* 81 (June 20, 1845): 972–99.

76. Perkins, "'Extraneous Government Business,'" 149–51; Parris, *Government and the Railways*, 101.

77. PP 699 (1846), iii.

78. PP 684 (1846), 4–21.

79. Simmons, *Railway in England and Wales*, 45–47; Marsden and Smith, *Engineering Empires*, 150–56.

80. Airy, *Autobiography of Sir George Biddell Airy*, 183. For George Airy's other personal reminiscences of this episode, see 171, 175–76, 180.

81. Gillin, "Mechanics and Mathematicians," gives a detailed analysis of these episodes.

82. For Airy's project to coordinate time, see Morus, "Nervous System of Britain," 464–70.

83. *Hansard* 93 (June 24, 1847): 841.

84. PP 1123 (1849), vii.

85. Airy, *Autobiography of Sir George Biddell Airy*, 185.

86. PP 1123 (1849), xiv.

87. PP 1123 (1849), 197–214.

88. PP 1123 (1849), xvi. Lewis and Gagg, "Aesthetics versus Function," analyze the cause of the bridge failure.

89. PP 1123 (1849), xii.

90. PP 1123 (1849), xvii.

91. George Stephenson had assistance from both William Fairbairn and George Airy in developing the tubular design of Britannia Bridge for the railway over the Menai Strait, which opened in 1850. When Airy was asked to submit a report and

charge a fee as a civil engineer, he refused; see Airy, *Autobiography of Sir George Biddell Airy*, 180.

92. Buchanan, "Rise of Scientific Engineering," 220.

93. Parris, *Government and the Railways*, 144.

94. Parris, *Government and the Railways*, 167.

95. Jackson, *Ascent of John Tyndall*, 5–7.

96. PP 1769 (1854), 3.

97. *Hansard* 135 (July 25, 1854): 696–99.

98. PP 1951 (1855), 63–64. Jackson, *Ascent of John Tyndall*, 281–82.

99. Michael Bailey, personal communication, n.d.

100. PP 684 (1846), 10–11.

101. PP 362 (1858), iii.

102. PP 3289 (1864).

103. Parris, *Government and the Railways*, 177–78. On safety systems and block working, see Simmons, *Railway in England and Wales*, 214–38.

104. PP 2498 (1859), 3.

105. PP 2498 (1859), 26.

106. PP 3844 (1867), xxxvii. The early part of the report gives an excellent summary of railway legislation since 1801; see vii–xxviii. A summary of the conclusions is at lxxxvii–xcii.

107. PP 3844 (1867), lxxx. On the issue of color blindness, see Jackson, *Case Studies in Scientific Advice*, no. 20.

108. PP 341 (1870).

109. *Hansard* 218 (April 27, 1874): 1150–72.

110. *Hansard* 218 (May 7, 1874): 1877–83.

111. *Hansard* 218 (May 7, 1874): 1883.

112. PP 341 (1870), 123.

113. PP C294 (1871), 38.

114. PP C2616 (1880), 32–33. Airy, *Autobiography of Sir George Biddell Airy*, 303, 331, 356–57.

115. Pinsdorf, "Engineering Dreams into Disaster," 492–501.

116. PP C2616 (1880), 44.

117. PP C2616 (1880), 5–16. Henry Rothery's report is at 17–49.

118. Lewis and Reynolds, "Forensic Engineering," reappraise the disaster, extending the report's conclusions about design and materials defects to include the effect of lateral oscillation on the high girders caused by trains passing over misaligned track, causing fatigue cracks in the cast iron lugs. They argue that the wind load added to the load on a bridge that was already defective; see Swinfen, *Fall of the Tay Bridge*.

119. PP C3000 (1881), 4–5.

120. *Hansard* 1 (March 1, 1892): 1610–13; *Hansard* 2 (March 4, 1892): 7.

121. PP 215 (1892), vi.

122. PP C8684 (1898), vii.

8. SHIPS, LIGHTHOUSES, AND THE BOARD OF TRADE

1. For the early history until the 1920s, see H. Smith, *Board of Trade*, and for an emphasis on the nineteenth-century transformation, see Prouty, *Transformation of the Board of Trade*.

2. On policy for merchant shipping, see Prouty, *Transformation of the Board of Trade*, 30–98; H. Smith, *Board of Trade*, 90–123.

3. On the origin of this strange formula, issued by the Navy Board, see Syrett, *Shipping and the American War*, 111–12.

4. PP 43 (1834).

5. PP 49 (1850).

6. On the importance of colonial expansion for the development of steamships, see Headrick, *Tools of Empire*, 129–41; Marsden and Smith, *Engineering Empires*, 88–128. Isambard Kingdom Brunel advised the Admiralty on screw propulsion; see A. Lambert, *Steam, Steel and Shellfire*, 33. See also D. Brown, *Before the Ironclad*, 61–72, 99–134; Hamilton, *Anglo-French Naval Rivalry*, 15–63.

7. PP 422 (1817).

8. Armstrong and Williams, *Impact of Technological Change*, 51–73.

9. Armstrong and Williams, *Impact of Technological Change*, 245–58.

10. PP 417 (1822), 180–83, 189–95. More than fifty years later, a 1,400-page Admiralty report exemplifies the difficulty, even relatively late in the nineteenth century, of coming to a clear judgement on the cause of a problem as simple to state as boiler corrosion; see Jackson, *Case Studies in Scientific Advice*, no. 5.

11. PP 417 (1822), 130. This was done, and revenue increased. For a detailed history of the Holyhead transport link to Ireland, see Watson, *Royal Mail to Ireland*.

12. PP 335 (1831).

13. PP 567 (1836).

14. Williams, "James Silk Buckingham."

15. Rose, "Military Background of John W. Pringle."

16. PP 273 (1839).

17. PP 273 (1839), 20.

18. PP 549 (1843); PP 581 (1843). The Select Committee on Smoke Prevention in 1843 also noted evidence that steamships could in principle be made smokeless, though nothing was done; see PP 583 (1843).

19. On factors influencing the development of iron ships, see Headrick, *Tools of Empire*, 142–49;

20. Jackson, *Case Studies in Scientific Advice*, no. 17.

21. For example, the *Amazon* in 1852. Thomas Graham gave evidence to the enquiry; see C. Smith, *Coal, Steam and Ships*, 192.

22. PP 370 (1870), 91–6.

23. Bartrip, "State and the Steam Boiler," discusses this for steam boilers.

24. PP 298 (1871), 30.

25. PP C1586 (1876), v.

26. PP C1586 (1876), xxvi–xxviii.

27. PP C1586 (1876), 109–10.

28. PP C1586 (1876), xxiv–xxv.

29. Jackson, *Case Studies in Scientific Advice*, no. 4.

30. R. Lambert, *Sir John Simon*, 375.

31. Prouty, *Transformation of the Board of Trade*, 33. See also Burn, *Age of Equipoise*, 137–78, as by this time the definition of *tonnage* had been settled. On Samuel Plimsoll and the government, see P. Smith, *Disraelian Conservatism*, 230–42.

32. PP C853 (1873); PP C1027 (1874).

33. PP C1027 (1874), xviii.

34. *Hansard* 225 (July 22, 1875): 1822–29.

35. Herd, "International Conference on Load Lines," 2.

36. PP C4577 (1885); PP C5227 (1887).

37. PP C5227 (1887), 14.

38. Herd, "International Conference on Load Lines," 2.

39. PP 591 (1822), 235–41, 319–20.

40. PP 591 (1822), 290–1.

41. PP 591 (1822), 3.

42. PP 591 (1822), 14–15.

43. J. Taylor, "Private Property, Public Interest," and Conway, "Illuminating Science," discuss David Brewster's role.

44. PP 590 (1834), iii.

45. PP 590 (1834), iv.

46. PP 590 (1834), xxvi.

47. *Hansard* 15 (February 21, 1833): 1074.

48. PP 590 (1834), xxiv, 172.

49. PP 590 (1834), 177.

50. Hannah Conway suggests that this demonstrates the committee's general lack of knowledge on how to properly incorporate scientific testing into questions of governmental policy; see Conway, "Illuminating Science," 62.

51. PP 590 (1834), 119–20.

52. Brewster, "Life-Boat, the Lightening Conductor, and the Lighthouse," 525.

53. *Hansard* 27 (March 25, 1835): 247.

54. *Hansard* 35 (July 12, 1836): 141.

55. *Hansard* 35 (July 12, 1836): 145.

56. Quoted in Churchill, *Twenty-One Years*, 127. This quote appears to originate

in the phrase *experts should be on tap and not on top*, written by the editor of the *Irish Homestead*, George Russell, "Notes of the Week," 1087.

57. James, "Michael Faraday and Lighthouses," 92. For Michael Faraday's work on lighthouses, see also James, "Civil-Engineer's Talent"; James, *Michael Faraday*, 50–55; James, *Correspondence of Michael Faraday, Volume 2*, xxxiv; James, *Correspondence of Michael Faraday, Volume 3*, xxv–xxvi, xxxiv–xxxvi; James, *Correspondence of Michael Faraday, Volume 4*, xxxviii–xl; James, *Correspondence of Michael Faraday, Volume 5*, xxx–xxxiv; James, *Correspondence of Michael Faraday, Volume 6*, xxxiii–xxxvii.

58. James, "Michael Faraday and Lighthouses," 97

59. Michael Faraday listed many of these tasks in his evidence to a royal commission in 1861; see PP 2793 (1861), 591–2.

60. James, "Civil-Engineer's Talent," 154–55.

61. James, *Correspondence of Michael Faraday, Volume 4*, xxxviii–xl.

62. James, *Correspondence of Michael Faraday, Volume 4*, xxxi–xxxii.

63. Courtney, "'Very Diadem of Light,'" 257.

64. Schiffer, "Electric Lighthouse in the Nineteenth Century," gives an analysis of why electric arc lights did not succeed in countries during the nineteenth century.

65. Schiffer, "Electric Lighthouse in the Nineteenth Century," 297.

66. Preece, "Address," 13.

67. PP 607 (1845), xxii–xxvii.

68. Prouty, *Transformation of the Board of Trade*, 46–49.

69. John Stuart Mill, in his *Principles of Political Economy* (1848), gave the building and operation of lighthouses visible at sea as a responsibility of government, as private interests would not develop them unless government indemnified them and enforced a compulsory levy. Effectively, they were a public-private partnership until they were nationalized; see Lindberg, "From Private to Public Provision."

70. Levitt, "When Lighthouses Became Public Goods," 147.

71. J. Taylor, "Private Property, Public Interest," 752.

72. J. Taylor, "Private Property, Public Interest," 750.

73. Conway, "Illuminating Science," i.

74. Schiffer, "Electric Lighthouse in the Nineteenth Century," 281.

75. James, *Correspondence of Michael Faraday, Volume 4*, xxxiii.

76. MacLeod, "Science and Government in Victorian England," 7–8.

77. PP 2793 (1861), 2–4, 71–102; Airy, *Autobiography of Sir George Biddell Airy*, 240–41.

78. PP 2793 (1861), 589–630. Faraday's evidence is at 591–93.

79. PP 2793 (1861), vi.

80. PP 2793 (1861), xxix.

81. James, *Correspondence of Michael Faraday, Volume 6*, xxxiv.

82. James, *Correspondence of Michael Faraday, Volume 6*, xxxiv–xxxvii, describes this episode.

83. For analysis of John Tyndall's involvement, see MacLeod, "Science and Government in Victorian England"; Jackson, *Ascent of John Tyndall*, 282–85, 316–21, 345–46, 406–13, 420–23, 433–34, 437.

84. MacLeod, "Science and Government in Victorian England," 15.

85. Jackson, *Ascent of John Tyndall*, 318–20, 327–28.

86. MacLeod, "Science and Government in Victorian England," 16–29; Jackson, *Ascent of John Tyndall*, 285–86, 316–17, 406–8, 411–13.

87. PP 2793 (1861), xxviii. Quoted by John Tyndall in a letter to the *Times*, January 12, 1885, 13.

88. PP 60 (1889), 4–12; John Wigham to Sir Michael Hicks Beach, May 19, 1888, PP 60 (1889), 13.

89. Sir Michael Hicks Beach to George Stokes, September 27, 1888, PP 60 (1889), 18. John Wigham objected on the grounds that although the council of the Royal Society contained "very distinguished men, geologists, botanists, naturalists, professional men associated with the medical profession &c., few of its members have taken up studies that have any bearing on the lighthouse question."

90. PP 2 (1890), 15–20, 22–36.

91. PP C5125 (1887), v.

92. PP C5763 (1889), 5–6.

93. PP C8675 (1897), 6.

9. FACTORIES, NUISANCES, AND THE HOME OFFICE

1. For graveyards, see Jackson, *Case Studies in Scientific Advice*, no. 22.

2. Pellew, *Home Office*, 6, 23, 30.

3. Pellew, *Home Office*, 122–82, describes the numerous Home Office inspectorates. She classifies the inspectorates as "industrial" (factory, mines, and explosives), "law and order" (prisons, police, and reformatory and industrial schools), "medical" (anatomy, burial grounds, and cruelty to animals), and "fisheries." Bartrip, "British Government Inspection," analyzes the roles, powers, and influence of inspectors. On factory acts, see Henriques, *Before the Welfare State*, 66–116.

4. For Edwin Chadwick's role, see S. Finer, *and Times of Sir Edwin Chadwick*, 50–68. For the factory inspectorate, see Pellew, *Home Office*, 123–28, 151–64; Bartrip, "Expertise and the Dangerous Trades," 89–109.

5. MacDonagh, *Early Victorian Government*, 1–77, explores factory reform up to 1844; see also Henriques, *Early Factory Acts*.

6. PP 397 (1816); PP 706 (1832). R. Gray, "Medical Men," 23–27, unpicks local and metropolitan views and tensions.

7. PP 397 (1816), 47–48.

8. PP 706 (1832); Hamlin, *Public Health and Social Justice*, 36–46.

9. PP 450 (1833); PP 519 (1833); PP 167 (1834); Jackson, *Case Studies in Scientific Advice*, no. 12.

10. Webb, "Southwood Smith"; Cook, "Thomas Southwood Smith."

11. Hamlin, *Public Health and Social Justice*, 49.

12. Lubenow, *Politics of Government Growth*, 137–79, outlines the factory acts and ten-hour issues in terms of government growth and intervention.

13. Bartrip, "British Government Inspection," discusses the origin and roles of the factory inspectors.

14. PP 628 (1845).

15. PP 1943 (1855), 12–13, 103–5.

16. PP 3678 (1866), vi. The full medical report is in PP 3416 (1864), 216–349.

17. For an overview of legislation, see MacDonagh, *Early Victorian Government*, 22–77.

18. Pellew, *Home Office*, 158.

19. Pontin, "Nuisance Law and the Industrial Revolution."

20. McLaren, "Nuisance Law and the Industrial Revolution," 158. Brenner, "Nuisance Law and the Industrial Revolution," explores reasons for the limited effect of nuisance law in the nineteenth century, especially in industrial towns, as does Hanley, *Healthy Boundaries*.

21. The evolution of nuisance laws from common law and community-oriented dimensions to state legislation prioritizing public health, is described in Hamlin, "Public Sphere to Public Health." On nuisances, private industry, and taxation, see Daunton, *State and Market in Victorian Britain*, 89–110.

22. PP 574 (1819); PP 244 (1820); Ashby and Anderson, *Politics of Clean Air*, 1–6; Beck, "History of Anti-Pollution Legislation," 476–79.

23. PP 244 (1820), 17.

24. Flick, "Movement for Smoke Abatement," 30.

25. Whitehead, *State, Science and the Skies*, 9–13. Ashby and Anderson, "Studies in the Politics of Environmental Protection I," follows Mackinnon's attempts to introduce legislation to abate smoke. See also Ashby and Anderson, *Politics of Clean Air*, 7–15. Thorsheim, *Inventing Pollution*, explores the changing cultural understanding of the effects of coal smoke in the nineteenth century and its comparison with miasma and extraparliamentary agitation. Wohl, *Endangered Lives*, 205–32, examines atmospheric pollution generally.

26. Flick, "Movement for Smoke Abatement," 39–50, describes and compares the methods used.

27. PP 583 (1843), iii–iv.

28. Farrar, "Andrew Ure," 315–16.

29. *Hansard* 76 (July 3, 1844): 285.

30. PP 289 (1845); Beck, "History of Anti-Pollution Legislation," 480–83.

31. *Report of the BAAS* (1844): 100–120.

32. PP 489 (1845).

33. PP 194 (1847), 2.

34. On the nuisance or sanitary inspectorate, see Crook, "Sanitary Inspection and the Public Sphere."

35. For Palmerston's sustained involvement, see Ashby and Anderson, *Politics of Clean Air*, 15–19.

36. PP 321 (1887); Ashby and Anderson, *Politics of Clean Air*, 50–64.

37. Ashby and Anderson, "Studies in the Politics of Environmental Protection II."

38. Thorsheim, "Paradox of Smokeless Fuels."

39. Ashby and Anderson, "Studies in the Politics of Environmental Protection III."

40. Dingle, "'Monster Nuisance of All,'" gives the background to the Alkali Act 1863. MacLeod, "Alkali Acts Administration," analyzes the acts in detail, and Ashby and Anderson, *Politics of Clean Air*, 20–54, 65–76, explore means of addressing the emissions. Garwood, "State and the Natural Environment," gives an extensive analysis of the alkali inspectorate, its operation, and impact. Wilmot, "Pollution and Public Concern," explores the approach of the chemical industry to pollution and regulation.

41. Reed, *Acid Rain and the Rise of the Environmental Chemist*.

42. *Hansard* 166 (May 9, 1862): 1452–67; Hawkins, *Forgotten Prime Minister*, 2:272–74.

43. PP 486 (1862), 50. The claim was not entirely true. The problems posed by the copper smelting industry are examined in Newell, "Atmospheric Pollution," who argues that the industry's escape from control was the result of a combination of lack of concern regarding the effect of copper smoke on health and the perceived high cost of imposing regulations on the smelting process.

44. PP 486 (1862), 99.

45. PP 486 (1862), ix.

46. Reed, "Robert Angus Smith," 156; Reed, *Acid Rain and the Rise of the Environmental Chemist*, 81–159; Gibson and Farrar, "Robert Angus Smith."

47. Eyler, "Conversion of Angus Smith."

48. Gorham, "Robert Angus Smith."

49. PP C2159 (1878), 2.

50. PP C2159 (1878), 36–37.

51. PP C2159 (1878), 38. Henry Roscoe was also conscious of the need not to interfere too much with the trade, see Roscoe, *Life & Experiences of Sir Henry Enfield Roscoe*, 162. He describes a variety of interventions in Parliament in scientific matters on 270–78, 285–88.

52. Sherard, *White Slaves of England*, 47–77.

53. The first two female factory inspectors were May Abraham and Mary

Paterson; see McFeeley, *Lady Inspectors*. On issues faced by female sanitary inspectors, see Crook, "Sanitary Inspection and the Public Sphere," 387–89.

54. Pellew, *Home Office*, 130–32, and Pellew, "Home Office and the Explosives Act," analyze the Explosives Act 1875 and Vivian Dering Majendie's role.

55. PP 267 (1865), 13–17.

56. Pellew, "Home Office and the Explosives Act," 180–82.

57. PP 977 (1874), 3–4.

58. PP 977 (1874), 3–74, 74, 75–81.

59. PP C586 (1872). The explosion killed twenty-four people and caused extensive damage a quarter of a mile from the site. This investigation illustrates the widespread involvement of chemists consulting for government. Vivian Dering Majendie could not call on Frederick Abel, as he was acting for the War Office, nor on William Odling, who was acting for the company. Charles Bloxam, Abel's brother-in-law, who had succeeded William Allen Miller as professor of chemistry at King's College London in 1870 after Miller's death, declined to act. So Majendie called on August Dupré, consulting chemist to the medical department of the Privy Council, and Thomas Keates, chemist to the Metropolitan Board of Works.

60. PP 243 (1874), iii–xi. Frederick Abel's classification of explosives into four categories of gunpowder, nitro-explosive, chlorate-explosive, and fulminate-explosive is at page 300. Vivian Dering Majendie extended this to include ammunition and fireworks; see 327–28.

61. PP C1364 (1875).

62. PP C7952 (1896), 22–24.

63. PP C977 (1874), 7.

64. Pellew, "Home Office and the Explosives Act," 191–2.

65. Jackson, *Case Studies in Scientific Advice*, no. 23.

66. French, *Anti-Vivisection and Medical Science*, 25–27. Richard French gives the classic account of the anti-vivisection movement in the context of the legislation in 1876. For the movement to prevent cruelty to animals, including wild birds, and the role of the Royal Society for the Prevention of Cruelty to Animals, see Hamilton, "On the Cruelty to Animals Act."

67. PP 667 (1832), 24–29. On rabies, see Jackson, *Case Studies in Scientific Advice*, no. 14. For Smithfield Market, see Jackson, *Case Studies in Scientific Advice*, no. 15.

68. Bynum, *Science and the Practice of Medicine*, 169; French, *Anti-Vivisection and Medical Science*, 46–60.

69. Finn and Stark, "Cruelty to Animals Act," 14–17.

70. Mitchell, *Frances Power Cobbe*, 229–66.

71. PP C1397 (1876); PP C1864 (1877); Harrison, "Animals and the State," 791–92, 804–5.

72. French, *Anti-Vivisection and Medical Science*, 91–111.

73. PP C1397 (1876), 183. Emmanuel Klein's amended evidence is at 328, and a

useful list of witnesses and summary of all the evidence is in PP C1397-I (1876). PP C1864 (1877) has a full analytical index; French, *Anti-Vivisection and Medical Science*, 103–6.

74. French, *Anti-Vivisection and Medical Science*, 70–71. For Huxley's views and role, see Desmond, *Evolution's High Priest*, 75–80; Huxley, *Life and Letters*, 1:427–41.

75. French, *Anti-Vivisection and Medical Science*, 112–58; Hamilton, "On the Cruelty to Animals Act," also describe the passage of the act.

76. The first annual report is PP 100 (1877), although the first with names of licensees, bland narrative explanation, and signed by George Busk is the third report PP 127 (1879).

77. Finn and Stark, "Cruelty to Animals Act," 17–22, detail the case of research into anthrax in Bradford in the years after 1876.

78. French, *Anti-Vivisection and Medical Science*, 200–219.

79. Lauriat, "'Examination of Everything,'" 36–37; French, *Anti-Vivisection and Medical Science*, 159–76.

80. R. Lambert, *Sir John Simon*, 582.

81. *Hansard* 277 (April 4, 1883): 1429, 1437.

82. Jackson, *Case Studies in Scientific Advice*, no. 28.

10. COAL, GAS, AND ELECTRICITY

1. Jackson, *Case Studies in Scientific Advice*, no. 24.

2. For a summary of nineteenth-century mining legislation, see PP Cd. 4820 (1909), 2–11. MacDonagh, "Coal Mines Regulation," and MacDonagh, *Early Victorian Government*, 78–95, offer an examination of coal mining in the context of early Victorian government. Bartrip, "British Government Inspection," and Pellew, *Home Office*, 128–32, explore the inspectorate.

3. PP 9 (1830), 32.

4. PP 603 (1835).

5. PP 603 (1835), 106 (Stephenson), 284–86 (Gurney), 286 (Birkbeck).

6. Jackson, *Case Studies in Scientific Advice*, no. 21.

7. PP 583 (1835), 5.

8. PP 603 (1835), 235.

9. *Hansard* 55 (August 4, 1840): 1260–79.

10. Martin, "Leonard Horner."

11. Frankel, "Blue Books and the Victorian Reader," 312.

12. PP 380 (1842), 255–61. For a perspective on the commissioners' views of women's positions in the mines, see E. Yeo, *Contest for Social Science*, 82–83.

13. *Hansard* 63 (June 7, 1842): 1320–64.

14. Webb, "Whig Inspector." For Seymour Tremenheere's first report, see PP 592 (1844).

15. Henriques, *Before the Welfare State*, 209.

16. The report is appended in PP 509 (1852), 155–227.

17. PP 613 (1849), 333.

18. For analysis of the Haswell colliery episode, see James, *Correspondence of Michael Faraday, Volume 3*, xxxi–xxxiv; James and Ray, "Science in the Pits"; James, *Michael Faraday*, 49–50.

19. Berman, *Social Change and Scientific Organization*, 177.

20. PP 232 (1845).

21. PP 232 (1845), 4.

22. Berman, *Social Change and Scientific Organization*, 177–86.

23. On the later "coal question," see Jackson, *Case Studies in Scientific Advice*, no. 24.

24. C. Russell, *Science and Social Change*, 201–2.

25. In 1883 the Mining Record Office was transferred from the Department of Science and Art to the Home Office.

26. PP 815 (1847), 4–12.

27. PP 815 (1847), 43.

28. PP 815 (1847), 51.

29. PP 815 (1847), 58.

30. PP 427 (1849).

31. PP 1051 (1849), 1.

32. PP 1214 (1850); PP 1222 (1850).

33. PP 613 (1849).

34. *Hansard* 113 (July 19, 1850): 3–4.

35. *Hansard* 113 (August 2, 1850): 759–60.

36. Bartrip, "British Government Inspection," 608.

37. Bartrip, "British Government Inspection," 622–23.

38. PP 509 (1852), iii–x.

39. PP 509 (1852), 27.

40. PP 325 (1854), 43–44.

41. PP 3889 (1864), 183–285.

42. Quoted in Eyler, "Conversion of Angus Smith," 222; Reed, *Acid Rain and the Rise of the Environmental Chemist*, 74–78.

43. PP 3889 (1864), 154–78.

44. *Hansard* 198 (August 9, 1869): 1517–22; *Hansard* 200 (April 12, 1870): 1708; *Hansard* 208 (July 31, 1871): 569–70; *Hansard* 209 (February 6, 1872): 57; *Hansard* 209 (February 12, 1872): 232–50.

45. PP C3036 (1881), iii.

46. PP C3036 (1881); PP C4699 (1886).

47. PP C2923 (1881), 5.

48. Bartrip, "British Government Inspection"; Bartrip, "State Intervention in Mid-Nineteenth-Century Britain."

49. PP 220 (1809); Tomory, *Progressive Enlightenment*, 169–237. On gas legislation, see Clifford, *History of Private Bill Legislation*, 203–31.

50. For Frederick Accum's work on gas, see Browne, "Life and Chemical Services of Frederick Accum," 1008–17. For account of the rapid expansion of gaslighting, see Schivelbusch, *Disenchanted Night*, 14–50. On William Murdoch and Frederick Winsor, see Tomory, *Progressive Enlightenment*, 71–83, 121–68.

51. PP 220 (1809), 45.

52. Tomory, "Environmental History the Early British Gas Industry," 38–39.

53. PP 267 (1828), 9–10.

54. Hutchinson, "Royal Society and the British Gas Industry"; RS CMO/9/52–54.

55. PP 193 (1823), 3–6; RS CMO/9/54–66.

56. PP 193 (1823), 6–10.

57. PP 193 (1823), 11–24.

58. On disputes about coal gas and oil gas, involving Thomas Brande in particular, see Berman, *Social Change and Scientific Organization*, 149–51.

59. PP 529 (1823), 3.

60. RS CMO/10/181–82, 201–2.

61. Hutchinson, "Royal Society and the British Gas Industry," 259; RS CMO/10/202. The committee consisted of Humphry Davy, Thomas Brande, Michael Faraday, Davies Gilbert, Charles Hatchett, John Herschel, Henry Kater, George Rennie, William Wollaston, and Thomas Young.

62. Daunton, *State and Market in Victorian Britain*, 111–27.

63. PP 2768 (1861), xxi. Thorsheim, "Paradox of Smokeless Fuels," explores pollution from gas and coke works.

64. PP 224 (1859).

65. Airy, *Autobiography of Sir George Biddell Airy*, 237.

66. C. Russell, *Edward Frankland*, 447.

67. Letters between Lord Monteagle, Charles Wheatstone, Edward Frankland, and George Lowe, as well as between George Airy and William Crosley, are in PP 547 (1860).

68. PP 100 (1860); PP 293 (1860); Perkins, "'Extraneous Government Business,'" 152.

69. Hutchinson, "Royal Society and the British Gas Industry," 263. On photometry, see Otter, *Victorian Eye*, 154–72.

70. PP 417 (1860), 169–83; PP 270 (1866), 46.

71. PP 270 (1866), v.

72. PP 270 (1866).

73. PP 270 (1866), 284. Henry Letheby's oral evidence is at 45–56, 71–81.

74. Hutchinson, "Royal Society and the British Gas Industry," 263.

75. PP 199 (1872), 3. The 1868 act also established three commissioners, of whom

William Odling was one, who sat for two years to review illuminating power and price.

76. PP 4156 (1869).

77. PP C394 (1871).

78. PP C270 (1870); PP 281 (1872).

79. PP C393 (1871).

80. PP 199 (1872), 24.

81. PP 260 (1876), 3.

82. PP 199 (1872), 52, 56.

83. "Photometric Standards," *Sanitary Engineer* 5 (1881): 10.

84. PP C7743 (1895).

85. PP C9164 (1899), x–xi.

86. Schivelbusch, *Disenchanted Night*, 50–78, describes its introduction.

87. The manner in which Victorians saw the telegraph as "annihilating time and space" is discussed in Morus, "Nervous System of Britain." On the history of the telegraph in Britain, see Marsden and Smith, *Engineering Empires*, 178–225.

88. PP 188 (1885), iv. New technology was not without its fears and opponents. In 1885 the select committee on telephone and telegraph wires decided that "the danger to the public from overhead wires has been greatly exaggerated." It found that a reported death was due "not to the breaking of the telegraph wire, but to a well-meant attempt by a member of the public to replace it."

89. Jackson, *Case Studies in Scientific Advice*, no. 18.

90. *Hansard* 244 (March 13, 1879): 803. Regulation of the electricity industry in this period is discussed in Stern, "Regulation and Contracts for Utility Services," 205–10. See also Byatt, *British Electrical Industry*, 197–209. Otter, *Victorian Eye*, 175–251, describes lighting technologies. On legislation, see Clifford, *History of Private Bill Legislation*, 231–47.

91. PP 224 (1879), 5.

92. Hughes "British Electrical Industry Lag," 30–33.

93. On the public perceptions and problems of the electricity industry in conjunction with gas in this period, see Otter, "Cleansing and Clarifying," 53–64. Victorian visions of electricity are explored in Gooday, "Electrical Futures Past."

94. PP 227 (1882), 171. Although it not was not mentioned, William Spottiswoode was also chairman of the English Edison Company.

95. PP 227 (1882), 152, 173, 184, 212.

96. PP 227 (1882), 230.

97. PP 252 (1886), 24 (Forbes), 46–47 (Bramwell), 128–29 (Lubbock).

98. T. Hughes, "British Electrical Industry Lag," 39.

99. H. Smith, *Board of Trade*, 178.

11. WATER, SANITATION, AND RIVER POLLUTION

1. Hanley, *Healthy Boundaries*; Luckin, *Death and Survival in Urban Britain*, and Hamlin, *Public Health and Social Justice* give extensive analysis. Crook, *Governing Systems*, takes a systems approach. Wohl, *Endangered Lives*, 166–204, explores the involvement of local government. See also Henriques, *Before the Welfare State*, 117–54.

2. Hassan, *History of Water*, 1–9, gives a historiography of related aspects of water supply and sewage treatment. Binnie, *Early Victorian Water Engineers*, examines the water engineers.

3. Stern, "Water Supply in Britain," describes the development of a public supply of water; Hassan, "Growth and Impact," gives an economic analysis; and Tynan, "Nineteenth Century London Water Supply," sets out the process of innovation by water companies. For metropolitan politics and water supply, see Luckin, *Death and Survival in Urban Britain*, and Tomory, *History of the London Water Industry*. On legislation, see Clifford, *History of Private Bill Legislation*, 100–198, 291–348. See also Jackson, *Case Studies in Scientific Advice*, no. 25.

4. This definition was inspired by words of Thomas Treadgold.

5. PP 537 (1821); PP 706 (1821). For the early history of London water provision, analysis, and its association with public health, see Hardy, "Water and the Search for Public Health"; Tomory, *History of the London Water Industry*. See also Hardy, "Parish Pump to Private Lives."

6. PP 537 (1821), 6.

7. PP 537 (1821), 7.

8. For a picture of the juxtaposition of the intake and the sewer outlet, see PP 267 (1828), 134.

9. PP 267 (1828).

10. PP 267 (1828), 77–83.

11. PP 267 (1828), 83.

12. Hamlin, *Science of Impurity*, 89–90.

13. PP 267 (1828), 136–40.

14. PP 267 (1828), 12.

15. PP 267 (1828), 12.

16. Hamlin, *Science of Impurity*, 1–98, describes the assumptions, practices, and disputes in detail.

17. PP 267 (1828), 8.

18. PP 267 (1828), 84–92.

19. PP 567 (1828), 5.

20. PP 176 (1834).

21. PP 571 (1834); Hardy, "Water and the Search for Public Health," 261n78.

22. Quoted in Peppercorne, *Supply of Water to the Metropolis*, 6.

23. Binnie, *Early Victorian Water Engineers*, 70–94, examines James Simpson's work.

24. Hamlin, *Public Health and Social Justice*, 13.

25. For background, see Crook, *Governing Systems*, 148–96; Pelling, *Cholera, Fever and English Medicine*, 1–33; MacDonagh, *Early Victorian Government*, 133–61; Wohl, *Endangered Lives*; Lubenow, *Politics of Government Growth*, 69–106; S. Finer, *Life and Times of Sir Edwin Chadwick*, 209–42, 293–482; Lewis, *Edwin Chadwick*.

26. See S. Finer, "Transmission of Benthamite Ideas," for an analysis of the influence of Jeremy Bentham and his circle, although D. Roberts, "Jeremy Bentham," argues that Bentham's impact in terms of specific policies and administrative structures was limited, and partial at best.

27. PP 384 (1840).

28. Cullen, *Statistical Movement in Early Victorian Britain*, 55–56.

29. PP 384 (1840), 3. On poverty and its practical implications, see Wohl, *Endangered Lives*, 43–79.

30. PP 474 (1846), 109.

31. PP 384 (1840), v.

32. Cullen, *Statistical Movement in Early Victorian Britain*, 56.

33. M. Brown, "From Foetid Air to Filth," traces the development of anticontagionism through Thomas Southwood Smith and Charles Maclean, who influenced him. For Maclean's prompting of the select committee on the validity of the doctrine of contagion in the plague in 1817, see Jackson, *Case Studies in Scientific Advice*, no. 27. Brown highlights the significance of Southwood Smith's Unitarian faith as a link between "the sin of filth and the punishment of pestilence." Ridding the world of the sin of filth would reduce suffering, in line with God's will.

34. S. Finer, *Life and Times of Sir Edwin Chadwick*, 160.

35. Hamlin, "Predisposing Causes and Public Health," 62–65.

36. Hamlin, *Public Health and Social Justice*, 114–20.

37. Hamlin, *Public Health and Social Justice*, 118.

38. Hamlin, *Public Health and Social Justice*, 84–120.

39. Cullen, *Statistical Movement in Early Victorian Britain*, 56–60, gives a statistical critique.

40. PP 187 (1839), 75.

41. S. Finer, *Life and Times of Sir Edwin Chadwick*, 221–24.

42. Hamlin, *Public Health and Social Justice*, 187.

43. PP 572 (1844), v; Binnie, *Early Victorian Water Engineers*, 8–30.

44. Jackson, *Case Studies in Scientific Advice*, no. 21.

45. Reid, *Memoirs and Correspondence of Lyon Playfair*, 63.

46. Owen to Chadwick, September 9, 1844, quoted in Lewis, *Edwin Chadwick*, 85.

47. Hamlin, *Public Health and Social Justice*, 222, referencing Edwin Chadwick to James Graham, March 15, 1843, Chadwick Papers, no. 849, University College London. For extensive analysis of the commission's activities, see Hamlin, *Public Health and Social Justice*, 221–42.

48. PP 572 (1844). Binnie, *Early Victorian Water Engineers*, 130–56, examines Thomas Hawksley and his skill as an expert witness.

49. PP 602 (1845).

50. PP 610 (1845), 59–60.

51. PP 572 (1844), 343–49; Goldman, "Statistics and the Science of Society," 423–25.

52. Reid, *Memoirs and Correspondence of Lyon Playfair*, 94.

53. PP 572 (1844), vi–xv.

54. PP 602 (1845), 1–76.

55. Hamlin, *Public Health and Social Justice*, 220.

56. Cullen, *Statistical Movement in Early Victorian Britain*, 60–62.

57. PP 556 (1846), iii–vii.

58. S. Finer, *Life and Times of Sir Edwin Chadwick*, 239.

59. Olien, *Morpeth*, gives his biography.

60. Hanley, *Healthy Boundaries*.

61. For the development of inspection in public health in the General Board of Health and subsequent bodies, see Hardy, "Water and the Search for Public Health," 39–51. See also Binnie, *Early Victorian Water Engineers*, 31–42, and Henriques, *Before the Welfare State*, 136–54. For local nuisance and sanitary inspection, see Otter, *Victorian Eye*, 101–9.

62. S. Finer, *Life and Times of Sir Edwin Chadwick*, 287.

63. On the overall politics around the General Board of Health, see Mandler, *Aristocratic Government in the Age of Reform*, 256–67.

64. PP 584 (1834).

65. PP 584 (1834), v.

66. Sunderland, "Monument to Defective Administration?"

67. Lewis, *Edwin Chadwick*, 149–57.

68. On the Metropolitan Board of Works, see Hanley, *Healthy Boundaries*, 89–110; Clifton, *Professionalism, Patronage and Public Service*.

69. Lewis, *Edwin Chadwick*, 370.

70. R. Lambert, "Central and Local Relations," 122–25.

71. R. Lambert, "Central and Local Relations," 131.

72. Cabinet "collective responsibility," introduced by Viscount Melbourne in 1838, also dates from this period; see Hawkins, *Victorian Political Culture*, 15, 115.

73. Schwartz, "John Stuart Mill and Laissez-Faire," 73–75. Hassan, "Growth and Impact," argues that the transfer of responsibility to municipal authorities with long-term horizons was a successful solution to the growing problems of water

supply, motivated by a desire to eliminate the avoidable costs of private ownership.

74. Hassan, *History of Water*, 25–30; Angus, "Cesspools, Sewage, and Social Murder," offers an eco-socialist account.

75. Clifton, *Professionalism, Patronage and Public Service*; Owen and MacLeod, *Government of Victorian London*, 23–46.

76. Tynan, "Nineteenth Century London Water Supply," 86. On river pollution, see Owen and MacLeod, *Government of Victorian London*, 47–73; Wohl, *Endangered Lives*, 233–56.

77. See Brock, *Justus von Liebig*, 203–14.

78. Hamlin, *Science of Impurity*, 127–40.

79. Lyndsay Blyth was born in Rye in 1823 and died in 1858, which accounts for his obscurity. I am grateful to Bill Brock for this information.

80. PP 2137 (1856), 4–6.

81. PP 2137 (1856), 93.

82. Porter, *Life and Times of Sir Goldsworthy Gurney*, 189–211.

83. Jackson, *Case Studies in Scientific Advice*, no. 21.

84. PP 442 (1858), iv–v.

85. Jackson, *Case Studies in Scientific Advice*, no. 21.

86. Hassan, *History of Water*, 30–50, explores the national context.

87. *Hansard* 177 (March 8, 1865): 1309–59.

88. Binnie, *Early Victorian Water Engineers*, 202–22, examines Robert Rawlinson's work.

89. PP 3634 (1866), 32–33; PP 3835 (1867), xxvi; PP 3850 (1867), liii–lvi.

90. PP 4169 (1869), xciv–xcv.

91. PP 3634 (1866), 71.

92. PP 3634 (1866), 50.

93. PP 307 (1868), 145 (Way), 161 (Rawlinson), 165 (Taylor).

94. Hamlin, *Science of Impurity*, 171.

95. PP 399 (1867), 352, 357.

96. PP C37 (1870); PP C180 (1870); PP C347 (1871); PP C603 (1872); PP C951 (1874); PP C1112 (1874); Hamlin, *Science of Impurity*, 172–74.

97. Hamlin, *Science of Impurity*, 174.

98. Jackson, *Case Studies in Scientific Advice*, no. 25.

99. Edward Frankland's rise and the challenges to him are discussed in Hamlin, *Science of Impurity*, 152–211. Barton, *X Club*, explores the X Club. On the importance of William Farr and Edward Frankland in improving water quality, see Luckin, *Death and Survival in Urban Britain*, 54.

100. PP C3337-I (1882), xvii–xxi, 127–65.

101. For details of the approach, assumptions, and criticisms, see Hamlin, "Edward Frankland's Early Career," 65–68.

102. C. Russell, *Edward Frankland*, 365–81.

103. For a comparison between the royal commissions on water supply in 1867–1869 and 1892–1893, see Hamlin, *Science of Impurity*.

104. Hamlin, "Edward Frankland's Early Career," 59–62.

105. C. Russell, *Edward Frankland*, 381–91.

106. For Frankland's work on water analysis, see C. Russell, *Edward Frankland*, 362–410; Hamlin, *Science of Impurity*, 152–210.

107. Rosenthal, *River Pollution Dilemma*, examines the operation of the nuisance laws.

108. *Hansard* 239 (May 9, 1878): 1596–97.

109. Thomson, "River Pollution," 355.

110. K. Waddington, "Vitriol in the Taff."

111. *Hansard* 11 (April 27, 1893): 1283–84.

112. On William Crookes and the ABC process, see Brock, *William Crookes*, 285–93.

113. PP C1464 (1876), iii–xxix.

114. PP C1410 (1876), xii–xiii.

115. Hamlin, "William Dibdin," 204–8.

116. Henry Roscoe describes his role and subsequent activity; see Roscoe, *Life & Experiences of Sir Henry Enfield Roscoe*, 303–6.

117. Hamlin, "William Dibdin," 189–91.

118. Sheail, "Town Wastes," 206–7.

119. PP Cd. 685 (1901), vii–xiii. Reports were published in PP Cd. 1178 (1902).

120. Sheail, "Town Wastes," 210.

12. INFECTION AND DISEASE

1. On miasmas and contagionism, see Worboys, *Spreading Germs*, 38–42.

2. Jackson, *Case Studies in Scientific Advice*, no. 27.

3. For example, the investigation of the Millbank penitentiary in 1823; see Jackson, *Case Studies in Scientific Advice*, no. 10.

4. Worboys, *Spreading Germs*. Fumigation was one early method; see PP 449 (1819); Jackson, *Case Studies in Scientific Advice*, no. 26.

5. Jackson, *Case Studies in Scientific Advice*, no. 27.

6. Brockington, "Public Health at the Privy Council, 1831–34"; Baldwin, *Contagion and the State in Europe*, 101–2, 114–18; Hanley, "Parliament, Physicians, and Nuisances," 712–19. For the relative lack of influence of the cholera issue on the Reform Act 1832, see Durey, *Return of the Plague*, 185–200.

7. Kidd and Wyke, "Cholera Epidemic in Manchester," describe the often tense situation in Manchester during this epidemic and the instigation of James Phillips Kay's notable report, *The Moral and Physical Condition of the Working Classes Employed in the Cotton Manufacture in Manchester* (London: Ridgway, 1832).

8. For the politics, see Morris, *Cholera 1932*, 21–37.

9. Hanley, "Parliament, Physicians, and Nuisances," 718–19.

10. For accounts of the 1832 epidemic, see Morris, *Cholera 1932*; Bynum, *Science and the Practice of Medicine*, 74–76; Baldwin, *Contagion and the State in Europe*, 37–122.

11. On the relationship between medicine and science around the 1830s, see Morris, *Cholera 1932*, 159–95; and for the mid-century period, see Pelling, *Cholera, Fever and English Medicine*.

12. Baldwin, *Contagion and the State in Europe*, 16–17.

13. Fullmer, "Technology, Chemistry, and the Law," 24–26.

14. On the medical profession in the nineteenth century, see Waddington, *Medical Profession in the Industrial Revolution*.

15. M. Brown, "Medicine, Reform," 1379–83.

16. Holloway, "Apothecaries Act." Loudon, *Medical Care and the General Practitioner*, 152–88, explores the generation of the act and its consequences.

17. C. Russell, *Science and Social Change*, 203. On medical reform and the general practitioner, see Loudon, *Medical Care and the General Practitioner*.

18. C. Russell, *Chemists by Profession*, 23–24.

19. Jackson, *Case Studies in Scientific Advice*, no. 28.

20. Jackson, *Case Studies in Scientific Advice*, no. 9.

21. Cullen, *Statistical Movement in Early Victorian Britain*, 29–43.

22. S. Finer, *Life and Times of Sir Edwin Chadwick*, 143.

23. For William Farr and the General Register Office, see Goldman, "Statistics and the Science of Society"; Eyler, *Victorian Social Medicine*, 37–65.

24. Higgs, "Some Forgotten Men." This paper highlights the contributions of the registrars-general, who have tended to be ignored in the context of Farr's work.

25. On William Farr, see Goldman, *Victorians and Numbers*, 211–22.

26. Higgs, "Struggle for the Occupational Census," 75–76. On the influence of statistics in public health, see Crook, *Governing Systems*, 63–105.

27. Szreter, "GRO and the Public Health Movement."

28. Higgs, "Struggle for the Occupational Census," 77–86.

29. On Justus Liebig's approach, see Pelling, *Cholera, Fever and English Medicine*, 113–45.

30. Chapman, "Science and the Public Good," 39.

31. Jackson, *Case Studies in Scientific Advice*, no. 6. On Florence Nightingale, William Farr, and Douglas Galton, see Goldman, *Victorians and Numbers*, 282–93.

32. On William Farr's approach to public health and medical science, see Pelling, *Cholera, Fever and English Medicine*, 81–112.

33. PP 888/895 (1847); PP 911 and PP921 (1848).

34. PP 979 (1848).

35. PP 888 (1847), iii.

36. Eyler, "William Farr on the Cholera," compares Farr's various studies and developing understanding of cholera in relation to John Snow's work and theory.

37. On typhus and typhoid, see Luckin, *Death and Survival in Urban Britain*, 35–55.

38. PP 888 (1847), 1–52.

39. PP 921 (1848), 7–12.

40. PP 921 (1848), 34–35; Gibson and Farrar, "Robert Angus Smith," 250–2.

41. PP 911 (1848), 19–21.

42. PP 911 (1848), 9.

43. Reid, *Memoirs and Correspondence of Lyon Playfair*, 103.

44. S. Finer, *Life and Times of Sir Edwin Chadwick*, 336–37.

45. Hanley, "Parliament, Physicians, and Nuisances," 728–29.

46. PP 1070 (1849), 5.

47. PP 1473 (1852), 1–7.

48. R. Lewis, *Edwin Chadwick*, 347–48. On changing policies in Britain toward quarantine in the nineteenth century, see McDonald, "History of Quarantine in Britain"; Hardy, "Cholera, Quarantine"; Maglen, *English System*; and for this period, see Baldwin, *Contagion and the State in Europe*, 149–55.

49. R. Lambert, *Sir John Simon*, gives a detailed biographical study. Sheard and Donaldson, *Nation's Doctor*, describe the chief medical officer's role.

50. R. Lambert, *Sir John Simon*, 99.

51. R. Lewis, *Edwin Chadwick and the Public Health Movement*, 206.

52. On John Bateman, see Binnie, *Early Victorian Water Engineers*, 157–201.

53. PP C1818 (1854); R. Lambert, *Sir John Simon*, 200; R. Lewis, *Edwin Chadwick and the Public Health Movement*, 356.

54. Brunton, *Politics of Vaccination*, 56–57.

55. PP 1893 (1855), 10.

56. PP 1980 (1855); Pelling, *Cholera, Fever and English Medicine*, 221–27.

57. PP 1980 (1855), 48.

58. PP 1980 (1855), 52.

59. Tynan, "Nineteenth Century London Water Supply," 85.

60. Quoted in Brunton, *Politics of Vaccination*, 55

61. For the impact of the Social Science Association, see Jackson, *Case Studies in Scientific Advice*, no. 11.

62. Goldman, *Science, Reform, and Politics in Victorian Britain*, 182–84.

63. R. Lewis, *Edwin Chadwick and the Public Health Movement*, 195–96.

64. Hanley, "Parliament, Physicians, and Nuisances," 729–31.

65. R. Lambert, *Sir John Simon*, 230.

66. R. Lambert, *Sir John Simon*, 234–36, describes John Simon's legislative program.

67. PP 2415 (1858).

68. R. Lambert, *Sir John Simon*, 263–67.

69. Brockington, "Public Health at the Privy Council, 1858–71," describes public health at the Privy Council between 1858 and 1871, with biographies of the medical protagonists and a list of surveys carried out.

70. R. Lambert, *Sir John Simon*, 276.

71. Goldman, *Science, Reform, and Politics in Victorian Britain*, 181.

72. Sheard and Donaldson, *Nation's Doctor*, 148–52.

73. Jackson, *Case Studies in Scientific Advice*, no. 7.

74. R. Lambert, *Sir John Simon*, 344–50, 376. For example, Simon was instrumental in establishing the provisions of the Sale of Poisons and Pharmacy Act 1868 with its requirement for the qualification of all dispensing druggists.

75. PP 2103 (1856), 6.

76. John Snow, "Cholera and the Water Supply in the South Districts of London," *British Medical Journal* 1 (1857), 864–65. Tynan, "Nineteenth Century London Water Supply," 83, notes that this finding was possible because of the differences in technology employed by the water companies; see Bynum, *Science and the Practice of Medicine*, 79–81.

77. Eyler, "Changing Assessments of John Snow," explains why Snow's ideas did not immediately prevail, despite William Farr's support for his analysis, and their different explanations of the data. Tulodziecki, "Case Study in Explanatory Power," examines the nature and explanatory power of Snow's theory of the "exciting cause" of cholera. See also Pelling, *Cholera, Fever and English Medicine*, 203–49, on John Snow and 250–94 on William Budd.

78. The first is PP 2512 (1859). John Simon issued thirteen annual reports in his position at the Privy Council Committee of Council on Health, and further reports under the Local Government Board.

79. On John Simon's team, see Goldman, *Victorians and Numbers*, 227–36.

80. On the importance of typhoid, and the Medical Department's approach to epidemiology, see Steere-Williams, *Filth Disease*.

81. PP 3645 (1866), 44.

82. PP 3484 (1865), 20.

83. PP 4072 (1868), lxxx. For a summary of the transition from atmospheric to waterborne understanding of cholera transmission, and William Farr's change of view, see Halliday, "Death and Miasma in Victorian London." Luckin, *Death and Survival in Urban Britain*, 56–68, stresses the roles of William Farr, Edward Frankland, and John Radcliffe. See also Worboys, *Spreading Germs*, 113–17.

84. PP 4072, 125. Indeed, Edward Frankland indicated that the East London Company's water seemed purer than usual even at the height of the cholera epidemic in August 1866. The tests also showed the water delivered in Manchester and Glasgow to be purer than Thames water, although the higher mortality rates in Manchester and Glasgow suggested otherwise; see Hamlin, *Science of Impurity*, 156–58.

85. Maglen, *English System*; Hardy, "Cholera, Quarantine," 255–61.

86. R. Lambert, "Central and Local Relations," 138–44.

87. Goldman, *Science, Reform, and Politics in Victorian Britain*, 189–97.

88. On disease theories around 1865, see Worboys, *Spreading Germs*, 28–42.

89. R. Lambert, *Sir John Simon*, 405–6; Jackson, *Case Studies in Scientific Advice*, no. 23.

90. R. Lambert, *Sir John Simon*, 448.

91. PP C281 (1871), 15–72, 73–177. See also PP 4218 (1870); PP C281-I (1871); PP C1109 (1874).

92. Quoted in R. Lambert, *Sir John Simon*, 452.

93. R. Lambert, *Sir John Simon*, 602.

94. Bellamy, *Administering Central-Local Relations*. For an overview of the medical activities of the Local Government Board and of the Poor Law medical officers, see Brand, *Doctors and the State*, 22–107. For the history from 1834 to 1871, see Hodgkinson, *Origins of the National Health Service*. On the developing system of local-central relations in public health, see Crook, *Governing Systems*, 23–62.

95. Brand, "John Simon," describes the conflict between John Simon and John Lambert. For the wider context, see Brand, *Doctors and the State*, 22–31.

96. R. Lambert, *Sir John Simon*, 545.

97. Coleman, *Conservatism and the Conservative Party*, 142, 146.

98. For example, see the reports he commissioned on the process of fever, infective inflammations, anatomical changes in enteric or typhoid fever, the aetiology of cancer, physiological chemistry, and disinfectants, published in 1875; see PP C1371 (1875). On typhoid in the 1870s, see Worboys, *Spreading Germs*, 132–7.

99. R. Lambert, *Sir John Simon*, 549.

100. For the work of the sanitary inspectors, see Crook, "Sanitary Inspection and the Public Sphere"; Crook, *Governing Systems*, 106–47. On Letheby's report and influence on Home Secretary Richard Cross, see P. Smith, *Disraelian Conservatism and Social Reform*, 220–21.

101. Mooney, "Public Health versus Private Practice," 245–46. Graham Mooney describes both the sanguine public acceptance of these powers and how compulsory notification was resisted in Liverpool by the local medical elite in private practice, who saw a threat to patient confidentiality and to their income, and a curtailment of their professional independence. On notification and "stamping out," see Crook, *Governing Systems*, 244–97.

102. Worboys, *Spreading Germs*, 193–276.

103. Goldman, *Science, Reform, and Politics in Victorian Britain*, 195.

104. R. Lambert, *Sir John Simon*, 571. On Simon's successor, see Jackson, *Case Studies in Scientific Advice*, no. 31. On influenza, see Jackson, *Case Studies in Scientific Advice*, no. 30.

105. McConaghey, "Notes of the Evolution of Medical Practice," describes

medical practice prior to 1858. Burn, *Age of Equipoise*, 202–11, describes the political process leading up to the act. Beard, "To What Extent"; M. Roberts, "Politics of Professionalization"; Loudon, *Medical Care and the General Practitioner*, 297–301, discuss the nature and implications of the 1858 act.

106. Beck, "British Medical Council," describes the period from 1858 to the passage of the 1886 act, especially with respect to medical education. For John Simon's role in the 1858 act and 1886 act, respectively, see R. Lambert, *Sir John Simon*, 461–77, 579–81. For Lyon Playfair's role in Parliament for the 1886 act, see Hardy, "Lyon Playfair and the Idea of Progress," 97–99.

107. Pellew, *Home Office*, 142.

108. Lawrence, "Incommunicable Knowledge," discusses the attitudes of the elite as expressed in a generalist culture of medicine, and its emphasis of gentlemanly characteristics and clinical experience over clinical science and specialist scientific knowledge. See also I. Waddington, *Medical Profession in the Industrial Revolution*.

109. Cited in French, *Anti-Vivisection and Medical Science*, 338.

110. Pattison, *British Veterinary Profession*, describes the development of the veterinary profession.

111. Jackson, *Case Studies in Scientific Advice*, no. 13.

112. Woods, "From Practical Men to Scientific Experts," discusses veterinary expertise in relation to its involvement in government and to laboratory-based research.

113. On comparisons between the veterinary and medical professions, including attitudes to research, see Worboys, "Germ Theories of Disease," 315.

114. Cooter, "Rise and Decline of the Medical Member," 64.

115. Jackson, *Case Studies in Scientific Advice*, no. 29 and appendix 2.

116. Bynum, *Science and the Practice of Medicine*, 109–14.

117. Sturdy and Cooter, "Science, Scientific Management," link the development of medical laboratories, from around 1870, to the increasingly specialized and corporate nature of medicine. F. Smith, *People's Health*, giving a perspective from patients and practitioners, argues that the medical profession became highly technical and lost sight of wider contributory factors to public health.

118. Weisz, "Emergence of Medical Specialization," 574–75.

13. CHEMICAL ANALYSIS, EXCISE, CUSTOMS, AND INLAND REVENUE

1. On Customs and Excise to 1845, see Ashworth, *Customs and Excise*.

2. PP 595 (1821), 16.

3. PP 499 (1846), 29.

4. Until their amalgamation in 1823 there were separate Boards of Excise and of Customs for England and Wales, Ireland, and Scotland.

5. PP 202 (1806), 118–19.

6. PP 220 (1819), 22–27.

7. PP 190 (1846), 9.

8. Later experiments by John Lawes in 1865 were consistent with these findings; see PP 3451 (1865), 6.

9. PP 411 (1846), 559.

10. For the history of the chemical laboratory, see Hammond, "150 Years of the Laboratory"; Hammond and Egan, *Weighed in the Balance*.

11. Farrar, "Andrew Ure," gives a delightful picture of Ure's colorful life.

12. For the growth of chemists and analysts in industry, see C. Russell, *Chemists by Profession*, 96–103.

13. PP 590 (1833).

14. PP 297 (1831), iv.

15. PP 297 (1831), 240–43.

16. PP 297 (1831), iii–xii. The early evidence is in PP 109 (1831).

17. Hammond and Egan, *Weighed in the Balance*, 70.

18. Hammond and Egan, *Weighed in the Balance*, 36.

19. Smoking involved cigars and pipes until rolled cigarettes were invented in the 1860s.

20. For the work on tobacco, see Hammond and Egan, *Weighed in the Balance*, 5–8, 13–32.

21. For the analyses, see PP 565 (1844), 588–92.

22. PP 565 (1844), 467–74.

23. PP 565 (1844), xxxix–xli.

24. Hammond and Egan, *Weighed in the Balance*, 19.

25. PP 253 (1840).

26. Hammond and Egan, *Weighed in the Balance*, 34–35.

27. PP 379 (1856), 9–14, 266. John Lindley and Joseph Hooker had also made a report to the government about distinguishing coffee from chicory using a microscope; see PP 379 (1856), 13.

28. Hammond and Egan, *Weighed in the Balance*, 44–45.

29. On Clarke's and the Sikes hydrometers, see Ashworth, *Customs and Excise*, 265–79.

30. Usselman, *Pure Intelligence*, 284–89.

31. Hammond and Egan, *Weighed in the Balance*, 3–4.

32. RS CMP/1/3, 26, 45–46, 98–101, 187–89; RS MS/411/3/3. For the Royal Society's Excise Committee, see RS MS/411 (1835–1836); RS MM/13/84–154.

33. PP 26 (1847), 1–2.

34. PP 426 (1847); PP 529 (1847).

35. Hammond and Egan, *Weighed in the Balance*, 53–55.

36. Hammond and Egan, *Weighed in the Balance*, 62–63.

37. Hammond and Egan, *Weighed in the Balance*, 103–18.

38. Hammond and Egan, *Weighed in the Balance*, 62.

39. Rowlinson, "Food Adulteration," describes the development and operation of the legislation, on which this account is based. See also Ashworth, *Customs and Excise*, 307–15.

40. For Accum's crusade against food adulteration, see Browne, "Life and Chemical Services of Frederick Accum," 1027–34.

41. Brian, "Food and Drugs Act," 51–52.

42. Gray, *By Candlelight*, 100–110. Arthur Hassall and Henry Letheby receive a brief mention in Charles Kingsley's *Water Babies* for this work.

43. R. Lambert, *Sir John Simon*, 243.

44. Rowlinson, "Food Adulteration," 66; *Hansard* 139 (June 26, 1855): 218–19; PP 432 (1855); PP 480 (1855); PP 379 (1856).

45. The select committee's report states that Henry Letheby gave evidence, which is referred to but does not appear to have been printed in the minutes.

46. PP 379 (1856), iii–iv.

47. PP 379 (1856), 103–6.

48. Jackson, *Case Studies in Scientific Advice*, no. 13.

49. R. Lambert, *Sir John Simon*, 375.

50. PP 3484 (1865), 22–25.

51. PP 353 (1873), 715.

52. PP 262 (1874), 284.

53. PP 262 (1874), iii–viii.

54. Brian, "Food and Drugs Act," gives an extensive analysis of the history of the Sale of Food and Drugs Act 1875 and its local administration. Charles Cameron was later a strong supporter of germ theories of disease; see Pemberton and Worboys, *Rabies in Britain*, 104.

55. For its history, see Chirnside and Hamence, "*Practising Chemists.*"

56. Brian, "Food and Drugs Act," 151–250.

57. Brian, "Food and Drugs Act," 128.

58. PP C6742 (1892), viii, xi.

59. For the Government Laboratory, see Morris, *Matter Factory*, 269–90.

60. Hammond and Egan, *Weighed in the Balance*, 121–26.

14. WEIGHTS, MEASURES, AND COINAGE

1. Schaffer, "Metrology, Metrication and Victorian Values," gives an overview of the scientific, cultural, and political significance of metrology in the Victorian period. On earlier history, see Ashworth, *Customs and Excise*, 280–88.

2. Adell, "British Metrological Standardization Debate," gives an extensive analysis. Hoppit, "Reforming Britain's Weights and Measures," describes the eighteenth-century background.

3. *Hansard* 27 (May 10, 1814): 810.

4. RS EC/1819/06.

5. PP 290 (1814).

6. PP 290 (1814), 4.

7. Todd, *Beyond the Blaze*, 162.

8. Lyon, *Royal Society*, 221–23; Waring, "Thomas Young," 82–102. For the minutes, see RS CMB 1/1 Pendulum Committee (May 1817–June 1819 and June 1829).

9. RS CMO/9/104–5.

10. PP 361 (1818).

11. PP 565 (1819), 3–4. On the metric revolution, see Vincent, *Beyond Measure*, 151–80.

12. Hoppit, "Reforming Britain's Weights and Measures," 99.

13. PP 314 (1820), 4.

14. PP 383 (1821), 3–4.

15. Waring, "Thomas Young," 45–46, 91.

16. PP 565 (1819), 3.

17. PP 571 (1821).

18. PP 94 (1824).

19. For background on Patrick Kelly, see Ashworth, "Calculating Eye," 422–23.

20. PP 94 (1824), 7–8.

21. PP 94 (1824), 15–16.

22. PP 464 (1834).

23. Waring, "Thomas Young," 139.

24. PP 356 (1841), 3.

25. PP 356 (1841), 10–11.

26. PP 356 (1841), 6.

27. PP 356 (1841), 6.

28. G. Turner, *Nineteenth-Century Scientific Instruments*, 49. The 1841 report of the commissioners was reprinted by the House of Commons in 1855 to inform this process; see PP 177 (1855).

29. PP 1786 (1854), 18.

30. An example is in PP 347 (1845).

31. Marples, "Science of Money."

32. See Craig, *Mint*, 256–77.

33. For these two episodes: Craig, *History of the London Mint*, 269, 293.

34. PP 465 (1837); PP 1026 (1849).

35. PP 1026 (1849), x.

36. PP 1026 (1849), 85–99.

37. Brock, "Brock Science Patronage," 178–79.

38. Dyer, "Master of the Royal Mint," describes John Herschel's period in office.

39. As described by Hofmann in "Obituary Notices of the Fellows Deceased," *Proceedings of the Royal Society of London* 18 (1870): xxv–xxvi.

40. PP C7 (1870).

41. Quoted in Craig, *History of the Mint*, 327.

42. Craig, *History of the Mint*, 312.

43. PP 1786 (1854), 14.

44. PP 851 (1853).

45. PP 851 (1853), 68.

46. "Third Annual Conference," *Journal of the Society of Arts* 2 (1854): 587.

47. PP 2212 (1857).

48. PP 2529 (1859), 4.

49. PP 2259 (1859), 16.

50. PP 411 (1862).

51. PP 411 (1862), 45–46.

52. PP 411 (1862), 84.

53. PP 165 (1864).

54. For their first annual report, see PP 3883 (1867).

55. PP 4077 (1868); PP 4186 (1869); PP C30 (1870); PP C147 (1870).

56. PP 4186 (1869), 4.

57. PP 4073 (1868).

58. PP 4073 (1868), xiv.

59. PP 346 (1895), iii. Britain had signed the Metre Convention in 1884, a price of Greenwich becoming the prime meridian.

60. "The Metric System of Weights and Measures," *Nature* 53 (December 5, 1895): 84–86.

61. Frederick Bramwell, an outlier as an ultraconservative engineer, wanted to stick with vulgar fractions; see Roscoe, *Life & Experiences of Sir Henry Enfield Roscoe*, 285.

CONCLUSION. CONSTRAINTS ON INFLUENCE

1. Peter Bartrip takes a particularly stark view, based on an analysis of the white lead trade and the use of phosphorus in "lucifer" matches, when he argues that the expert did not possess "the capacity to carry through proposals based on rigorous scientific and technological analysis, for scientific and medical goals tended to be subordinate to, and even the servant of, political and economic considerations"; see Bartrip, "Expertise and the Dangerous Trades," 108–9. I find it unsurprising that they were subordinate.

2. Joyce, *State of Freedom*, 33. See also the analysis in Scott, *Seeing like a State*, and note 10 in this book's introduction (on pp. 321–22).

3. Eastwood, "Making Public Policy," 13, 21.

4. Rhodes, *Inspectorates in British Government*, 10.

5. Hartley, "Inspectorates in British Central Government," offers a typology of inspectors; Harris, *British Government Inspection*, addresses the development of central government inspection; see also Henriques, *Before the Welfare State*, 250–52.

6. Bellamy, *Administering Central-Local Relations*.

7. MacDonagh, *Early Victorian Government*, 5–6.

8. Parry, *Rise and Fall of Liberal Government*, 180.

9. F. Smith, "Contagious Diseases Acts," 200; Jackson, *Case Studies in Scientific Advice*, no. 23.

10. See Henriques, *Before the Welfare State*, 249–50.

11. Albu, "Member of Parliament," 18–20, gives a commentary of the need for parliamentary as well as executive access to scientific advice.

12. Hennock, "Central/Local Government Relations," 40–41; Eastwood, "Making Public Policy," 20.

13. R. Yeo, *Defining Science*, 117.

14. F. Turner, "Public Science in Britain," 592–96.

15. Marsden and Smith, *Engineering Empires*, 221.

16. Odling, "Science in Courts of Law," in response to criticism by others, including Robert Angus Smith.

17. Hamlin, "Scientific Method and Expert Witnessing," 495–96.

18. Collingridge and Reeve, *Science Speaks to Power*, x. Weinberg, "Science and Trans-Science," tackles science, "trans-science," and politics. He argues that the inherent uncertainties surrounding complex and socially divisive problems lead to questions being asked of science that it cannot answer.

19. Lewis, *Edwin Chadwick*, 134–38.

20. Gooday, "Liars, Experts and Authorities," 432–34.

21. Charles Darwin's lobbying of John Lubbock, MP, to get a clause inserted about marriages between cousins in a forthcoming bill, effectively to test his own theories, is described in Patton, *Science, Politics and Business*, 93. It was rejected by Parliament.

22. Hawkins, *Victorian Political Culture*, 368.

23. Smith, *Disraelian Conservatism and Social Reform*, 19.

24. Hawkins, *Victorian Political Culture*, 342–45.

25. Nevertheless, Mandler, *Aristocratic Government in the Age of Reform*, 281–82, argues for the importance of Whig attitudes and government in driving legislation on social reform in the 1830s and 1840s.

26. Coleman, *Conservatism and the Conservative Party*, 125–30; and Smith, *Disraelian Conservatism and Social Reform*, 4–5.

27. Parris, *Government and the Railways*, 213–14. On the "independent member," see Beales, "Parliamentary Parties and the 'Independent' Member," 1–19.

28. Smith, *Disraelian Conservatism and Social Reform*, 157–61.

29. On the importance and changing nature of local government, see, for example, Waller, *Town, City, and Nation*; Keith-Lucas, *English Local Government*; H. Finer, *Theory and Practice of Modern Government*, 755–59.

30. J. Taylor, "Private Property, Public Interest," 750.

31. Bartrip, "State and the Steam Boiler," 79.

32. Charles Babbage gave one early example of such an analysis in his "Combinations of Masters against the Public," in *On the Economy of Machinery and Manufactures* (London: John Murray, 1846), 312–35.

33. For a comprehensive list of inspectorates, and their dates of introduction in legislation, see Bartrip, "British Government Inspection," 607. D. Roberts, *Victorian Origins of the British Welfare State*, 106, lists sixteen government "departments" in 1854 that had nationwide supervisory powers over local government and private institutions.

34. Kearns, "Private Property and Public Health Reform."

35. Sherard, *White Slaves of England.*

36. Ashby and Anderson, "Studies in the Politics of Environmental Protection III," 204–5.

37. Millward, *Private and Public Enterprise in Europe*, 33–41.

38. Byatt, *British Electrical Industry*, 197–98, identifies the turning point in moves away from laissez-faire in reports on gas bills in 1847–1848.

39. Jackson, *Case Studies in Scientific Advice*, no. 29.

40. *Hansard* 212 (July 10, 1872): 931.

41. PP C1123 (1849), xvii–xviii.

42. For example, Clifton, *Professionalism, Patronage and Public Service*, explores the metropolis from mid-century; while Hassan, *History of Water in Modern England and Wales*, 10–50, explores water policy throughout Britain.

43. The Crown Estates Paving Commissioners still exist, and are responsible, among others, for the gardens outside the Athenaeum.

44. On central/local relations, see Ogborn, "Local Power and State Regulation"; Henriques, *Before the Welfare State*, 252–54. A prime example in the later nineteenth century is the operation of the Local Government Board; see Bellamy, *Administering Central-Local Relations.*

45. Questions to Siemens and Odling are typical; see PP 224 (1879), 17; PP C3842 (1884), 632.

46. PP 322 (1887), 3.

47. Barton, *X Club.*

48. Haigh, "William Brande," argues both for his neglected significance and his influence on a scientific basis for medicine.

49. For the history, see Chambers, *Register of the Associates and Old Students.*

50. Leggett and Davey, "Expertise and Authority in the Royal Navy," 11.

51. Pielke, *Honest Broker.*

52. Hamlin, *Science of Impurity*, 161.

53. C. Russell, *Edward Frankland*, 398–403. Considering Frankland's expertise in analysis and his character, I am inclined to side with Russell over Hamlin.

54. Crook, "Sanitary Inspection and the Public Sphere," 393.

55. PP C408 (1871), 636.

56. PP C408 (1871), 640.

57. PP C5413 (1888), 6–20.

58. Sutherland, *Studies in the Growth of Nineteenth-Century Government*, 10.

59. For Treasury control, see E. Hughes, "Sir Charles Trevelyan"; Wright, "Treasury Control"; MacLeod, "Science and the Treasury."

60. Garwood, "British State and the Natural Environment," 134–39. Garwood suggests that the reasons include the lower status of the Local Government Board compared to the Home Office, the focus of factory and mines inspectors on the protection of human life and morals rather than private property, and the low status of men of science.

61. PP C281 (1871), 35–36.

62. Hart, "Genesis of the Northcote-Trevelyan Report," 81. See E. Hughes, "Sir Charles Trevelyan."

63. For an overview, see MacDonagh, *Early Victorian Government*, 197–213; and Pellew, *Home Office*, 183–205, on the professionalization of the civil service.

64. Kitson Clark, "'Statesmen in Disguise,'" 23; Fry, *Statesmen in Disguise*.

65. Berman, *Social Change and Scientific Organization*, 123.

66. Snyder, *Reforming Philosophy*, 4–8.

67. R. Yeo, *Defining Science*, and R. Yeo, "Idol of the Market-Place." Snyder, *Reforming Philosophy*, and Gillin, *Victorian Palace of Science*, 69–73, describe the philosophies of John Stuart Mill and William Whewell. Bord, *Science and Whig Manners*, addresses science and the Whigs. On the wider place of knowledge in Victorian Britain, see Daunton, *Organisation of Knowledge*, 1–27.

68. Worboys, *Spreading Germs*, 109–10. Worboys argues that tuberculosis was catalytic in advancing the importance and acceptance of bacteriology; see Worboys, *Spreading Germs*, 193–276.

69. For example, John Simon's success in persuading the Treasury to make regular the temporary grant for laboratory research for the medical department was a consequence of his close relationship with the chancellor of the exchequer, Robert Lowe. See Wright, "Treasury Control," 210.

70. MacDonagh, "Government, Industry and Science," 504. The United Alkali Company, for example, did establish a central laboratory in 1890.

71. Foster, "State and Scientific Research."

72. PP C8976 (1898), 5. See Meadows, *Science and Controversy*, 270–77; Macdonald, *Kew Observatory*, 165–202.

73. Albu, "Member of Parliament," 3.

74. MacLeod, "Science for Imperial Efficiency," analyzes the life of the British Science Guild.

75. Hull, "War of Words," 464. See MacLeod, "Science for Imperial Efficiency," 172–73.

76. Vig, *Science and Technology in British Politics*, gives background to this.

77. *Hansard* 149 (March 1, 1858): 41.

Bibliography

PARLIAMENTARY PAPERS

For ease of referencing, the papers are referred to by their year of publication and their unique number, which, with the abbreviated title, should make them easy to find.

The following guides provide useful starting points for parliamentary papers on particular topics, although papers referenced in this book extend beyond their coverage: See P. Ford and G. Ford, eds., *Hansard's Catalogue and Breviate of Parliamentary Papers, 1696–1834* (Oxford: Basil Blackwell, 1953), and P. Ford and G. Ford, eds., *Select List of British Parliamentary Papers 1833–1899* (Oxford: Basil Blackwell, 1953). The original publication of the first is also available as PP 626 (1834).

Other primary sources—for example, from *Hansard*, the Royal Society (RS), the Royal Institution (RI), and contemporary journals and newspapers—are referenced directly in notes.

PP 65 (1805)	Report from Committee on . . . Swedish Herrings
PP 149 (1806)	Copy of Treasury Minute . . . Swedish Herrings
PP 202 (1806)	Experiments . . . to ascertain the relative Qualities of Malt made from Barley and Scotch Bigg
PP 212 (1806)	First Report from Committee on the Use of Broad Wheels . . .
PP 321 (1806)	Second Report from Committee on the Use of Broad Wheels . . .
PP 179 (1809)	Report from the Committee on the Acts . . . regarding the Use of Broad Wheels . . .
PP 220 (1809)	Minutes of Evidence . . . Coke, Oil, Tar, Ammoniacal Liquor, essential Oils, and inflammable Air, from Coal . . .
PP 238 (1809)	Second Report from Committee on the Acts . . . Use of Broad Wheels . . .
PP 271 (1809)	Third Report from Committee on the Acts . . . Use of Broad Wheels . . .
PP 166 (1810)	Report from Committee on Holyhead Roads and Harbour

PP 203 (1810)   Report from Committee on Howth Harbour

PP 240 (1811)   Report from Committee on the Highways . . .

PP 57 (1813)   Report from Select Committee . . . Corn Trade . . .

PP 26 (1814)   Reports respecting Grain . . .

PP 290 (1814)   Report from Select Committee on Weights and Measures

PP 339 (1814)   Report from Select Committee . . . Corn Laws

PP 397 (1816)   Report of Select Committee on the State of the Children
                  employed in the Manufactories . . .

PP 462 (1816)   Report from Committee on Holyhead Harbour

PP 275 (1817)   Third Report from Select Committee on Finance: Ordnance

PP 311 (1817)   Report from Committee on Petitions relating to Machinery for
                  Manufacturing of Flax

PP 422 (1817)   Report from Select Committee on Steam Boats

PP 462 (1817)   Report from Select Committee on the Poor Laws

PP 144 (1818)   Report of the Committee . . . Experiments of Dr. Sickler . . .

PP 361 (1818)   Experiments Relating to the Pendulum Vibrating Seconds of
                  Time in the Latitude of London

PP 2 (1819)   Report of the Commissioners . . . Forgery of Bank Notes.

PP 220 (1819)   Minutes taken . . . the high price and inferior quality of Beer . . .

PP 449 (1819)   Report from Select Committee . . . Validity of the Doctrine of
                  Contagion in the Plague

PP 565 (1819)   First report of Commissioners . . . Weights and Measures

PP 574 (1819)   Report from Select Committee on Steam Engines and Furnaces
                  . . .

PP 64 (1820)   Final Report of Commissioners . . . Forgery of Bank Notes

PP 244 (1820)   Report from Select Committee on Steam Engines and Furnaces
                  . . .

PP 301 (1820)   Report from Select Committee on the Turnpike Roads and
                  Highways . . .

PP 314 (1820)   Second report of Commissioners . . . Weights and Measures

PP 383 (1821)   Third report of Commissioners . . . Weights and Measures

PP 537 (1821)   Report from Select Committee on the Supply of Water to the
                  Metropolis

PP 571 (1821)	Report from Select Committee on Weights and Measures
PP 595 (1821)	Report from Committee on the Adulteration of Clover and Trefoil Seeds
PP 706 (1821)	Minutes of Evidence . . . Select Committee on the Supply of Water to the Metropolis
PP 417 (1822)	Fifth Report of Select Committee on the Roads from London to Holyhead . . .
PP 591 (1822)	Report from Select Committee . . . Foreign Trade of the Country. Lights, Harbour Dues, and Pilotage
PP 193 (1823)	Gas-Light Establishments . . . Report of the Royal Society . . .
PP 370 (1823)	Mr. Babbage's Invention
PP 529 (1823)	Report from Select Committee on Gas-Light Establishments
PP 94 (1824)	Report from Select Committee of the House of Lords . . . Petition of the Chamber of Commerce and Manufactures . . . Glasgow
PP 305 (1824)	First Report of Commissioners . . . Improvement of the Road from London to Holyhead
PP 427 (1824)	Report from Select Committee . . . Salmon Fisheries . . .
PP 173 (1825)	Report from Select Committee . . . Salmon Fisheries . . .
PP 393 (1825)	Second Report from Select Committee . . . Salmon Fisheries . . .
PP 267 (1828)	Supply of Water in the Metropolis . . .
PP 567 (1828)	Report from Select Committee on the Supply of Water in the Metropolis
PP 332 (1829)	Report from Select Committee . . . Patents for Inventions
PP 9 (1830)	Report from Select Committee . . . State of the Coal Trade . . .
PP 109 (1831)	Report from Select Committee on the Use of Molasses in Breweries and Distilleries
PP 297 (1831)	Report from Select Committee on the Use of Molasses in Breweries and Distilleries
PP 324 (1831)	Report from Select Committee on Steam Carriages
PP 335 (1831)	Report from Select Committee on Steam Navigation
PP 380 (1832)	Report from Select Committee . . . Patent granted for Morton's Slip
PP 667 (1832)	Report from Committee . . . cruel and improper Treatment of Animals . . .

PP 706 (1832)   Report from Committee. . . . Labour of Children in the Mills and Factories . . .

PP 450 (1833)   Factories Inquiries Commission. First Report . . .

PP 519 (1833)   Factories Inquiries Commission. Second Report . . .

PP 590 (1833)   Sugar Refining . . .

PP 612 (1833)   Report from Select Committee on Agriculture

PP 676 (1833)   Report from Select Committee on British Channel Fisheries

PP 43 (1834)   Admeasurement of Shipping: Report of Committee . . . Tonnage of Ships

PP 167 (1834)   Factories Inquiries Commission. Supplementary Report . . .

PP 176 (1834)   Metropolis Water Supply. Report of Thomas Telford . . .

PP 464 (1834)   Minutes of Evidence . . . relating to Weights and Measures

PP 571 (1834)   Report from Select Committee on Metropolis Water

PP 584 (1834)   Report from Select Committee on Metropolis Sewers

PP 590 (1834)   Report from Select Committee on Lighthouses

PP 626 (1834)   Catalogue of Parliamentary Reports 1696–1834

PP 367 (1835)   Mr Kyan's Patent . . . Report from the Committee . . .

PP 370 (1835)   Report from Select Committee . . . State of the Harbours of Leith and Newhaven

PP 583 (1835)   Report from Select Committee on the Ventilation of the Houses of Parliament

PP 603 (1835)   Report from Select Committee on Accidents in Mines

PP 334 (1836)   Report from Select Committee on Harbours of Refuge

PP 393 (1836)   Report from Select Committee on Salmon Fisheries, Scotland

PP 398 (1836)   Report from Select Committee on Dover Harbour

PP 547 (1836)   Report from Select Committee on Turnpike Trusts and Tolls

PP 567 (1836)   Report from Select Committee . . . Causes of Shipwrecks

PP 465 (1837)   Report from Select Committee on the Royal Mint

PP 187 (1839)   First Annual Report of Registrar-General of Births, Deaths, and Marriages . . .

PP 273 (1839)   Report of Steam-Vessel Accidents

PP 517 (1839)   Second Report from Select Committee on Railways

PP 50 (1840)   First Report from Select Committee on Railway Communication

PP 253 (1840)   Soap. Reports . . . on the Specific Gravity of Soap

PP 256 (1840)   Report of Commissioners for inquiring into the State of the Roads in England and Wales

PP 299 (1840)   Third Report from Select Committee on Railway Communication

PP 368 (1840)   Harbours (South Eastern Coast) . . . report of the Commissioners . . .

PP 384 (1840)   Report from Select Committee on the Health of Towns

PP 437 (1840)   Fourth Report from Select Committee on Railway Communication

PP 474 (1840)   Fifth Report from Select Committee on Railway Communication

PP 287 (1841)   Report of Officers of the Railway Department . . .

PP 354 (1841)   Report from Select Committee on Railways

PP 356 (1841)   Report of Commissioners . . . Restoration of the Standards of Weight & Measure

PP 380 (1842)   Children's Employment Commission. First Report . . .

PP 549 (1843)   First Report from Select Committee on Shipwrecks

PP 581 (1843)   Second Report from Select Committee on Shipwrecks

PP 583 (1843)   Report from Select Committee on Smoke Prevention

PP 166 (1844)   Railways. Third Report from Select Committee on Railways

PP 318 (1844)   Railways. Fifth Report from Select Committee on Railways

PP 565 (1844)   Report from Select Committee on Tobacco Trade

PP 572 (1844)   First Report of Commissioners . . . State of Large Towns and Populous Districts

PP 592 (1844)   Report of Commissioner . . . State of the Population in the Mining Districts

PP 232 (1845)   Haswell Collieries. Copy of the Report of Messrs. Lyell and Faraday . . .

PP 289 (1845)   Report from Select Committee on Smoke Prevention

PP 347 (1845)   Coinage. . . . Report of the last Pix Jury on the Coinage at the Mint . . .

PP 489 (1845)   Second Report from Select Committee on Smoke Prevention

PP 602 (1845)   Second Report of Commissioners . . . State of Large Towns and
                   Populous Districts

PP 607 (1845)   Report from Select Committee on Lighthouses

PP 610 (1845)   Second Report of Commissioners . . . State of Large Towns and
                   Populous Districts, Appendix, Part II

PP 611 (1845)   Report of Commissioners . . . Harbours of Refuge

PP 628 (1845)   Report on the Fall of the Cotton Mill at Oldham, and part of the
                   Prison at Northleach

PP 665 (1845)   Tidal Harbours Commission. First Report of Commissioners

PP 28 (1846)   Potatoes (Ireland)

PP 33 (1846)   Potato Crop

PP 190 (1846)   Cows and Bullocks. . . . Experiments to determine the Effect of
                   Barley and Malt on the Milk of Cows and the fattening of
                   Bullocks . . .

PP 308 (1846)   Report on Metropolitan Sewage Manure Company Bill

PP 411 (1846)   Report from Select Committee of the House of Lords on the
                   Burdens affecting Real Property

PP 474 (1846)   Report from Select Committee on Metropolitan Sewage Manure

PP 499 (1846)   Burdens on Land. . . . a Communication made by Lord Monteagle
                   to the Board of Trade . . .

PP 556 (1846)   Report from Select Committee on Private Bills

PP 684 (1846)   Report of Gauge Commissioners

PP 692 (1846)   Tidal Harbours Commission. Second Report of Commissioners

PP 699 (1846)   Minutes of Evidence . . . Gauge Commissioners

PP 26 (1847)   Breweries and Distilleries. . . . Report . . . on the use of Barley,
                   Malt, Sugar, and Molasses, in Breweries and Distilleries

PP 194 (1847)   Smoke Prohibition. Report upon the Means of obviating the
                   Evils arising from the Smoke occasioned by Factories . . .

PP 411 (1847)   Harbours of Refuge. . . . Report of Harbour of Refuge Commis-
                   sion of 1846 . . .

PP 426 (1847)   Spirit from Sugar. . . . Reports . . . on the Production of Spirit
                   from Sugar and Molasses

PP 529 (1847)   Spirit from Sugar. . . . Report . . . on the Production of Spirit
                   from Sugar and Molasses.

PP 815 (1847)   Collieries. Reports on the Gases and Explosions in Collieries . . .

PP 821 (1847)   Report on the Harbour of Refuge to be constructed in Dover Bay

PP 874 (1847)   Tidal Harbours Commission . . .

PP 888/895 (1847) Metropolitan Sanitary Commission. First Report of Commissioners . . .

PP 911 (1848)   Metropolitan Sanitary Commission. Second Report of Commissioners . . .

PP 915 (1848)   Museum of Practical Geology. First Report on the Coals suited to the Steam Navy . . .

PP 921 (1848)   Metropolitan Sanitary Commission. Second Report. Minutes of Evidence . . .

PP 943 (1848)   Tidal Harbours Commission. Supplement II to Appendix C. to Second Report of Commissioners

PP 979 (1848)   Metropolitan Sanitary Commission. Third Report of Commissioners . . .

PP 427 (1849)   Collieries and Mines. Instructions issued . . . to Professor Phillips F.R.S., and J.R. Blackwell, Esq. . . . .

PP 613 (1849)   Report from Select Committee of the House of Lords . . . best Means of preventing the Occurrence of Dangerous Accidents in Coal Mines

PP 1026 (1849)   Report of Commissioners . . . Constitution, Management, and Expense of the Royal Mint

PP 1051 (1849)   Reports on the Explosion in Darley Main Colliery

PP 1070 (1849)   General Board of Health. Report on Quarantine

PP 1086 (1849)   Museum of Practical Geology. Second Report on the Coals suited to the Steam Navy . . .

PP 1123 (1849)   Report of Commissioners . . . Application of Iron to Railway Structures

PP 49 (1850)   Tonnage: Report of Committee . . . defects of the method of measuring ships for tonnage

PP 1214 (1850)   Report on the Ventilation of Mines . . .

PP 1222 (1850)   Report on the Ventilation of Mines and Collieries . . .

PP 390 (1851)   Victoria (Redcar) Harbour and Docks Bill. . . . Reports . . . on the Victoria (Redcar) Harbour and Docks Bill . . .

PP 1345 (1851)   Museum of Practical Geology. Third Report on the Coals suited to the Steam Navy . . .

PP 509 (1852)   Report from Select Committee on Coal Mines

PP 1472 (1852)   General Board of Health. . . . Practical Application of Sewer
                  Water and Town Manures to Agricultural Production

PP 1473 (1852)   General Board of Health. Second Report on Quarantine. Yellow
                  Fever

PP 747 (1853)   Report from Select Committee on Communication between
                  London and Dublin

PP 851 (1853)   Report from Select Committee on Decimal Coinage

PP 236 (1854)   Report from Select Committee on Small Arms

PP 325 (1854)   Fourth Report from Select Committee on Accidents in Coal
                  Mines

PP 1713 (1854)   Report on the Organisation of the Permanent Civil Service . . .

PP 1769 (1854)   Report of Captain Wynne . . . Professor Gluckman's Invention . . .

PP 1786 (1854)   Report of Commissioners . . . Construction of New Parliamen-
                  tary Standards of Length and Weight

PP 1818 (1854)   Report of Commissioners . . . Outbreak of Cholera in the Towns
                  of Newcastle-upon-Tyne, Gateshead, and Tynemouth

PP 177 (1855)   Weights and Measures. . . . Report of Commissioners . . . Steps
                  to be taken for Restoration of the Standards of Weight and
                  Measures

PP 432 (1855)   First Report of Select Committee on Adulteration of Food . . .

PP 480 (1855)   Second Report of Select Committee on Adulteration of Food . . .

PP 1870 (1855)   Papers on the Re-Organisation of the Civil Service

PP 1893 (1855)   General Board of Health. Letter . . . Report from Dr. Sutherland
                  on Epidemic Cholera in the Metropolis in 1854

PP 1943 (1855)   Report of the Commissioner . . . Provision of the Acts for the
                  better Regulation of Mills and Factories to Bleaching Works

PP 1951 (1855)   Two reports . . . on the explosion of boilers of locomotive engines
                  . . .

PP 1980 (1855)   General Board of Health. Medical Council. Report of Committee
                  for Scientific Inquiries in relation to the Cholera-Epidemic of
                  1854

PP 379 (1856)   Report of Select Committee on Adulteration of Food . . .

PP 2103 (1856)   Report on the Last Two Cholera-Epidemics of London . . .

PP 2137 (1856)   General Board of Health. Reports . . . Metropolis Water Supply

PP 233 (1857)   Metropolitan Drainage. . . . Plans for the Main Drainage of the
                  Metropolis . . .

PP 262 (1857)   Report from Select Committee on Harbours of Refuge

PP 2212 (1857)   Preliminary Report of Decimal Coinage Commissioners

PP 2234 (1857)   Report of Meteorological Department of the Board of Trade

PP 2262 (1857)   Report on the Means of Deodorizing and Utilizing the Sewage of Towns . . .

PP 344 (1858)   Report from Select Committee on Harbours of Refuge

PP 362 (1858)   Report from Select Committee on Accidents on Railways

PP 442 (1858)   Report from Select Committee on the River Thames

PP 2372 (1858)   Sewage of Towns. Preliminary Report of Commission . . .

PP 2396 (1858)   Report of Ordnance Survey Commission

PP 2415 (1858)   General Board of Health. Papers relating to the Sanitary State of the People of England

PP 224 (1859)   Report from Select Committee on Gas (Metropolis)

PP 2474 (1859)   Report of Commissioners . . . Harbours of Refuge

PP 2498 (1859)   Report . . . Accidents which have occurred on Railways during the Year 1858

PP 2512 (1859)   Public Health. Report of the Medical Officer of the Privy Council, 1858

PP 2529 (1859)   Final Report of Decimal Coinage Commissioners

PP 100 (1860)   Sale of Gas. . . . Report of the Astronomer Royal . . .

PP 293 (1860)   Sale of Gas Act. . . . Correspondence between the Lords Commissioners of Her Majesty's Treasury, the Board of Trade, the Comptroller of the Exchequer, and the Astronomer Royal . . .

PP 417 (1860)   Gas (Metropolis) Bill. . . . Evidence taken before Select Committee on the Gas (Metropolis) Bill

PP 456 (1860)   Report from Select Committee of the House of Lords . . . Salmon Fishings on the Sea Coasts and in Rivers and Estuaries in Scotland . . .

PP 547 (1860)   Sale of Gas Act. . . . Correspondence . . . on the subject of the Sale of Gas Act . . .

PP 187 (1861)   Enfield Whitworth Committee. . . . report for the Enfield Whitworth Committee . . .

PP 2768 (1861)   Report of Commissioners appointed to inquire into Salmon Fisheries (England and Wales)

PP 2793 (1861)   Report of Commissioners . . . Condition and Management of Lights, Buoys, and Beacons

PP 2882 (1861)   Sewage of Towns. Second Report of Commission . . .

PP 160 (1862)   First Report from Select Committee on Sewage of Towns

PP 411 (1862)   Report from Select Committee on Weights and Measures

PP 448 (1862)   Report from Select Committee on Ordnance

PP 469 (1862)   Second Report from Select Committee on Sewage of Towns

PP 486 (1862)   Report from Select Committee of the House of Lords, on Injury from Noxious Vapours

PP 139 (1863)   Army (Rifles). . . . Report of Committee on Small Bore Rifles . . .

PP 3106 (1863)   Report of Royal Commission . . . Trawling for Herring

PP 165 (1864)   Weights and Measures (Metric System). A Bill to render permissive the use of the Metric System of Weights and Measures . . .

PP 487 (1864)   Report from Select Committee on Sewage (Metropolis)

PP 543 (1864)   Metford's Explosive Bullet. . . . Correspondence between the Authorities at the War Office and Mr. Metford . . .

PP 3289 (1864)   Railway Department. Board of Trade. Report by Mr. Fairbairn . . . of his Experiments for ascertaining the Strength of Iron Structures

PP 3416 (1864)   Public Health. Sixth Report of the Medical Officer of the Privy Council . . .

PP 3889 (1864)   Report of Commissioners . . . Condition of All Mines in Great Britain . . . with Reference to the Health and Safety of Persons employed . . .

PP 267 (1865)   Gunpowder. . . . Reports of Lieutenant Colonel Boxer . . .

PP 459 (1865)   Rifled Guns. . . . Report . . . by the Ordnance Select Committee on the Trials . . .

PP 462 (1865)   Enfield Rifles. . . . Report of Committee . . . on the Trials at Woolwich of Enfield Rifles converted to Breech-Loaders

PP 3419 (1865)   Report of Commissioners . . . Working of the Law relating to Letters Patent for Inventions

PP 3451 (1865)   Preliminary Abstract report . . . Experiments of the Comparative Qualities of Unmalted and Malted Barley as Food for Stock

PP 3472 (1865)   Sewage of Towns. Third Report and Appendices of Commission . . .

PP 3484 (1865)   Public Health. Seventh Report of the Medical Officer of the Privy Council . . .

PP 270 (1866)   Report from Select Committee on the London (City) Corpora-
                  tion Gas, &c. Bills

PP 435 (1866)   Enfield Rifles.... Report of Committee ... on the Trials at
                  Woolwich of Enfield Rifles converted to Breech-Loaders

PP 3596 (1866)   Report of Commissioners ... Sea Fisheries of the United
                  Kingdom

PP 3605 (1866)   Report of the Special Armstrong and Whitworth Committee

PP 3634 (1866)   First Report of Commissioners ... Best Means of Preventing the
                  Pollution of Rivers. (River Thames.)

PP 3645 (1866)   Public Health. Eighth Report of the Medical Officer of the Privy
                  Council ...

PP 3678 (1866)   Children's Employment Commission (1862). Fifth Report of
                  Commissioners

PP 394 (1867)   Iron.... Report of Committee on Metals ...

PP 399 (1867)   Report from Select Committee on East London Water Bills ...

PP 3835 (1867)   Second Report of Commissioners ... Best Means of Preventing
                  the Pollution of Rivers. (River Lee.)

PP 3844 (1867)   Royal Commission on Railways. Report of Commissioners

PP 3850 (1867)   Third Report of Commissioners ... Best Means of Preventing the
                  Pollution of Rivers. (Rivers Aire and Calder.)

PP 3883 (1867)   First Report of the Warden of the Standards of the Proceedings
                  and Business of the Standard Weights and Measures Depart-
                  ment of the Board of Trade

PP 307 (1868)   ... Evidence taken before Select Committee on the Lee Water
                  Conservancy Bill

PP 3998 (1868)   Ordnance.... Report of Ordnance Select Committee on Coiled
                  Wrought-Iron Inner Tubes for Ordnance

PP 4003 (1868)   Report of a Special Committee on the "Gibraltar" Shields

PP 4072 (1868)   Report of the Cholera Epidemic of 1866 in England

PP 4073 (1868)   Report from Royal Commission on International Coinage

PP 4077 (1868)   Standards Commission. First Report of Commissioners ... Con-
                  dition of the Exchequer Standards

PP 179 (1869)   Gun Cotton.... Reports ... to guide the Employment of such
                  Substances for Mining Purposes

PP 361 (1869)   Report from Select Committee on Salmon Fisheries

PP 4156 (1869)   Report . . . by the Referees appointed under the "City of London
                     Gas Act 1868," together with the Instructions for Testing
                     issued to the Gas Examiners . . .

PP 4169 (1869)   Royal Commission on Water Supply. Report of Commissioners

PP 4186 (1869)   Standards Commission. Second Report of Commissioners . . .
                     on the Question of the Introduction of the Metric System of
                     Weights and Measures into the United Kingdom

PP 308 (1870)   Ordnance. . . . Report of Ordnance Council on the proposed 35-
                     Ton Gun Competition

PP 341 (1870)   Report from Select Committee on Railway Companies

PP 368 (1870)   Report from Select Committee on Salmon Fisheries

PP 370 (1870)   Report from Select Committee on Steam Boiler Explosions

PP 4218 (1870)   First Report of Royal Sanitary Commission . . .

PP C7 (1870)   Mint: . . . Reports on the Mint

PP C30 (1870)   Standards Commission. Third Report of Commissioners . . . on
                     the Abolition of the Troy Weight

PP C37 (1870)   Rivers Pollution Commission (1868). First Report of Commis-
                     sioners . . . Best Means of Preventing the Pollution of Rivers.
                     (Mersey and Ribble Basins)

PP C147 (1870)   Standards Commission. Fourth Report of Commissioners . . .
                     Inspection of Weights and Measures . . .

PP C180 (1870)   Rivers Pollution Commission (1868). Second Report of Commis-
                     sioners . . . Best Means of Preventing the Pollution of Rivers.
                     The A.B.C. Process of Treating Sewage

PP C270 (1870)   Sulphur Question. Report of the Gas Referees upon the Sulphur
                     Question

PP 298 (1871)   Report from Select Committee on Steam Boiler Explosions

PP C258 (1871)   Report of a Committee on the Education of Artillery Officers

PP C281 (1871)   Second Report of Royal Sanitary Commission. Vol. I. The Report

PP C281-I (1871)   Second Report of Royal Sanitary Commission. Vol. II . . .

PP C294 (1871)   General Report by Captain Tyler . . . upon the Accidents which
                     have occurred on Railways during the Year 1870

PP C299 (1871)   Reports of Special Committee on Martini-Henry Breech-
                     Loading Rifles

PP C347 (1871)   Rivers Pollution Commission (1868). Third Report of Commis-
                     sioners . . . Best Means of Preventing the Pollution of Rivers.
                     Pollution arising from the Woollen Manufacture . . .

PP C393 (1871)   Gas (Ammonia Impurity). . . . Report . . . by the Gas Referees on the Ammonia Impurity in Gas . . .

PP C394 (1871)   Gas Burners. . . . First Report . . . by the Gas Referees on the Construction of Gas Burners . . .

PP C408 (1871)   Report of Royal Commission . . . Administration and Operation of the Contagious Diseases Acts

PP 199 (1872)   Gas Referees.—Gas Referees' Report (Beckton Gasworks). . . . Correspondence . . . relative to the Appointment of Gas Referees . . .

PP 281 (1872)   Gas Referees. (Beckton and Bow Gasworks). . . . Reports by the Gas Referees on the Sulphur-Purification at the Becton and Bow Gasworks . . .

PP 371 (1872)   Lithofracteur. . . . report . . . by the Special Committee on Gun Cotton and Lithofracteur . . .

PP C477 (1872)   Report of Committee . . . to examine the Designs upon which Ships of War have recently been constructed

PP C586 (1872)   Report on the Explosion of Gun-Cotton at Stowmarket . . .

PP C603 (1872)   Rivers Pollution Commission (1868). Fourth Report . . . Pollution of Rivers of Scotland.

PP 353 (1873)   Report from Select Committee on Contagious Diseases (Animals)

PP C853 (1873)   Royal Commission on Unseaworthy Ships. Preliminary Report . . .

PP 243 (1874)   Report from Select Committee on Explosive Substances

PP 262 (1874)   Report from the Select Committee on Adulteration of Food Act (1872).

PP C951 (1874)   Rivers Pollution Commission (1868). Fifth Report . . . Pollution arising from Mining Operations and Metal Manufactures.

PP C977 (1874)   Reports on the Necessity for the Amendment of the Law relating to Gunpowder and other Explosives . . .

PP C1027 (1874)   Royal Commission on Unseaworthy Ships. Final Report . . .

PP C1109 (1874)   Second Report of Royal Sanitary Commission. Vol. III. Part 2 . . .

PP C1112 (1874)   Rivers Pollution Commission (1868). Sixth Report . . . The Domestic Water Supply of Great Britain

PP 428 (1875)   Fisheries (Norfolk). Report on the Fisheries of Norfolk, especially Crabs, Lobsters, Herrings, and the Broads . . .

PP C1364 (1875)   Report of the Explosion on Gunpowder in the Regent's Park . . .

PP C1371 (1875)   Public Health. . . . Report . . . on Scientific Investigations . . . in Aid of Pathology and Medicine.

PP 260 (1876)   Gas Purification (Patterson's Patent) . . .

PP C1397 (1876)   Report of Commission on the Practice of Subjecting Live Ani-
                     mals to Experiments for Scientific Purposes

PP C1410 (1876)   Sewage Disposal. Report of Committee . . . Modes of Treating
                     Town Sewage

PP C1464 (1876)   Report of Sir John Hawkshaw, . . . Purification of the River
                     Clyde.

PP C1586 (1876)   Report of Royal Commissioners . . . Spontaneous Combustion of
                     Coal in Ships

PP 100 (1877)   Cruelty to Animals. Return of Licences granted under the Act .
                     . .

PP C1637 (1877)   Royal Commission on Railway Accidents. Report . . .

PP C1695 (1877)   Reports on the Crab and Lobster Fisheries . . .

PP C1819 (1877)   Report on the Use of Dynamite for Killing Fish

PP C1864 (1877)   Royal Commission on the Practice of Subjecting Live Animals to
                     Experiments for Scientific Purposes . . .

PP C1917 (1878)   Report of Committee on the "Inflexible"

PP C1979 (1878)   Report on the Herring Fisheries of Scotland

PP C2159 (1878)   Noxious Vapours Commission. Report . . .

PP 127 (1879)   Experiments on Animals. Report . . . Number of Experiments
                     performed on Living Animals . . .

PP 224 (1879)   Report from Select Committee on Lighting by Electricity

PP 274 (1880)   Report from Select Committee on Potato Crop

PP C2616 (1880)   Tay Bridge Disaster. Report of Court of Inquiry . . .

PP C2660 (1880)   Report on the Disease which has recently prevailed among the
                     Salmon in the Tweed, Eden, and other Rivers . . .

PP C2778 (1881)   Preliminary Report from her Majesty's Commissioners on
                     Agriculture

PP C2923 (1881)   Report on the Results of Experiments made, with Samples of
                     Dust collected at Seaham Colliery . . .

PP C3000 (1881)   Report of Committee . . . Question of Wind Pressure on Railway
                     Structures

PP C3036 (1881)   Accidents in Mines. Preliminary Report . . .

PP C3096 (1881)   Minutes of Evidence . . . Commissioners on Agriculture. Vol. II

PP 227 (1882)   Report from Select Committee on Electric Lighting Bill

PP C3309 (1882)   Agricultural Commission. Report . . .

PP C3337-I (1882) Eleventh Annual Report of the Local Government Board. . . .
                  Report of the Medical Officer . . .

PP 255 (1883)   Report from Select Committee on Harbour Accommodation

PP C3741 (1883)   First Annual Report of the Fishery Board for Scotland

PP 290 (1884)   Report from Select Committee on Harbour Accommodation

PP C3842 (1884)   Royal Commission on Metropolitan Sewage Discharge. First
                  Report . . .

PP C3977 (1884)   Annual Report of the Agricultural Department . . .

PP 188 (1885)   Report from Select Committee on Telephone and Telegraph
                  Wires

PP 271 (1885)   Report from Select Committee on Salmon Fisheries (Ireland)

PP C4328 (1885)   Trawl Net and Beam Trawl Fishing. Report of the Commission-
                  ers . . .

PP C4543 (1885)   Reports in Insects Injurious to Hop Plants, Corn Crops, and
                  Fruit Crops . . . (No. I.—Insects Injurious to Hop Plants.)

PP C4577 (1885)   Loss of Life at Sea. First Report of the Royal Commission . . .

PP 252 (1886)   Report from Select Committee of the House of Lords on Electric
                  Lighting . . .

PP C4680 (1886)   Reports in Insects Injurious to Hop Plants, Corn Crops, and
                  Fruit Crops . . . (No. II.—Insects Injurious to Corn, Grass,
                  Pea, Bean, and Clover Crops.)

PP C4699 (1886)   Accidents in Mines. Final Report . . .

PP C4836 (1886)   Reports in Insects Injurious to Hop Plants, Corn Crops, and
                  Fruit Crops . . . (No. III.—Insects Injurious to Fruit Crops.)

PP C4843 (1886)   Report on Swine-Fever

PP 321 (1887)   Report from Select Committee of the House of Lords on the
                  Smoke Nuisance Abatement (Metropolis) Bill

PP 322 (1887)   Report from Select Committee of the House of Lords on Rabies
                  in Dogs

PP C4944 (1887)   Report on the Hessian Fly . . .

PP C4986 (1887)   Reports on Insects Injurious to Root and certain other Crops . . .

PP C5062-I (1887) Report of Royal Commission . . . Patterns of Warlike Stores . . .

PP C5125 (1887)   Light Vessels &c., Electrical Communication Committee. Report
                  . . .

PP C5217 (1887)   Report of Commissioners appointed by the Government . . .
                   Present Visitation of the Hessian Fly on Corn Crops in Great
                   Britain

PP C5227 (1887)   Loss of Life at Sea. Final Report of Royal Commission . . .

PP C5275 (1888)   First Annual Report of the Agricultural Adviser to the Lords of
                   the Committee of Council for Agriculture. 1887. (Insects and
                   Fungi Injurious to the Crops of the Farm, the Orchard, and
                   the Garden.)

PP C5305 (1888)   Report of a Committee . . . Plans proposed for the Fortification
                   and Armament of our Military and Home Mercantile Ports .
                   . .

PP C5413 (1888)   Warlike Stores. Second Report of Royal Commission . . .

PP C5481 (1888)   Report on Eruptive Diseases of the Teats and Udders of Cows in
                   Relation to Scarlet Fever in Man . . .

PP 60 (1889)   Lighthouse Illuminants (South Foreland experiments) . . .

PP C5679 (1889)   Annual Report of the Agricultural Department Privy Council
                   Office on the Contagious Diseases Inspection and Transit of
                   Animals . . .

PP C5763 (1889)   Light Vessels &c., Electrical Communication Committee. Second
                   and Final Report . . .

PP 2 (1890)   Lighthouse illuminants (South Foreland Experiments) . . .

PP C5979 (1890)   Preliminary and Further Reports . . . of Royal Commissioners . . .
                   Civil and Professional Administration of the Naval and Mili-
                   tary Departments . . .

PP C5995 (1890)   Board of Agriculture. Annual Report of the Veterinary Depart-
                   ment . . .

PP 215 (1892)   Report from joint Select Committee . . . on the Electric and
                   Cable Railways (Metropolis)

PP C6647 (1892)   Board of Agriculture. Report on recent Experiments on checking
                   Potato Disease . . .

PP C6742 (1892)   Adulteration of Artificial Manures . . . Report of the Departmen-
                   tal Committee . . .

PP C7018 (1893)   Board of Agriculture. Report of the Intelligence Department on
                   Rust or Mildew in Wheat Plants . . .

PP C7138 (1893)   Board of Agriculture. Report on further Experiments in check-
                   ing Potato Disease . . .

PP 346 (1895)   Report from Select Committee on Weights and Measures

PP C7743 (1895)   Photometric Standards. Report made to the Board of Trade . . .

PP C7862 (1895)   Royal Commission on Secondary Education. Volume 1 . . .

PP C7952 (1896)   Committee on the Manufacture of Compressed Gas Cylinders.
                  Report of the Committee . . .

PP C8540 (1897)   Royal Commission on Agriculture. Final Report . . . Agricultural
                  Depression

PP C8675 (1897)   Royal Commission on Electrical Communication with Light-
                  houses and Light-Vessels. Fifth and Final Report . . .

PP C8684 (1898)   Metropolitan Railway (Ventilation of Tunnels) . . .

PP C8787 (1898)   Fishery Board for Scotland . . . Report of Investigations on the
                  Life History of Salmon

PP C8976 (1898)   National Physical Laboratory. Report of Committee appointed
                  by the Treasury . . .

PP C9164 (1899)   Water Gas Committee. Report of Departmental Committee . . .

PP Cd. 685 (1901) Royal Commission on Sewage Disposal. Interim Report . . .

PP Cd. 1178 (1902) Royal Commission on Sewage Disposal. Second Report . . .

PP Cd. 4820   Royal Commission on Mines. Second Report . . .
   (1909)

SECONDARY SOURCES

Adell, Rebecca. "The British Metrological Standardization Debate, 1756–1824: The Importance of Parliamentary Sources in its Reassessment." *Parliamentary History* 22 (2003): 165–82.

Airy, Wilfred, ed. *Autobiography of Sir George Biddell Airy*. Cambridge: Cambridge University Press, 1896.

Albu, Austin. "The Member of Parliament, the Executive, and Scientific Policy." *Minerva* 2 (1963): 1–20.

Allibone, Thomas. *The Royal Society and Its Dining Clubs*. Oxford: Pergamon, 1976.

Alter, Peter. *The Reluctant Patron: Science and the State in Britain, 1850–1920*. Oxford: Berg, 1987.

Anderson, Katharine. *Predicting the Weather: Victorians and the Science of Meteorology*. Chicago: University of Chicago Press, 2005.

Angus, Ian. "Cesspools, Sewage, and Social Murder: Environmental Crisis and Metabolic Rift in Nineteenth-Century London." *Monthly Review* 70 (2018): 32–68.

Armstrong, John, and David Williams. *The Impact of Technological Change: The Early Steamship in Britain*. Research in Maritime History no. 47. Liverpool: Liverpool University Press, 2017.

Ashby, Eric, and Mary Anderson. *The Politics of Clean Air*. Oxford: Clarendon Press, 1981.

Ashby, Eric, and Mary Anderson. "Studies in the Politics of Environmental Protection: The Historical Roots of the British Clean Air Act, 1956: I. The Awakening of Public Opinion over Industrial Smoke, 1843–1853." *Interdisciplinary Science Reviews* 1 (1976): 279–90.

Ashby, Eric, and Mary Anderson. "Studies in the Politics of Environmental Protection: The Historical Roots of the British Clean Air Act, 1956: II. The Appeal to Public Opinion over Domestic Smoke, 1880–1892." *Interdisciplinary Science Reviews* 2 (1977): 9–26.

Ashby, Eric, and Mary Anderson. "Studies in the Politics of Environmental Protection: The Historical Roots of the British Clean Air Act, 1956: III. The Ripening of Public Opinion, 1898–1952." *Interdisciplinary Science Reviews* 2 (1977): 190–206.

Ashworth, William. "The Calculating Eye: Baily, Herschel, Babbage and the Business of Astronomy." *British Journal for the History of Science* 27 (1994): 409–41.

Ashworth, William. *Customs and Excise: Trade, Production, and Consumption in England, 1640–1845.* Oxford: Oxford University Press, 2003.

Ashworth, William. "John Herschel, George Airy, and the Roaming Eye of the State." *History of Science* 36 (1998): 151–78.

Ashworth, William. "'System of Terror': Samuel Bentham, Accountability and Dockyard Reform during the Napoleonic Wars." *Social History* 23 (1998): 63–79.

Ashworth, William. "Expertise and Authority in the Royal Navy, 1800–1945." *Journal for Maritime Research*, 16 (2014): 103–116.

Atkinson, Norman. *Joseph Whitworth: The World's Best Mechanician.* Sutton 1997.

Bailey, Mark. "The 1844 Railway Act: A Violation of Laissez-Faire Political Economy?." *History of Economic Ideas* 12 (2004): 7–24.

Baldwin, Peter. *Contagion and the State in Europe, 1830–1930.* Cambridge: Cambridge University Press, 1999.

Barrow, John. *Sketches of Royal Society and Royal Society Club.* London: Frank Cass, 1849.

Barton, Ruth. "'An Influential Set of Chaps': The X-Club and Royal Society Politics, 1864–85." *British Journal for the History of Science* 23 (1990): 53–81.

Barton, Ruth. "'Men of Science': Language, Identity and Professionalization in the Mid-Victorian Scientific Community." *History of Science* 41 (2003): 73–119.

Barton, Ruth. *The X Club: Power and Authority in Victorian Science.* Chicago: University of Chicago Press, 2018.

Bartrip, Peter. "British Government Inspection, 1832–1875: Some Observations." *Historical Journal* 25 (1982): 605–26.

Bartrip, Peter. 1988. "Expertise and the Dangerous Trades, 1875–1900." In *Government and Expertise: Specialists, Administrators and Professionals, 1860–1919*, edited by Roy MacLeod, 89–109. 1988. Reprint, Cambridge: Cambridge University Press, 2003.

Bartrip, Peter. "Food for the Body and Good for the Mind: The Regulation of Freshwater Fisheries in the 1870s." *Victorian Studies* 28 (1985): 285–304.

Bartrip, Peter. "The State and the Steam Boiler in Nineteenth-Century Britain." *International Review of Social History* 25 (1980): 77–105.

Bartrip, Peter. "State Intervention in Mid-Nineteenth-Century Britain: Fact or Fiction?." *Journal of British Studies* 33 (1983): 63–83.

Bastable, Marshall. *Arms and the State: Sir William Armstrong and the Remaking of British Naval Power, 1854–1914.* Farnham, UK: Ashgate, 2004.

Beales, Derek. 1967. "Parliamentary Parties and the 'Independent' Member, 1810–1860." In *Ideas and Institutions of Victorian Britain: Essays in Honour of George Kitson-Clark,* edited by Robert Robson, 1–19. London: G. Bell & Sons, 1967.

Beard, John. "To What Extent Did the 1858 Medical Act Bring Unity to the British Medical Profession?," *Journal of Medical Biography* 21 (2013): 95–99.

Beck, Ann. "The British Medical Council and British Medical Education in the Nineteenth Century." *Bulletin of the History of Medicine* 30 (1956): 150–62.

Beck, Ann. "Some Aspects of the History of Anti-Pollution Legislation in England, 1819–1954." *Journal of the History of Medicine* 14 (1959): 475–89.

Bellamy, Christine. *Administering Central-Local Relations, 1871–1919: The Local Government Board in its Fiscal and Cultural Context.* Manchester: Manchester University Press, 1988.

Ben-David, Joseph. *Centers of Learning.* 1977. Reprint, New Brunswick, NJ: Transaction Publishers, 1992.

Ben-David, Joseph. *The Scientist's Role in Society: A Comparative Analysis.* Hoboken, NJ: Prentice-Hall, 1971.

Berman, Morris. *Social Change and Scientific Organization: The Royal Institution, 1799–1844.* London: Heinemann Educational, 1978.

Binnie, Geoffrey. *Early Victorian Water Engineers.* London: Thomas Telford Ltd., 1981.

Bogart, Dan. "The Turnpike Roads of England and Wales." n.d. https://www.campop.geog.cam.ac.uk/research/projects/transport/onlineatlas/britishturnpiketrusts.pdf.

Bompass, George. *Life of Frank Buckland.* London: Smith, Elder, 1888.

Bonar, J., and H. Macrosty. *Annals of the Royal Statistical Society, 1834–1934.* London, 1934.

Bord, Joe. *Science and Whig Manners: Science and Political Style in Britain, c. 1790–1850.* London: Palgrave Macmillan, 2009.

Bourne, J. *Patronage and Society in Nineteenth-Century England.* London: Edward Arnold, 1986.

Brand, Jeanne. "John Simon and the Local Government Board Bureaucrats, 1871–1876." *Bulletin of the History of Medicine* 37 (1963): 184–94.

Brand, Jeanne. *Doctors and the State.* Baltimore, MD: Johns Hopkins University Press, 1965.

Brassley, Paul. "Agricultural Research in Britain, 1850–1914: Failure, Success and Development." *Annals of Science* 52 (1995): 465–80.

Brenner, Joel. "Nuisance Law and the Industrial Revolution." *Journal of Legal Studies* 3 (1974): 403–33.

Brian, Janet. "The Local Implementation of the Sale of Food and Drugs Act, 1875." PhD diss., Open University, 2006.

Brock, William. *Justus von Liebig: The Chemical Gatekeeper.* Cambridge: Cambridge University Press, 1997.

Brock, William. "The Spectrum of Science Patronage." In *The Patronage of Science in the Nineteenth Century,* edited by Gerard Turner, 173–206. Leiden: Noordhoff International Publishing, 1976.

Brock, William. *William Crookes (1832–1919) and the Commercialization of Science.* Farnham, UK: Ashgate, 2008.

Brockington, Fraser. "Public Health at the Privy Council, 1831–34." *Journal of the History of Medicine and Allied Sciences* 16 (1961): 161–85.

Brockington, Fraser. "Public Health at the Privy Council, 1858–71." *Medical Officer* 101 (1959): 173–77, 185–90, 197–200, 211–15, 243–46, 259–60, 278–80, 287–90.

Brown, David. *Before the Ironclad: The Development of Ship Design, Propulsion and Armament in the Royal Navy, 1815–1860.* London: Conway Maritime Press, 1990.

Brown, Lucy. *The Board of Trade and the Free Trade Movement, 1830–1842.* Oxford: Clarendon Press, 1958.

Brown, Michael. "From Foetid Air to Filth: The Cultural Transformation of British Epidemiological Thought." *Bulletin of the History of Medicine* 82 (2008): 515–44.

Brown, Michael. "Medicine, Reform and the 'End' of Charity in Early Nineteenth-Century England." *English Historical Review* 124 (2009): 1353–88.

Browne, Charles. "The Life and Chemical Services of Frederick Accum." *Journal of Chemical Education* 2 (1925): 829–51, 1008–35, 1140–48.

Brunton, Deborah. *The Politics of Vaccination: Practice and Policy in England, Wales, Ireland, and Scotland, 1800–1874.* Rochester NY: University of Rochester Press, 2008.

Buchanan, Robert. *The Engineers: A History of the Engineering Profession in Britain, 1750–1914.* London: J. Kingsley, 1989.

Buchanan, Robert. "Engineers and Government in Nineteenth-Century Britain." In *Government and Expertise: Specialists, Administrators and Professionals, 1860–1919,* edited by Roy MacLeod, 41–58. 1988. Reprint, Cambridge: Cambridge University Press, 2003.

Buchanan, Robert. "The Rise of Scientific Engineering in Britain." *British Journal for the History of Science* 18 (1985): 218–33.

Bud, Robert, and Gerrylynn Roberts. *Science versus Practice: Chemistry in Victorian Britain.* Manchester University Press, 1984.

Buffery, Caroline. "Changing Landscapes: A Legal Geography of the River Severn." PhD diss., University of Birmingham, 2015.

Burn, William. *The Age of Equipoise: A Study of the Mid-Victorian Generation.* London: George Allen and Unwin Ltd., 1968.

Butterworth, Harry. "The Science and Art Department 1853–1900." PhD diss., University of Sheffield, 1968.

Byatt, Ian. *The British Electrical Industry, 1875–1914*. Oxford: Clarendon Press, 1979.

Bynum, W. *Science and the Practice of Medicine in the Nineteenth Century*. Cambridge: Cambridge University Press, 1994.

Cahan, David. *An Institute for an Empire: The Physikalisch-Technische Reichsanstalt, 1871–1918*. Cambridge: Cambridge University Press, 1989.

Cardwell, Donald. *The Organisation of Science in England*. 2nd ed. London: Heinemann, 1972.

Chambers, Theodore. *Register of the Associates and Old Students of the Royal College of Chemistry and the Royal College of Science*. London: Hazell, 1896.

Chapman, Allan. "Science and the Public Good: George Biddell Airy (1801–92) and the Concept of a Scientific Civil Servant." In *Science, Politics and the Public Good*, edited by Nicolaas Rupke, 36–62. London: Macmillan, 1988.

Checkland, S. *The Rise of Industrial Society in England, 1815–1885*. London: Longmans, 1964.

Chirnside, Ralph, and J. Hamence. *The "Practising Chemists": A History of the Society for Analytical Chemistry, 1874–1974*. London: Society for Analytical Chemistry, 1974.

Churchill, Randolph. *Twenty-One Years*. London: Weidenfeld & Nicolson, 1965.

Clark, John. "Eleanor Ormerod (1828–1901) as an Economic Entomologist: 'Pioneer of Purity Even More than of Paris Green.'" *British Journal for the History of Science* 25 (1992): 431–52.

Clarke, Ernest. *History of the Board of Agriculture, 1793–1822*. London: Royal Agricultural Society of England, 1898.

Clifford, Frederick. *A History of Private Bill Legislation* 2 vols. London: Butterworths, 1885–1887.

Clifton, Gloria. *Professionalism, Patronage and Public Service in Victorian London: The Staff of the Metropolitan Board of Works, 1856–1889*. London: Athlone Press, 1992.

Clokie, Hugh, and J. William Robinson. *Royal Commissions of Inquiry: The Significance of Investigations in British Politics*. Stanford, CA: Stanford University Press, 1937.

Coleman, Bruce. *Conservatism and the Conservative Party in Nineteenth-Century Britain*. London: Edward Arnold, 1988.

Collingridge, David, and Colin Reeve. *Science Speaks to Power: The Role of Experts in Policy-Making*. New York: St. Martin's Press, 1986.

Cook, G. "Thomas Southwood Smith FRCP (1788–1861): Leading Exponent of Diseases of Poverty and Pioneer of Sanitary Reform in the Mid-Nineteenth Century." *Journal of Medical Biography* 10 (2002): 194–205.

Cooter, Roger. "The Rise and Decline of the Medical Member: Doctors and Parliament in Edwardian and Interwar Britain." *Bulletin of the History of Medicine* 78 (2004): 59–107.

Conway, Hannah. "Illuminating Science: The Lighthouse as Public Good and the Role of the Scientific Expert in Nineteenth-Century British Lighthouse Reform." Master's thesis, Citadel Graduate College, 2015

Coull, James. "The Development of Marine Superintendence in Scotland under the Fishery Boards." *International Journal of Maritime History* 10 (1998): 41–59.

Courtney, Stephen. "'A Very Diadem of Light': Exhibitions in Victorian London, the Parliamentary Light and the Shaping of the Trinity House Lighthouses." *British Journal of the History of Science* 50 (2017): 249–65.

Craig, John. *Mint: A History of the London Mint from A.D. 287 to 1948*. Cambridge: Cambridge University Press, 2010.

Crook, Tom. *Governing Systems: Modernity and the Making of Public Health in England, 1830–1910*. Berkeley: University of California Press, 2016.

Crook, Tom. "Sanitary Inspection and the Public Sphere in Late Victorian and Edwardian Britain: A Case Study in Liberal Governance." *Social History* 33 (2007): 369–93.

Crosland, Maurice. *Science under Control: The French Academy of Sciences 1795–1914*. Cambridge: Cambridge University Press, 1992.

Crowther, J. *Statesmen of Science*. London: Cresset Press, 1965.

Cullen, Michael. *The Statistical Movement in Early Victorian Britain: The Foundations of Empirical Social Research*. 1975. Reprint, Brighton, UK: Edward Everett Root, 2016.

Daunton, Martin. *State and Market in Victorian Britain: War, Welfare and Capitalism*. Martlesham, UK: Boydell Press, 2008.

Daunton, Martin, ed. *The Organisation of Knowledge in Victorian Britain*. Oxford: Oxford University Press, 2005.

Dearce, Miguel. "Correspondence of Charles Darwin on James Torbitt's Project to Breed Blight-Resistant Potatoes." *Archives of Natural History* 35 (2008): 208–22.

DeJager, Timothy. "Pure Science and Practical Interests: The Origins of the Agricultural Research Council, 1930–1937." *Minerva* 31 (1993): 129–50.

Desmond, Adrian. *Huxley: The Devil's Disciple*. London: Michael Joseph, 1994.

Desmond, Adrian. *Huxley: Evolution's High Priest*. London: Michael Joseph, 1997.

Dicey, Albert. *Lectures on the Relation between Law and Public Opinion in England during the Nineteenth Century*. London: Macmillan, 1905.

Dingle, A. "'The Monster Nuisance of All': Landowners, Alkali Manufacturers, and Air Pollution, 1828–64." *Economic History Review* 35 (1982): 529–48.

Donnelly, James. "Representations of Applied Science: Academics and Chemical Industry in Late Nineteenth-Century England." *Social Studies of Science* 16 (1986): 195–234.

Douglas, Mary. *Purity and Danger: An Analysis of Concepts of Pollution and Taboo*. London: Routledge, 1966.

Dunn, Richard. 2012. "'Their Brains Over-Taxed': Ships, Instruments and Users." In *Reinventing the Ship: Science, Technology and the Maritime World, 1800–1918*, edited by Don Leggett and Richard Dunn, 131–55. Farnham, UK: Ashgate, 2012.

Dunn, Richard and Rebekah Higgitt. *Finding Longitude*. London: Collins, 2014.

Durey, Michael. *The Return of the Plague: British Society and the Cholera of 1831–32.* London: Macmillan, 1979.

Dyer, Graham P. "'One of the Best Men of Business': Master of the Royal Mint." In *John Herschel, 1792–1871: A Bicentennial Commemoration,* edited by Desmond King-Hele, 105–13. London: Royal Society, 1992.

Eastwood, David. "Making Public Policy in 19th Century Britain." University of Wales Swansea, 1997. https://collections.swansea.ac.uk/s/mitchell-welsh-arts-archive /item/2571#?c=0&m=0&s=0&cv=0.

Egerton, David, and John Pickstone. "United Kingdom." In *The Cambridge History of Science, Volume 8: Modern Science in National, Transnational, and Global Context,* edited by Hugh Slotten, Ronald Numbers, and David Livingstone, 151–91. Cambridge, Cambridge University Press, 2020.

Emmerson, George. *John Scott Russell: A Great Victorian Engineer and Naval Architect.* London: John Murray, 1977.

Eyler, John. "The Changing Assessments of John Snow's and William Budd's Cholera Studies." *Sozial- und Präventivmedizin* 46 (2001): 225–32.

Eyler, John. "The Conversion of Angus Smith: The Changing Role of Chemistry and Biology in Sanitary Science." *Bulletin of the History of Medicine* 54 (1980): 216–34.

Eyler, John. *Victorian Social Medicine: The Ideas and Methods of William Farr.* Baltimore, MD: Johns Hopkins University Press, 1979.

Eyler, John. "William Farr on the Cholera: The Sanitarian's Disease Theory and the Statistician's Method." *Journal of the History of Medicine,* 28 (1973): 79–100.

Farrar, Wilfred. "Andrew Ure, F.R.S., and the Philosophy of Manufactures." *Notes and Records* 27 (1973): 299–324.

Finer, Herman. *The Theory and Practice of Modern Government.* 4th ed. London: Methuen Publishing, 1961.

Finer, Samuel. "The Individual Responsibility of Ministers." *Public Administration* 34 (1956): 377–96.

Finer, Samuel. *The Life and Times of Sir Edwin Chadwick.* London: Methuen Publishing, 1952.

Finer, Samuel. "The Transmission of Benthamite Ideas, 1820–1850." In *Studies in the Growth of Nineteenth-Century Government,* edited by Gillian Sutherland, 11–32. Lanham, MD: Rowman & Littlefield, 1972.

Finn, Michael, and James Stark. 2015. "Medical Science and the Cruelty to Animals Act 1876: A Re-Examination of Anti-Vivisectionism in Provincial Britain." *Studies in History and Philosophy of Biological and Biomedical Sciences* 49 (2015): 12–23.

Fletcher, T. "The Great Depression of English Agriculture, 1873–1896." *Economic History Review* 13 (1961): 417–32.

Flick, Carlos. "The Movement for Smoke Abatement in 19th-Century Britain." *Technology and Culture* 21 (1980): 29–50.

Foote, George. "Science and Its Function in Early Nineteenth-Century England." *Osiris* 11 (1954): 438–54.

Forbes, Arthur. *History of the Army Ordnance Services: Volume Two, Modern History*. 1929. Reprint, Uckfield, UK: Naval & Military Press, 2020.

Ford, Matthew. "Towards a Revolution in Firepower? Logistics, Lethality, and the Lee-Metford." *War in History* 20 (2013): 273–99.

Foster, Michael. "The State and Scientific Research." *Nineteenth Century and After* 55 (1904): 741–51.

Frankel, Oz. "Blue Books and the Victorian Reader." *Victorian Studies* 46 (2004): 308–18.

Fraser, Derek. *The Evolution of the British Welfare State: A History of Social Policy since the Industrial Revolution*. London: Macmillan, 1973.

French, Richard. *Anti-Vivisection and Medical Science in Victorian Society*. Princeton, NJ: Princeton University Press, 1975.

Fries, Russell. "British Response to the American System: The Case of the Small-Arms Industry after 1850." *Technology and Culture* 16 (1975): 377–403.

Fry, Geoffrey. *Statesmen in Disguise: The Changing Role of the Administrative Class of the British Home Civil Service 1853–1966*. London: Macmillan, 1969.

Fulford, Tim. "The Role of Patronage in Early Nineteenth-Century Science, as Evidenced in Letters from Humphry Davy to Joseph Banks." *Notes and Records* 73 (2019): 457–75.

Fullmer, June. "Technology, Chemistry, and the Law in Early 19th-Century England." *Technology and Culture* 21 (1980): 1–28.

Fussell, G. "Impressions of Sir John Sinclair, Bart., First President of the Board of Agriculture." *Agricultural History* 25 (1951): 162–69.

Garwood, Christine. "The British State and the Natural Environment: With Special Reference to the Alkali Inspectorate, circa 1860–1906." PhD diss., University of Leicester, 1998.

Gascoigne, John. "The Royal Society and the Emergence of Science as an Instrument of State Policy." *British Journal for the History of Science* 32 (1999): 171–84.

Gascoigne, John. *Science and the State*. Cambridge: Cambridge University Press, 2019.

Gascoigne, John. *Science in the Service of Empire: Joseph Banks, the British State and the Uses of Science in the Age of Revolution*. Cambridge: Cambridge University Press, 1998.

Gibson, A., and W. Farrar. "Robert Angus Smith, FRS, and 'Sanitary Science.'" *Notes and Records* 28 (1974): 241–62.

Gillin, Edward. "Mechanics and Mathematicians: George Biddell Airy and the Social Tensions in Constructing Time at Parliament, 1845–1860." *History of Sciences* 58 (2020): 301–25.

Gillin, Edward. *The Victorian Palace of Science: Scientific Knowledge and the Building of the Houses of Parliament*. Cambridge: Cambridge University Press, 2017.

Gladstone, John. *Michael Faraday*. London: Macmillan & Co., 1874.

Goldman, Lawrence. "The Origins of British 'Social Science': Political Economy, Natural Science and Statistics, 1830–35." *Historical Journal* 26 (1983): 587–616.

Goldman, Lawrence. *Science, Reform, and Politics in Victorian Britain: The Social Science Association, 1857–1886*. Cambridge: Cambridge University Press, 2002.

Goldman, Lawrence. "Statistics and the Science of Society in Early Victorian Britain: An Intellectual Context for the General Register Office." *Social History of Medicine* 4 (1991): 415–34.

Goldman, Lawrence. *Victorians and Numbers: Statistics and Society in Nineteenth-Century Britain*. Oxford: Oxford University Press, 2022.

Gooday, Graeme. "Electrical Futures Past." *Endeavour* 29 (2005): 150–55.

Gooday, Graeme. "Liars, Experts and Authorities." *History of Science* 46 (2008): 431–56.

Gosnell, Harold. "British Royal Commissions of Enquiry." *Political Science Quarterly* 49 (1934): 84–118.

Gorham, Eville. "Robert Angus Smith, F.R.S., and 'Chemical Climatology.'" *Notes and Records* 36 (1982): 267–72.

Gratwick, Marion, ed. *Crop Pests in the UK: Collected Edition of MAFF Leaflets*. Dordrecht, Netherlands: Springer Science, 1992.

Gray, Ernest. *By Candlelight: The Life of Dr Arthur Hill Hassall, 1817–94*. London: Robert Hale, 1983.

Gray, Robert. "Medical Men, Industrial Labour and the State in Britain, 1830–1850." *Social History* 16 (1991): 19–43.

Guldi, Jo. *Roads to Power: Britain Invents the Infrastructure State*. Cambridge, MA: Harvard University Press, 2012.

Haigh, Elizabeth. 1991. "William Brande and the Chemical Education of Medical Students." In *British Medicine in an Age of Reform*, edited by Roger French and Andrew Wear, 186–202. London: Routledge, 1991.

Hall, Marie. *All Scientists Now: The Royal Society in the Nineteenth Century*. Cambridge: Cambridge University Press. 1984.

Hall, Marie. "Public Science in Britain: The Role of the Royal Society." *ISIS* 72 (1981): 627–69.

Halliday, S. "Death and Miasma in Victorian London: An Obstinate Belief." *British Medical Journal* 323 (2001): 1469–71.

Hamer, W. *The British Army: Civil-Military Relations, 1885–1905*. Oxford: Clarendon Press, 1970.

Hamilton, Charles. *Anglo-French Naval Rivalry, 1840–1870*. Oxford: Oxford University Press, 1993.

Hamilton, Charles. "Three Cultures at the Admiralty, c.1800–1945: Naval Staff, the Secretariat and the Arrival of Scientists." *Journal for Maritime Research* 16 (2014): 89–102.

Hamilton, Susan. "On the Cruelty to Animals Act, 15 August 1876." *BRANCH: Britain, Representation and Nineteenth-Century History*, n.d. https://branchcollective.org/?ps_articles=susan-hamilton-on-the-cruelty-to-animals-act-15-august-1876.

Hamlin, Christopher. "Edward Frankland's Early Career as London's Official Water Analyst, 1865–1876: The Context of 'Previous Sewage Contamination.'" *Bulletin of the History of Medicine* 56 (1982): 56–76.

Hamlin, Christopher. "Predisposing Causes and Public Health in Early Nineteenth-Century Medical Thought." *Social History of Medicine* 5 (1992): 43–71.

Hamlin, Christopher. "Providence and Putrefaction: Victorian Sanitarians and the Natural Theology of Health and Disease." *Victorian Studies* 78 (1985): 385–412.

Hamlin, Christopher. *Public Health and Social Justice in the Age of Chadwick; Britain, 1800–1954.* Cambridge: Cambridge University Press, 1998.

Hamlin, Christopher. 2002. "Public Sphere to Public Health: The Transformation of 'Nuisance.'" In *Medicine, Health and the Public Sphere in Britain, 1600–2000*, edited by Steve Sturdy, 189–204. London: Routledge, 2002.

Hamlin, Christopher. *A Science of Impurity: Water Analysis in Nineteenth-Century Britain.* Berkeley: University of California Press, 1990.

Hamlin, Christopher. "Scientific Method and Expert Witnessing: Victorian Perspectives on a Modern Problem." *Social Studies of Science* 16 (1986): 485–513.

Hamlin, Christopher. "William Dibdin and the Idea of Biological Sewage Treatment." *Technology and Culture* 29 (1988): 189–218.

Hammond, P. "150 Years of the Laboratory of the Government Chemist." *Analytical Proceedings* 29 (1992): 311–14.

Hammond, P., and Harold Egan. *Weighed in the Balance: A History of the Laboratory of the Government Chemist.* London: HMSO, 1992.

Hanley, James. *Healthy Boundaries: Property, Law, and Public Health in England and Wales, 1815–1872.* Rochester, NY: University of Rochester Press, 2016.

Hanley, James. "Parliament, Physicians, and Nuisances: The Demedicalization of Nuisance Law, 1831–1855." *Bulletin of the History of Medicine* 80 (2006): 702–32.

Hardy, Anne. "Cholera, Quarantine and the English Preventive System, 1850–1895." *Medical History* 37 (1993): 250–69.

Hardy, Anne. "Lyon Playfair and the Idea of Progress: Science and Medicine in Victorian Parliamentary Politics." In *Doctors, Politics and Society: Historical Essays*, edited by Dorothy Porter and Roy Porter, 81–106. Amsterdam: Rodopi, 1993.

Hardy, Anne. "Parish Pump to Private Pipes: London's Water Supply in the Nineteenth Century." *Medical History* Supplement No. 11 (1991): 76–93.

Hardy, Anne. "Water and the Search for Public Health in London in the Eighteenth and Nineteenth Centuries." *Medical History* 28 (1984): 250–82.

Harris, Jose. *British Government Inspection: The Local Services and the Central Departments.* London: Stevens & Sons, 1955.

Harrison, Andrew. "Scientific Naturalists and the Government of the Royal Society." PhD diss., Open University, 1989.

Harrison, Brian. "Animals and the State in Nineteenth-Century England." *English Historical Review* 88 (1973): 786–820.

Harrison, Elaine. "Women Members and Witnesses on British Government Ad Hoc Committees of Inquiry 1850–1930, with Special Reference to Royal Commissions of Inquiry." PhD diss., LSE, 1998.

Hart, Jenifer. 1972. "The Genesis of the Northcote-Trevelyan Report." In *Studies in the Growth of Nineteenth-Century Government*, edited by Gillian Sutherland, 63–81. Lanham, MD: Rowman & Littlefield, 1972.

Hart, Jenifer. "Nineteenth-Century Social Reform: A Tory Interpretation of History." *Past and Present* 31 (1965): 39–61.

Hartley, Owen. "Inspectorates in British Central Government." *Public Administration* 50 (1972): 447–66.

Hassan, John. "The Growth and Impact of the British Water Industry in the Nineteenth Century." *Economic History Review* 38 (1985): 531–47.

Hassan, John. *A History of Water in Modern England and Wales*. Manchester: Manchester University Press, 1998.

Hawkins, Angus. *The Forgotten Prime Minister: The 14th Earl of Derby*. 2 vols. Oxford: Oxford University Press, 2008.

Hawkins, Angus. *Victorian Political Culture*. Oxford: Oxford University Press, 2015.

Headrick, Daniel. *The Tools of Empire: Technology and Imperialism in the Nineteenth Century*. Oxford: Oxford University Press, 1981.

Heald, Henrietta. *William Armstrong: Magician of the North*. Newcastle-upon-Tyne: Northumbria Press, 2010.

Hennock, Peter. "Central/Local Government Relations in England: An Outline 1800–1950." *Urban History Yearbook* 9 (1982): 38–49.

Henriques, Ursula. *The Early Factory Acts and Their Enforcement*. London: Historical Association, 1971.

Henriques, Ursula. *Before the Welfare State: Social Administration in Early Industrial Britain*. London: Longman, 1979.

Herd, R. "Technical Background to International Conference on Load Lines, 1966." Royal Institution of Naval Architects. https://www.rina.org.uk/res/1967-5%20Herd %20RJ%20-%20Technical%20Background%20to%20the%20International%20 Conference%20on%20Load%20Lines%201966.pdf.

Higgitt, Rebekah. "The 1818 Longitude Act: A Watershed in Government Funding for Science?" Lecture delivered at the British Society for the History of Science Annual Conference, Exeter, England, July 16, 2011.

Higgitt, Rebekah, and Charles Withers. "Science and Sociability: Women as Audience at the British Association for the Advancement of Science, 1831–1901." *ISIS* 99 (2008): 1–27.

Higgs, Edward. "Some Forgotten Men: The Registrars General of England and Wales and the History of State Demographic and Medical Statistics, 1837–1920." Paper presented at the conference "Birth Pains and Death Throes: The Creation of Vital Statistics in Scotland and England," University of Glasgow, September 17, 2004.

Higgs, Edward. "The Struggle for the Occupational Census, 1841–1911," In *Government and Expertise: Specialists, Administrators and Professionals, 1860–1919*, edited by Roy MacLeod, 73–86. 1988. Reprint, Cambridge: Cambridge University Press, 2003.

Hill, I. "Statistical Society of London—Royal Statistical Society: The First 100 Years: 1834–1934." *Journal of the Royal Statistical Society A* 147 (1984): 130–39.

Hilts, Victor. "*Aliis Exterendum*, Or, the Origins of the Statistical Society of London." *ISIS* 69 (1978): 21–43.

Hodgkinson, Ruth. *The Origins of the National Health Service: The Medical Services of the New Poor Law, 1834–1871*. London: Wellcome Historical Medical Library, 1967.

Holloway, S. "The Apothecaries Act, 1815: A Reinterpretation." *Medical History* 10 (1960): 107–29, 221–36.

Hoppit, Julian. "Reforming Britain's Weights and Measures, 1660–1824." *English Historical Review* 108 (1993): 82–104.

Howarth, Oliver. *The British Association for the Advancement of Science: A Retrospect, 1831–1931*. London: British Association, 1931.

Howes, Anton. *Arts and Minds: How the Royal Society of Arts Changed a Nation*. Princeton, NJ: Princeton University Press, 2020.

Hughes, Edward. "Sir Charles Trevelyan and Civil Service Reform 1853–5." *English Historical Review* 64 (1949): 53–88.

Hughes, Thomas. "British Electrical Industry Lag: 1882–1888." *Technology and Culture* 3 (1962): 27–44.

Hull, Andrew. "War of Words: The Public Science of the British Scientific Community and the Origins of the Department of Scientific and Industrial Research, 1914–16." *British Journal for the History of Science* 32 (1999): 461–81.

Hutchinson, Kenneth. "The Royal Society and the Foundation of the British Gas Industry." *Notes and Records* 39 (1985): 245–70.

Huxley, Leonard. *Life and Letters of Thomas Henry Huxley*. 2 vols. London: Macmillan, 1900.

Jackson, Roland. *The Ascent of John Tyndall*. Oxford: Oxford University Press, 2018.

Jackson, Roland, *Case Studies in Scientific Advice to the Nineteenth-Century British State*. Welwyn, UK: Welwyn Press, 2023.

Jackson, Roland. "John Tyndall and the Early History of Diamagnetism." *Annals of Science* 72 (2015): 435–89.

Jackson, Roland. "John Tyndall and the Royal Medal That Was Never Struck." *Notes and Records* 68 (2014): 151–64.

James, Frank. "'Agricultural Chymistry Is at Present in It's Infancy': The Board of Agriculture, The Royal Institution and Humphry Davy." *Ambix* 62 (2015): 363–85.

James, Frank. "'The Civil-Engineer's Talent': Michael Faraday, Science, Engineering and the English Lighthouse Service, 1836–1865." *Transactions of the Newcomen Society* 70 (1999): 153–60.

James, Frank. "Davy in the Dockyard: Humphry Davy, the Royal Society and the Electro-Chemical Protection of the Copper Sheeting of His Majesty's Ships in the Mid-1820s." *Physis* 29 (1992): 205–25.

James, Frank. "Michael Faraday and Lighthouses." In *The Golden Age: Essays in British Social and Economic History 1850–1870*, edited by Ian Inkster, Colin Griffith, Geoff Hill, and Judith Rowbotham, 92–104. Farnham, UK: Ashgate, 2000.

James, Frank. *Michael Faraday: A Very Short Introduction*. Oxford: Oxford University Press, 2010.

James, Frank. "Michael Faraday's Work on Optical Glass." *Physics Education* 26 (1991): 296–300.

James, Frank. "The Military Context of Chemistry: The Case of Michael Faraday." *Bulletin of the History of Chemistry* 11 (1991): 36–40.

James, Frank. "Time, Tide and Michael Faraday." *History Today* (1991): 28–34.

James, Frank, ed. *'The Common Purposes of Life': Science and Society at the Royal Institution of Great Britain*. Farnham, UK: Ashgate, 2002.

James, Frank, ed. *The Correspondence of Michael Faraday, Volume 1: 1811–1831*. London: Institute of Electrical Engineering, 1991.

James, Frank, ed. *The Correspondence of Michael Faraday, Volume 2: 1832–1840*. London: Institute of Electrical Engineering, 1993.

James, Frank, ed. *The Correspondence of Michael Faraday, Volume 3: 1841–1848*. London: Institute of Electrical Engineering, 1996.

James, Frank, ed. *The Correspondence of Michael Faraday, Volume 4: 1849–1855*. London: Institute of Electrical Engineering, 1996.

James, Frank, ed. *The Correspondence of Michael Faraday, Volume 5: 1855–1860*. London: Institute of Electrical Engineering, 2008.

James, Frank, ed. *The Correspondence of Michael Faraday, Volume 6: 1861–1867*. London: Institute of Electrical Engineering, 2011.

James, Frank, and Margaret Ray. "Science in the Pits: Michael Faraday, Charles Lyell and the Home Office Enquiry into the Explosion at Haswell Colliery, Country Durham, in 1844." *History and Technology* 15 (1999): 213–31.

Johnson, Peter. "The Board of Longitude, 1714–1828." *Journal of the British Astronomical Association* 99 (1989): 63–9.

Joyce, Patrick. *The State of Freedom: A Social History of the British State since 1800*. Cambridge: Cambridge University Press, 2014.

Kearns, Gerry. "Private Property and Public Health Reform in England 1830–1870." *Social Science & Medicine* 26 (1988): 187–99.

Keith-Lucas, Bryan. *English Local Government in the Nineteenth and Twentieth Centuries.* London: Historical Association, 1977.

Kidd, Alan, and Terry Wyke. "The Cholera Epidemic in Manchester, 1831–32." *Bulletin of the John Rylands Library* 87 (2005): 43–56.

Kitson Clark, George. "'Statesmen in Disguise': Reflexions on the History of the Neutrality of the Civil Service." *Historical Journal* 2 (1959): 19–39.

Knight, David. *Humphry Davy: Science and Power.* Cambridge: Cambridge University Press, 1992.

Knight, Roger. *Britain against Napoleon: The Organization of Victory.* London: Allen Lane, 2013.

Knight, Roger. "The Introduction of Copper Sheathing into the Royal Navy, 1779–1786." *Mariner's Mirror* 59 (1973): 299–309.

Lambert, Andrew. *Battleships in Transition: The Creation of the Steam Battleship.* London: Conway Maritime Press, 1984.

Lambert, Andrew. *The Crimean War: British Grand Strategy, 1853–56.* Manchester: Manchester University Press, 1990.

Lambert, Andrew. "Science and Sea Power: The Navy Board, the Royal Society and the Structural Reforms of Sir Robert Seppings." *Transactions of the Naval Dockyards Society* 1 (2006): 9–19.

Lambert, Andrew. *Steam, Steel and Shellfire: The Steam Warship, 1815–1905.* London: Conway Maritime Press, 1992.

Lambert, Royston. "Central and Local Relations in Mid-Victorian England: The Local Government Act Office, 1858–71." *Victorian Studies* 6 (1962): 121–50.

Lambert, Royston. *Sir John Simon, 1816–1904, and English Social Administration.* London: MacGibbon & Kee, 1963.

Lauriat, Barbara. "'The Examination of Everything': Royal Commissions in British Legal History." *Statute Law Review* 31 (2010): 24–46.

Lawrence, Christopher. "Incommunicable Knowledge: Science, Technology and the Clinical Art in Britain 1850–1914." *Journal of Contemporary History* 20 (1985): 503–20.

Layton, David. "Lord Wrottesley, F.R.S., Pioneer Statesman of Science." *Notes and Records* 23 (1968): 230–46.

Leggett, Don. 2012. "Neptune's New Clothes: Actors, Iron and the Identity of the Mid-Victorian Warship." In *Reinventing the Ship: Science, Technology and the Maritime World, 1800–1918*, edited by Don Leggett and Richard Dunn, 71–92. Farnham, UK: Ashgate, 2012.

Leggett, Don. "Naval Architecture, Expertise and Navigating Authority in the British Admiralty, c. 1885–1906." *Journal for Maritime Research* 16 (2014): 73–88.

Leggett, Don. *Shaping the Royal Navy: Technology, Authority and Naval Architecture, c. 1830–1906.* Manchester: Manchester University Press, 2015.

Leggett, Don, and James Davey. "Introduction: Expertise and Authority in the Royal Navy, 1800–1945." *Journal for Maritime Research*, 16 (2014): 1–13.

Leggett, Don, and Richard Dunn, eds. *Reinventing the Ship: Science, Technology and the Maritime World, 1800–1918*. Farnham, UK: Ashgate, 2012.

Levitt, Theresa. "When Lighthouses Became Public Goods: The Role of Technological Change." *Technology and Culture* 61 (2020): 144–72.

Lewis, Peter, and Colin Gagg. "Aesthetics versus Function: The Fall of the Dee Bridge, 1847." *Interdisciplinary Science Reviews* 29 (2004): 177–91.

Lewis, Peter, and Ken Reynolds. "Forensic Engineering: A Reappraisal of the Tay Bridge Disaster." *Interdisciplinary Science Reviews* 27 (2002): 287–98.

Lewis, Richard. *Edwin Chadwick and the Public Health Movement, 1832–1854*. London: Longmans, 1952.

Lidwell-Durnin, John. "Cultivating Famine: Data, Experimentation and Food Security, 1795–1848." *British Journal for the History of Science* 53 (2020): 159–81.

Lightman, Bernard. *Evolutionary Naturalism in Victorian Britain: The "Darwinians" and their Critics*. Farnham, UK: Ashgate/Variorum, 2009.

Lindberg, Erik. "From Private to Public Provision of Public Goods: English Lighthouses between the Seventeenth and Nineteenth Centuries." *Journal of Policy History* 25 (2013): 538–56.

Lloyd, Harriet. "Rulers of Opinion: Women at the Royal Institution of Great Britain, 1799–1812." PhD diss., University College London, 2019.

Loudon, Irvine. *Medical Care and the General Practitioner, 1750–1850*. Oxford: Oxford University Press, 1986.

Luckin, William. *Death and Survival in Urban Britain: Disease, Pollution and Environment, 1800–1950*. London: I. B. Tauris, 2015.

Lubenow, William. *The Politics of Government Growth: Early Victorian Attitudes Towards State Intervention, 1833–1848*. Newton Abbott: David & Charles, 1971.

Lyons, Henry. *The Royal Society, 1660–1940*. Cambridge: Cambridge University Press, 1944.

MacDonagh, Oliver. "Coal Mines Regulation: The First Decade, 1842–1852. " In *Ideas and Institutions of Victorian Britain: Essays in honour of George Kitson-Clark*, edited by Robert Robson, 58–86. London: G. Bell & Sons, 1967.

MacDonagh, Oliver. *Early Victorian Government, 1830–1870*. London: Weidenfeld & Nicolson, 1977.

MacDonagh, Oliver. "Government, Industry and Science in 19th Century Britain: A Particular Study." *Historical Studies* 16 (1975): 503–17.

MacDonagh, Oliver. "The Nineteenth-Century Revolution in Government: A Reappraisal." *Historical Journal* 1 (1958): 52–67.

Macdonald, Lee. *Kew Observatory and the Evolution of Victorian Science, 1840–1910*. Pittsburgh: University of Pittsburgh Press, 2018.

MacLeod, Roy. "The Alkali Acts Administration, 1863–84: The Emergence of the Civil Scientist." *Victorian Studies* 9 (1965): 85–112.

MacLeod, Roy. *The Creed of Science in Victorian England*. Farnham, UK: Ashgate, 2000.

MacLeod, Roy. "Government and Resource Conservation: The Salmon Acts Administration, 1860–86." *Journal of British Studies* 7 (1968): 114–50.

MacLeod, Roy. "Of Medals and Men: A Reward System in Victorian Science, 1826–1914." *Notes and Records* 26 (1971): 81–105.

MacLeod, Roy. *Public Science and Public Policy in Victorian England*. Farnham, UK: Ashgate Variorum, 1996.

MacLeod, Roy. "The Royal Society and the Government Grant: Notes on the Administration of Scientific Research, 1849–1914." *Historical Journal* 14 (1971): 323–58.

MacLeod, Roy. "Science and Government in Victorian England: Lighthouse Illumination and the Board of Trade, 1866–1886." *ISIS* 60 (1969): 5–38.

MacLeod, Roy. "Science and the Civil List, 1824–1914." *Technology and Society* 6 (1970): 1–20.

MacLeod, Roy. 1976. "Science and the Treasury: Principles, Personalities and Politics, 1870–85." In *The Patronage of Science in the Nineteenth Century*, edited by Gerard Turner, 115–72. Leiden: Noordhoff International Publishing, 1976.

MacLeod, Roy. "Science for Imperial Efficiency: Reflections on the British Science Guild, 1905–1936." *Public Understanding of Science* 3 (1994): 155–94.

MacLeod, Roy. "Statesmen Undisguised." *American Historical Review* 78 (1973): 1386–1405.

MacLeod, Roy. 1983. "Whigs and Savants: Reflections on the Reform Movement in the Royal Society, 1830–48." In *Metropolis and Province: Science in British Culture 1780–1850*, edited by Ian Inkster and Jack Morrell, 55–90. Philadelphia: University of Pennsylvania Press, 1983.

MacLeod, Roy, ed. *Government and Expertise: Specialists, Administrators and Professionals, 1860–1919*. 1988. Reprint, Cambridge: Cambridge University Press, 2003.

MacLeod, Roy, and Peter Collins, eds. *The Parliament of Science*. Northwood, UK: Science Reviews, 1981.

Maglen, Krista. *The English System: Quarantine, Immigration and the Making of Port Sanitary Zone*. Manchester: Manchester University Press, 2014.

Mandler, Peter. *Aristocratic Government in the Age of Reform: Whigs and Liberals, 1830–1852*. Oxford: Clarendon Press, 1990.

Mårald, Erland. "Everything Circulates: Agricultural Chemistry and Recycling Theories in the Second Half of the Nineteenth Century." *Environment and History* 8 (2002): 65–84.

Marples, Alice. "The Science of Money: Isaac Newton's Mastering of the Mint." *Notes and Records* 76 (2022): 507–25.

Marsden, Ben, and Crosbie Smith. *Engineering Empires: A Cultural History of Technology in Nineteenth-Century Britain*. London: Palgrave Macmillan, 2005.

Martin, Bernice. "Leonard Horner: A Portrait of an Inspector of Factories." *International Review of Social History* 14 (1969): 412–43.

Mauskopf, Seymour. "'From an Instrument of War to an Instrument of the Laboratory: The Affinities Certainly Do Not Change' Chemists and the Development of Munitions, 1785–1885." *Bulletin of the History of Chemistry* 24 (1999): 1–15.

Mauskopf, Seymour. "Long Delayed Dream: Frederick Abel and Smokeless Powder." *Royal Society of Chemistry, Occasional Papers* 3 (2003): 1–27.

McDonald, J. "The History of Quarantine in Britain during the 19th Century." *Bulletin of the History of Medicine* 25 (1951): 22–44.

McConaghey, R. "Notes of the Evolution of Medical Practice Prior to 1858." *Journal of the College of General Practitioners and Research Newsletter* 1 (1958): 267–78.

McFeeley, Mary. *Lady Inspectors: The Campaign for a Better Workplace, 1893–1921.* Oxford: Blackwell, 1988.

McGregor, Oliver. "Social Research and Social Policy in the Nineteenth Century." *British Journal of Sociology* 8 (1957): 146–57.

McLaren, John. "Nuisance Law and the Industrial Revolution—Some Lessons from Social History." *Oxford Journal of Legal Studies* 3 (1983): 159–61.

McLean, Iain. "The Origin and Strange History of Regulation in the UK: Three Case Studies in Search of a Theory." Paper prepared for European Science Foundation Exploratory Workshop, "The Politics of Regulation," Barcelona, 2002.

Meadows, Jack. *Science and Controversy: A Biography of Sir Norman Locker, Founder Editor of Nature.* 2nd ed. London: Macmillan, 2008.

Miller, David. "Between Hostile Camps: Sir Humphry Davy's Presidency of the Royal Society of London, 1820–1827." *British Journal for the History of Science* 16 (1983): 1–47.

Miller, David. "The Royal Society of London, 1800–1835: A Study in the Cultural Politics of Scientific Organization." PhD diss., University of Pennsylvania, 1981.

Millward, Robert. *Private and Public Enterprise in Europe: Energy, Telecommunications and Transport, 1830–1990.* Cambridge: Cambridge University Press, 2005.

Mitchell, Sally. *Frances Power Cobbe.* Charlottesville: University of Virginia Press, 2004.

Mitchison, Rosalind. "The Old Board of Agriculture (1793–1822)." *English Historical Review* 74 (1959): 41–69.

Mooney, Graham. "Public Health versus Private Practice: The Contested Development of Compulsory Infectious Disease Notification in Late-Nineteenth-Century Britain." *Bulletin of the History of Medicine* 73 (1999): 238–67.

Morrell, Jack. "Individualism and the Structure of British Science in 1830." *Historical Studies in the Physical Sciences* 3 (1971): 183–204.

Morrell, Jack, and Arnold Thackray. *Gentlemen of Science: Early Years of the British Association for the Advancement of Science.* Oxford: Clarendon Press, 1982.

Morris, Robert. *Cholera 1932. The Social Response to an Epidemic.* London: Croom Helm, 1976.

Morris, Peter. *The Matter Factory: A History of the Chemistry Laboratory*. London: Reaktion Books, 2015.

Morriss, Roger. *Science, Utility and British Naval Technology, 1793–1815*. London: Routledge, 2021.

Morus, Iwan Rhys. "The Nervous System of Britain: Space, Time, and the Electric Telegraph in the Victorian Age." *British Journal for the History of Science* 33 (2000): 455–75.

Newell, Edmund. "Atmospheric Pollution and the British Copper Industry, 1690–1920." *Technology and Culture* 38 (1997): 655–89.

Odling, William. "Science in Courts of Law." *Journal of the Society of Arts* 7 (1860): 167–68.

Ogborn, Miles. "Local Power and State Regulation in Nineteenth Century Britain." *Transactions of the Institute of British Geographers* 17 (1992): 215–26.

O'Gorman, Frank. *The Emergence of the British Two-Party System, 1760–1832*. London: Hodder Arnold, 1982.

Olien, Diana Davids. *Morpeth: A Victorian Public Career*. Washington, DC: University Press of America, 1963.

Ogawa, Mariko. "Liebig and the Royal Agricultural Society Meeting at Bristol, 1842." *Ambix* 55 (2008): 136–52.

Orange, A. "The Beginnings of the British Association, 1831–1851." In *The Parliament of Science*, edited by Roy MacLeod and Peter Collins, 43–64. Northwood, UK: Science Reviews, 1981.

Otter, Christopher. "Cleansing and Clarifying: Technology and Perception in Nineteenth-Century London." *Journal of British Studies* 43 (2004): 40–64.

Otter, Christopher. "Making Liberal Objects: British Techno-Social Relations 1800–1900." *Cultural Studies* 21 (2007): 570–90.

Otter, Christopher. *The Victorian Eye: A Political History of Light and Vision in Britain, 1800–1910*. Chicago: University of Chicago Press, 2008.

Owen, David, and Roy MacLeod, eds. *The Government of Victorian London, 1855–1889: The Metropolitan Board of Works, the Vestries, and the City Corporation*. Cambridge, MA: Harvard University Press, 1982.

Parker, Charles, ed. *Sir Robert Peel: From His Private Papers*. Vol. 3. London: John Murray, 1899.

Parris, Henry. *Government and the Railways in Nineteenth-Century Britain*. London: Routledge and Kegan Paul, 1965.

Parris, Henry. "The Nineteenth-Century Revolution in Government: A Reappraisal Reappraised." *Historical Journal* 3 (1960): 17–37.

Parry, Jonathan. *The Rise and Fall of Liberal Government in Victorian Britain*. New Haven, CT: Yale University Press, 1993.

Pattison, Iain. *The British Veterinary Profession, 1791–1948*. London: J. A. Allen, 1984.

Patton, Mark. *Science, Politics and Business in the Work of Sir John Lubbock*. Farnham, UK: Ashgate, 2007.

Peel, George, ed. *The Private Letters of Sir Robert Peel*. London: John Murray, 1920.

Pellew, Jill. "The Home Office and the Explosives Act of 1875." *Victorian Studies* 18 (1974): 175–94.

Pellew, Jill. *The Home Office, 1848–1914: From Clerks to Bureaucrats*. London: Heinemann, 1982.

Pelling, Margaret. 1978. *Cholera, Fever and English Medicine, 1825–1865*. Oxford: Oxford University Press.

Pemberton, Neil, and Michael Worboys. *Rabies in Britain: Dogs, Disease and Culture, 1830–2000*. 2007. Reprint, London: Palgrave Macmillan, 2012.

Peppercorne, Frederick. *Supply of Water to the Metropolis*. London: John Weale, 1840.

Perkin, Harold. "Individualism versus Collectivism in Nineteenth-Century Britain: A False Antithesis." *Journal of British Studies* 17 (1977): 105–18.

Perkins, Adam. "'Extraneous Government Business': The Astronomer Royal as Government Scientist: George Airy and His Work on the Commissions of State and Other Bodies, 1838–1880." *Journal of Astronomical History and Heritage* 4 (2001): 143–54.

Pickstone, John. "Science in Nineteenth-Century England: Plural Configurations and Singular Politics." In *The Organisation of Knowledge in Victorian Britain*, edited by Martin Daunton, 87–114. Oxford: Oxford University Press, 2005.

Pielke, Roger. *The Honest Broker: Making Sense of Science in Policy and Politics*. Cambridge: Cambridge University Press, 2007.

Pinsdorf, Marion. "Engineering Dreams into Disaster: History of the Tay Bridge." *Business and Economic History* 26 (1997): 491–504.

Pontin, Ben. "Nuisance Law and the Industrial Revolution: A Reinterpretation of Doctrine and Institutional Competence." *Modern Law Review* 75 (2012): 1010–36.

Poole, John, and Kay Andrews, eds. *The Government of Science in Britain*. London: Weidenfeld & Nicolson, 1972.

Porter, Dale. *The Life and Times of Sir Goldsworthy Gurney: Gentleman Scientist and Inventor, 1793–1875*. Bethlehem, PA: Lehigh University Press, 1998.

Prouty, Roger. *The Transformation of the Board of Trade, 1830–55: A Study of Administrative Reorganisation in the Heyday of Laissez-Faire*. London: William Heinemann, 1957.

Preece, William. "Address." *Journal of the Society of Arts* 50 (1901): 8–22.

Reed, Peter. *Acid Rain and the Rise of the Environmental Chemist in Nineteenth-Century Britain: The Life and Work of Robert Angus Smith*. Farnham, UK: Ashgate, 2014.

Reed, Peter. "Robert Angus Smith and the Alkali Inspectorate." In *The Chemical Industry in Europe, 1850–1914: Industrial Growth, Pollution, and Professionalization*, edited by Ernst Homburg, Anthony Travis, and Harm Schröter, 149–63. Dordrecht, Netherlands: Kluwer Academic. 1998.

Reid, Robert. *Tongues of Conscience: War and the Scientist's Dilemma*. London: Constable, 1969.

Reid, Wemyss. *Memoirs and Correspondence of Lyon Playfair*. London: Cassell, 1900.

Rhodes, Gerald. *Inspectorates in British Government: Law Enforcement and Standards of Efficiency*. London: George Allen & Unwin, 1981.

Ringer, Fritz. *The Decline of the German Mandarins: The German Academic Community, 1890–1933*. Cambridge, MA: Harvard University Press, 1969.

Roberts, David. "Jeremy Bentham and the Victorian Administrative State." *Victorian Studies* 2 (1959): 193–210.

Roberts, David. *Victorian Origins of the British Welfare State*. New Haven, CT: Yale University Press, 1960.

Roberts, Michael. "The Politics of Professionalization: MPs, Medical Men, and the 1858 Medical Act." *Medical History* 53 (2009): 37–56.

Rolt, Lionel. *Thomas Telford*. London: Longmans, 1958.

Roscoe, Henry. *The Life & Experiences of Sir Henry Enfield Roscoe*. London: Macmillan, 1906.

Rose, Edward. "The Military Background of John W. Pringle, in 1826 Founding Superintendent of the Geological Survey of Ireland." *Irish Journal of Earth Sciences* 17 (1998/1999): 61–70.

Rosenthal, Leslie. *The River Pollution Dilemma in Victorian England: Nuisance Law versus Economic Efficiency*. London: Routledge, 2014.

Ross, S. "*Scientist*: The Story of a Word." *Annals of Science* 18 (1962): 65–85.

Rowlinson, P. "Food Adulteration Its Control in 19th Century Britain." *Interdisciplinary Science Reviews*, 7 (1982): 63–72.

Russell, Colin. *Chemists by Profession*. London: Royal Society of Chemistry, 1977.

Russell, Colin. *Edward Frankland: Chemistry, Controversy and Conspiracy in Victorian England*. London: Macmillan, 1996.

Russell, Colin. *Science and Social Change, 1700–1900*. London: Macmillan, 1983.

Russell, George. "Notes of the Week: Fair Play in Legislation." *Irish Homestead* 17 (1910): 1087.

Schaffer, Simon. "Metrology, Metrication and Victorian Values." In *Victorian Science in Context*, edited by Bernard Lightman, 438–74. Chicago: University of Chicago Press, 1997.

Schiffer, Michael. "The Electric Lighthouse in the Nineteenth Century: Aid to Navigation and Political Technology." *Technology and Culture* 46 (2005): 275–305.

Schivelbusch, Wolfgang. *Disenchanted Night: The Industrialization of Light in the 19th Century*. Berkeley: University of California Press, 1988.

Schwartz, Pedro. "John Stuart Mill and Laissez-Faire: London Water." *Economica* 33 (1966): 71–83.

Scott, James. *Seeing like a State*. New Haven, CT: Yale University Press, 2020.

Sheail, John. "Town Wastes, Agricultural Sustainability and Victorian Sewage." *Urban History* 23 (1996): 189–210.

Sheard, Sally, and Liam Donaldson. *The Nation's Doctor: The Role of the Chief Medical Officer, 1855–1998*. London: Nuffield Trust, 2005.

Sheffield, Suzanne. *Revealing New Worlds: Three Victorian Women Naturalists*. London: Routledge, 2001.

Sherard, Robert. *The White Slaves of England*. London: James Bowden, 1897.

Shoen, Harriet. "Prince Albert and the Application of Statistics to Problems of Government." *Osiris* 5 (1938): 276–318.

Simmons, Jack. *The Railway in England and Wales, 1830–1914*. Leicester: Leicester University Press, 1978.

Skentelbery, Norman. *A History of the Ordnance Board*. London: Ordnance Board Press, 1967.

Smiles, Samuel. *The Life of Thomas Telford, Civil Engineer*. London: John Murray, 1867.

Smith, Crosbie. *Coal, Steam and Ships: Engineering, Enterprise and Empire on the Nineteenth-Century Seas*. Cambridge: Cambridge University Press, 2018.

Smith, Crosbie. "'The Great National Undertaking': John Scott Russell, the Master Shipwrights and the Royal Mail Steam Packet Company." In *Reinventing the Ship: Science, Technology and the Maritime World, 1800–1918*, edited by Don Leggett and Richard Dunn, 25–52. Farnham, UK: Ashgate, 2012.

Smith, Francis. "The Contagious Diseases Acts Reconsidered." *Social History of Medicine* 3 (1990): 197–215.

Smith, Frances. *The People's Health, 1830–1910*. Canberra: Australia National University Press, 1979.

Smith, Hubert. *Board of Trade*. London: Putnam, 1928.

Smith, Paul. *Disraelian Conservatism and Social Reform*. London: Routledge and Kegan Paul, 1967.

Snyder, Laura. *Reforming Philosophy: A Victorian Debate on Science and Society*. Chicago: University of Chicago Press, 2006.

Southward, Alan, and E. Roberts. "100 Years of Marine Research at Plymouth." *Journal of the Marine Biological Association* 67 (1987): 465–506.

Stafford, Robert. *Scientist of Empire: Sir Roderick Murchison, Scientific Exploration and Victorian Imperialism*. Cambridge: Cambridge University Press, 2010.

Steere-Williams, Jacob. *The Filth Disease: Typhoid Fever and the Practices of Epidemiology in Victorian England*. Rochester, NY: University of Rochester Press, 2020.

Stern, Walter. "Water Supply in Britain: The Development of a Public Service." *Perspectives in Public Health* 74 (1954): 998–1004.

Stern, Jon. "Regulation and Contracts for Utility Services: Substitutes or Complements? Lessons from UK Railway and Electricity History." *Policy Reform* 6 (2003): 193–215.

Sturdy, Steve, and Roger Cooter. "Science, Scientific Management, and the Transformation of Medicine in Britain c. 1870–1950." *History of Science* 36 (1998): 421–66.

Sunderland, David. "A Monument to Defective Administration? The London Commissions of Sewers in the Early Nineteenth Century." *Urban History* 26 (1999): 349–72.

Sutherland, David. "The Scottish Continental Herring Trade 1810–1914." http://scottishherringhistory.uk/index.html.

Sutherland, Gillian, ed. *Studies in the Growth of Nineteenth-Century Government.* Lanham, MD: Rowman & Littlefield, 1972.

Swann, D. "The Engineers of English Port Improvements, 1600–1830." *Transport History* 1 (1968): 153–67, 260–76.

Swinfen, David. *The Fall of the Tay Bridge.* Edinburgh: Mercat Press, 1994.

Syrett, David. *Shipping and the American War 1775–1783: A Study of British Transport Organization.* London: Athlone Press, 1970.

Szreter, Simon. "The GRO and the Public Health Movement in Britain, 1837–1914." *Social History of Medicine* 4 (1991): 435–63.

Taylor, Arthur. *Laissez-Faire and State Intervention in 19th Century Britain.* London: Macmillan, 1972.

Taylor, James. "Private Property, Public Interest, and the Role of the State in Nineteenth-Century Britain: The Case of the Lighthouses." *Historical Journal* 44 (2001): 749–71.

Thomson, Gilbert. "River Pollution. With Special Reference to Present and Prospective Legislation." *Perspectives in Public Health* 46 (1925): 355–63.

Thorsheim, Peter. *Inventing Pollution: Coal, Smoke, and Culture in Britain since 1800.* Athens: Ohio University Press, 2006.

Thorsheim, Peter. "The Paradox of Smokeless Fuels: Gas, Coke and the Environment in Britain, 1813–1949." *Environment and History* 8 (2002): 381–401.

Todd, A. *Beyond the Blaze: A Biography of Davies Gilbert.* Truro, UK: D. Bradford Barton, 1967.

Todd, Alphaeus. *On Parliamentary Government in England: Its Origin, Development, and Practical operation.* Vol. 2. London: Longmans, 1869.

Tomory, Leslie. "The Environmental History of the Early British Gas Industry, 1812–1830." *Environmental History* 17 (2012): 29–54.

Tomory, Leslie. *The History of the London Water Industry, 1580–1820.* Baltimore, MD: Johns Hopkins University Press, 2017.

Tomory, Leslie. *Progressive Enlightenment: The Origins of the Gaslight Industry, 1780–1820.* Cambridge: MIT Press, 2012.

Tulodziecki, Dana. "A Case Study in Explanatory Power: John Snow's Conclusions about the Pathology and Transmission of Cholera." *Studies in History and Philosophy of Biological and Biomedical Sciences* 42 (2011): 306–16.

Turner, Frank. "Public Science in Britain, 1880–1919." *ISIS* 71 (1980): 589–608.

Turner, Frank. "Rainfall, Plague and the Prince of Wales: A Chapter in the Conflict of Religion and Science." *Journal of British Studies* 13 (1974): 46–65.

Turner, Gerard, ed. *Nineteenth-Century Scientific Instruments.* London: Philip Wilson, 1983.

Turner, Michael. "Output and Prices in UK Agriculture, 1867–1914, and the Great Agricultural Depression Reconsidered." *Agricultural History Review* 40 (1992): 38–51.

Tynan, Nicola. "Nineteenth Century London Water Supply: Processes of Innovation and Improvement." *Review of Austrian Economics* 26 (2013): 73–91.

Usselman, Melvyn. *Pure Intelligence: The Life of William Hyde Wollaston.* Chicago: University of Chicago Press, 2015.

Vig, Norman. *Science and Technology in British Politics.* Oxford: Pergamon Press, 1968.

Vincent, James. *Beyond Measure: The Hidden History of Measurement.* London: Faber, 2022.

Waddington, Ivan. *The Medical Profession in the Industrial Revolution.* Dublin: Humanities Press, 1984.

Waddington, Keir. "Vitriol in the Taff: River Pollution, Industrial Waste, and the Politics of Control in Late Nineteenth-Century Rural Wales." *Rural History* 29 (2018): 23–44.

Walker, Simon. "The Rise and Fall of the Second British Empire: The Evolution of British Small Arms and Rifles, 1850–1950." Bachelor's thesis, University of Strathclyde, 2013.

Waller, Philip. *Town, City, and Nation: England, 1850–1914.* Oxford: Oxford University Press, 1983.

Warren, Kenneth. *Armstrong: The Life and Mind of an Armaments Maker.* Northern Heritage Services, 2010.

Waring, Sophie. "The Board of Longitude and the Funding of Scientific Work: Negotiating Authority and Expertise in the Early Nineteenth Century." *Journal of Maritime Research* 16 (2014): 55–71.

Waring, Sophie. "Thomas Young, the Board of Longitude and the Age of Reform." PhD diss., University of Cambridge, 2014.

Warner, John. "The Idea of Science in English Medicine: The 'Decline of Science' and the Rhetoric of Reform, 1815–45." In *British Medicine in an Age of Reform*, edited by Roger French and Andrew Wear, 136–64. London: Routledge, 1991.

Watson, Edward. *The Royal Mail to Ireland.* London: Edward Arnold, 1917.

Webb, Robert. "Southwood Smith: The Intellectual Sources of Public Service." In *Doctors, Politics and Society: Historical Essays*, edited by Dorothy Porter and Roy Porter, 46–80. Amsterdam: Rodopi, 1993.

Webb, Robert, "A Whig Inspector." *Journal of Modern History* 27 (1955): 352–64.

Weinberg, Alvin. "Science and Trans-Science." *Minerva* 10 (1972): 209–22.

Weintraub, Stanley. *Uncrowned King: The Life of Prince Albert*. New York: Free Press, 1997.

Weisz, George. "The Emergence of Medical Specialization in the Nineteenth Century." *Bulletin of the History of Medicine* 77 (2003): 536–75.

Wheeler, Michael. *The Athenaeum: More than Just Another London Club*. New Haven, CT: Yale University Press, 2020.

Whitehead, Charles. *Retrospections*. Maidstone, UK: W. E. Thorpe, 1913.

Whitehead, Mark. *State, Science and the Skies: Governmentalities of the British Atmosphere*. London: John Wiley, 2009.

Williams, David. "James Silk Buckingham: Sailor, Explorer and Maritime Reformer." In *Merchants and Mariners*, Research in Maritime History, no. 18, 109–26. Liverpool: Liverpool University Press, 2000.

Wilmot, Sarah. *"The Business of Improvement": Agriculture and Scientific Culture in Britain, c.1700–c.1870*. Historical Geography Research Group, 1990.

Wilmot, Sarah. "Pollution and Public Concern: The Response of the Chemical Industry in Britain to Emerging Environmental Issues, 1860–1901." In *The Chemical Industry in Europe, 1850–1914: Industrial Growth, Pollution, and Professionalization*, edited by Ernst Homburg, Anthony Travis, and Harm Schröter, 121–47. Dordrecht, Netherlands: Kluwer Academic, 1998.

Wohl, Anthony. *Endangered Lives: Public Health in Victorian Britain*. Cambridge, MA: Harvard University Press, 1983.

Woods, Abigail. "From Practical Men to Scientific Experts: British Veterinary Surgeons and the Development of Government Scientific Expertise, c. 1878–1919." *History of Science* 51 (2013): 457–80.

Worboys, Michael. "Germ Theories of Disease and British Veterinary Medicine, 1860–1890." *Medical History* 35 (1991): 308–27.

Worboys, Michael. *Spreading Germs: Disease Theories and Medical Practice in Britain, 1865–1900*. Cambridge: Cambridge University Press, 2000.

Wright, Maurice. "Treasury Control, 1854–1914." In *Studies in the Growth of Nineteenth-Century Government*, edited by Gillian Sutherland, 196–226. Lanham, MD: Rowman & Littlefield, 1972.

Yeo, Eileen. *The Contest for Social Science: Relations and Representations of Gender and Class*. London: Rivers Oram Press, 1996.

Yeo, Richard. *Defining Science: William Whewell, Natural Knowledge and Public Debate in Early Victorian Britain*. Cambridge: Cambridge University Press, 1993.

Yeo, Richard. "An Idol of the Market-Place: Baconianism in Nineteenth Century Britain." *History of Science* 61 (1985): 251–98.

Yeo, Richard. "Science and Intellectual Authority in Mid-Nineteenth-Century Britain: Robert Chambers and *Vestiges of the Natural History of Creation*." *Victorian Studies* 28 (1984): 5–31.

Index